I0064595

Intelligent Control Systems

Intelligent Control Systems

Edited by **Alfred Silva**

NY RESEARCH
P R E S S

New York

Published by NY Research Press,
23 West, 55th Street, Suite 816,
New York, NY 10019, USA
www.nyresearchpress.com

Intelligent Control Systems
Edited by Alfred Silva

© 2016 NY Research Press

International Standard Book Number: 978-1-63238-477-5 (Hardback)

This book contains information obtained from authentic and highly regarded sources. Copyright for all individual chapters remain with the respective authors as indicated. All chapters are published with permission under the Creative Commons Attribution License or equivalent. A wide variety of references are listed. Permission and sources are indicated; for detailed attributions, please refer to the permissions page and list of contributors. Reasonable efforts have been made to publish reliable data and information, but the authors, editors and publisher cannot assume any responsibility for the validity of all materials or the consequences of their use.

The publisher's policy is to use permanent paper from mills that operate a sustainable forestry policy. Furthermore, the publisher ensures that the text paper and cover boards used have met acceptable environmental accreditation standards.

Trademark Notice: Registered trademark of products or corporate names are used only for explanation and identification without intent to infringe.

Printed in the United States of America.

Contents

Permissions

List of Contributors

Preface

Intelligent control systems use artificial intelligence to manage and control other systems such as production or manufacturing for machines or controlling equipment. These systems utilize fuzzy logic, neural networks, and fuzzy neural networks in principle. Adaptive controls, advanced control algorithms and applications, artificial intelligence, knowledge engineering, automation system, autonomous systems, complex systems and intelligent robots, hybrid systems, industrial automation and online monitoring, integrated and complex automation systems, intelligent automation and manufacturing are some of the important topics discussed in this book. It strives to provide a fair idea about this discipline and to help develop a better understanding of the latest advances within this field. While understanding the long-term perspectives of the topics, the book makes an effort in highlighting their impact as a modern tool for the growth of the discipline. Researchers and students actively engaged in this field will find this book full of crucial and unexplored concepts.

The researches compiled throughout the book are authentic and of high quality, combining several disciplines and from very diverse regions from around the world. Drawing on the contributions of many researchers from diverse countries, the book's objective is to provide the readers with the latest achievements in the area of research. This book will surely be a source of knowledge to all interested and researching the field.

In the end, I would like to express my deep sense of gratitude to all the authors for meeting the set deadlines in completing and submitting their research chapters. I would also like to thank the publisher for the support offered to us throughout the course of the book. Finally, I extend my sincere thanks to my family for being a constant source of inspiration and encouragement.

Editor

Research on Workpiece Sorting System Based on Machine Vision Mechanism

Juan Yan, Huibin Yang

College of Mechanical Engineering, Shanghai University of Engineering Science, Shanghai, China
Email: aliceyan_shu@126.com

Abstract

This paper describes industrial sorting system, which is based on robot vision technology, introduces main image processing methodology used during development, and simulates algorithm with Matlab. Besides, we set up image processing algorithm library via C# program and realize recognition and location for regular geometry workpiece. Furthermore, we analyze camera model in vision algorithm library, calibrate the camera, process the image series, and resolve the identify problem for regular geometry workpiece with different colours.

Keywords

Machine Vision, Industrial Robot, Target Recognition, Image Processing

1. Introduction

With the development of society, enterprises have become increasingly demanding on production automation, so that more and more industrial robots have been used in automated production lines. For example, in order to enhance operational efficiency in modern logistics distribution centers and to improve the efficiency of large-scale centralized distribution to reduce operating costs, as the core equipment distribution center and the main operating procedures, sorting systems and sorting operation efficiency and technology are increasingly by theorists and engineering concern. Another example is the tool change robot in the machining center. Compared to manual tool change method, this tool works more quickly and more accurately, greatly reduces tool change time and increases production efficiency. In the automotive industry, due to the auto parts more complex vehicle bodies, in many cases on the body cannot use manual welding. Currently, automobile manufacturers utilize a lot of body welding robots to improve product quality and to reduce the labor intensity. Visibly, doing the work on the production line with industrial robots is an inevitable trend.

Machine vision technology uses cameras and computers to simulate human vision features, which are widely used in electronics, aerospace, automotive and automotive parts manufacturing industry, the pharmaceutical industry and electronics fields [1]. In the traditional production line, sorting artifacts with industrial robots usually uses teaching methods or offline programming operations. All the actions and placements, etc. should be strictly preset. Once the working environment or condition changes, it will affect the efficiency and accuracy of crawling robot, unable to meet the high-speed production of large quantities. Introduction of machine vision in industrial production can greatly improve productivity and reduce costs. Industrial machine vision technology based on sorting system with high detection speed, high reliability, and high real-time, compared to traditional mechanical sorting, is more intelligent, more efficient and has other irreplaceable advantages.

This paper researches the sorting system based on the machine vision and the methods of camera calibration. The straight linear method is used and the system is simulated by Matlab. It also studied the correlation of image processing algorithms, combined thresholding proposed centroid location method to solve the different colors of regular geometric shape of the workpiece recognition problem. Threshold segmentation and edge detection method can lock the workpiece area; centroid location can accurately locate the geometric center of the workpiece. And this integrated approach can effectively solve the regular geometric shape of the workpiece sorting problem.

2. Sorting System Based on Machine Vision Hardware Architecture Artifacts

In this paper, the mechanical axis motion platform, is built on the work of sorting machine vision systems. Overall system structure is shown in **Figure 1**.

The hardware platform consists of mechanical three-axis motion platform, camera platforms, PC and motion control unit, four major components. Hardware platform package sorting system is shown in **Figure 2**.

Mechanical three-axis motion platform: Using screw-nut mechanism, while the linear guide linear motion needs as a guide and support. The system uses a servo motor drive: although the control unit is complex, the high repeat accuracy, no step phenomenon and no cumulative error [2].

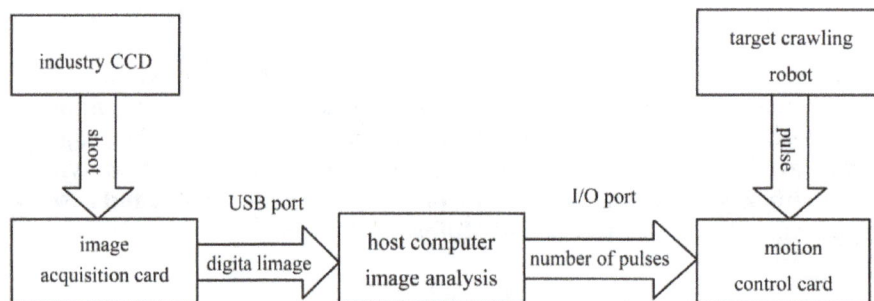

Figure1. Piece sorting system structure based on machine vision.

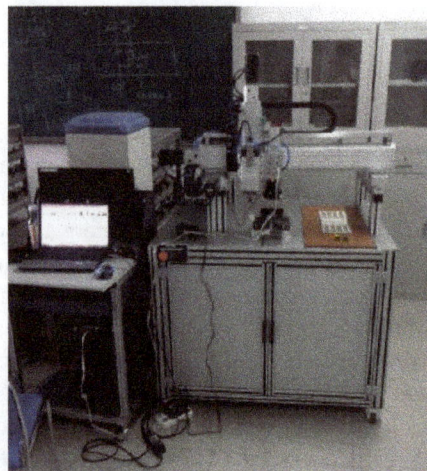

Figure 2. Workpiece sorting system based on machine vision hardware platform.

Camera Platform: Camera platform unit is mainly constructed by MV-2000UC camera and light source components. Hanging bracket role monocular camera is to obtain the experimental stage of the video image of the workpiece. White LED surface light source provides light for digital cameras captured images. The light will be fixed at the top of the workpiece in order to eliminate the shadow of the workpiece itself.

PC: Using existing laboratories as a PC, since the development environment is good, fast, easy to operate and low cost. At that time with respect to the IPC, its anti-interference ability and stability is poor, such as teaching or laboratory for stable work environment occasions. PC camera platform receives the experimental stage to obtain a video image of the workpiece using the vision system to identify the target species, the centroid of the workpiece is then calculated, and finally the image according to the relationship between the coordinate system and the object coordinate system, the calculated position of the target relative and direction, and then passes the information parameter to the control unit.

Motion Control unit: mainly control the composition of the cabinet and robots. It is responsible for the parameters analyzing, then the robot-related operations and finally the completion of the target workpiece sorting grab and place.

3. Camera Calibration

Camera calibration is one of the important parts in machine vision [3]. Calibration includes calibration camera and visual systems. First of all based model of the camera, and then to establish the relationship between the coordinate system, and finally through the calibration can establish the correspondence between image pixels and the spatial three-dimensional coordinates [4]. Workpiece sorting system is based on machine vision. The camera calibration as a key link to other related work is a prerequisite for the smooth conduct of the direct impact on subsequent sorting processes. Camera calibration process is based on the model of the camera imaging process to solve internal and external camera parameters, and these parameters determine the correspondence between the image pixel point and three-dimensional coordinate space.

There are many ways in camera calibration, the current methods commonly used in machine vision are direct linear method, perspective projection matrix method, Zhang Zhengyou France, the two-step camera calibration method. For the facts of relatively stable laboratory and low interference environment, this paper chooses the direct linear method, which is characterized by convenient computing but less accuracy. It is difficult to be precise shooting environment complex self-correcting. DLT method called direct linear method, by Abdel-Aziz and Karara and was first proposed in 1971 [5]. This method does not consider the non-linear distortion of the camera, getting the relevant parameters of the camera by directly solving a set of linear equations. It acquires pinhole imaging model for the study, ignoring specific intermediate imaging process and using a 3×4 order matrix to represent a direct correspondence between the point and the two-dimensional space object image point. The linear transformation matrix is with only a difference of a perspective matrix scale factor. Providing space point $P\left(X, Y, Z\right)$ coordinates in the camera coordinate system of $\left(x, y, z\right)$, the camera will capture the image on the CCD imaging plane, which is provided for the image point coordinates $\left(x, y\right)$ corresponding to the image pixel coordinates $\left(u, v\right)$.

By the world coordinate system to the camera coordinate system, using the following formula rotation—translation conversion formula described:

$$\begin{bmatrix} x \\ y \\ z \\ 1 \end{bmatrix} = \begin{bmatrix} R & T \\ 0 & 1 \end{bmatrix} \begin{bmatrix} X \\ Y \\ Z \\ 1 \end{bmatrix} \tag{1}$$

where in R is a rotation matrix, T is the translation matrix. In the case where distortion is not considered (*i.e.*, an ideal perspective projection), the image coordinates of the camera coordinate system is described by the following formula:

$$s\begin{bmatrix} X \\ Y \\ 1 \end{bmatrix} = \begin{bmatrix} f & 0 & 0 \\ 0 & f & 0 \\ 0 & 0 & 1 \end{bmatrix} \begin{bmatrix} x \\ y \\ z \end{bmatrix} \tag{2}$$

In the image processing, is often used in image coordinate of the pixel as a unit, so to convert the image coordinate system to the coordinate system of the pixel, *i.e.*

$$
\begin{bmatrix} u \\ v \\ 1 \end{bmatrix} = \begin{bmatrix} \dfrac{1}{d_X} & 0 & 0 \\ 0 & \dfrac{1}{d_Y} & 0 \\ 0 & 0 & 1 \end{bmatrix} \begin{bmatrix} X \\ Y \\ 1 \end{bmatrix}
\tag{3}
$$

Without considering the various types of the imaging by the specific process, can be obtained:

$$
s \begin{bmatrix} u \\ v \\ 1 \end{bmatrix} = \begin{bmatrix} r_{11} & r_{12} & r_{13} & r_{14} \\ r_{21} & r_{22} & r_{23} & r_{24} \\ r_{31} & r_{32} & r_{33} & r_{34} \end{bmatrix} \begin{bmatrix} X \\ Y \\ Z \\ 1 \end{bmatrix}
\tag{4}
$$

where, (X, Y, Z) of the three-dimensional world coordinate space points, (u, v) coordinates of the corresponding pixel, r_{ij} is a perspective view of elements of the transformation matrix. It contains three equations, finishing eliminates, obtain the following two linear equations on people r_{ij}:

$$
r_{11}X + r_{12}Y + r_{13}Z + r_{34} - uXr_{31} - uYr_{32} - uZr_{33} = ur_{34}
\tag{5}
$$

$$
r_{21}X + r_{22}Y + r_{23}Z + r_{24} - vXr_{31} - vYr_{32} - uZr_{33} = ur_{34}
\tag{6}
$$

This equation describes the relationship between the two three-dimensional world and the corresponding points between image points. If the three-dimensional world coordinates and the corresponding image coordinates are known, the transformation matrix regarded as unknown, then there will be a total of 12 unknowns. For two equations above each object point has, in general, can be set $r_{34} = 1$, then there will be a total of 11 unknowns. Taking six goals can get 12 points a overdetermined equation Using the least squares method described above can easily find solution of linear equations [6]. Using more points, we make much more than the number of equations number of unknowns, using the least squares method for solving can reduce the impact of errors caused by the perspective transformation matrix obtained after its decomposition can get cameras inside and outside the parameters.

Based on the direct linear method was calibrated to the camera. Put the calibration paper on a suitable position in the robot coordinate system, as shown in **Figure 3**. Cross figure crosshairs for the world coordinate system. Only a simple translation of the relationship is needed between it and the robot base coordinate system, *i.e.*, calculating the target position in the world coordinate system, then subtracting a certain shift amount and finally obtaining the target based on the robot base in the standard position.

Figure 3. Camera calibration chart based DLT.

Corner extraction algorithm using corner point data is shown in **Table 1**.

On the Matlab model with a blue asterisk plotted coordinates of each point (real value), the table on each corner point coordinates a left multiplication result (calculated) M-1 obtained plotted with a green plus sign, as shown in **Figure 4**.

The verification result from **Figure 4** is that, by using direct linear method in this system, the calibration results is accurate, easy to calculate and close to actual value. Besides, the calibration results are also good, which provides a reliable foundation for accurate positioning of the workpiece in a sorting system.

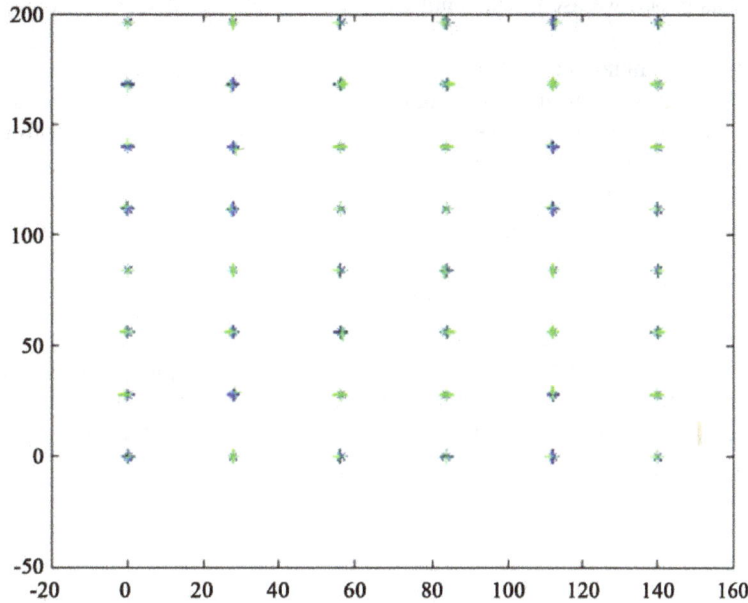

Figure 4. Camera calibration results based DLT.

Table 1. Angular point data table.

(u,v)	Column 1	Column 2	Column 3	Column 4
		Angular point data table		
Line 1	(234.5, 168)	(261.5, 166.5)	(289.5, 166.5)	(316.5, 166.5)
Line 2	(235.5, 194.5)	(262.5, 194.5)	(289.5, 193.6)	(316.5, 193.5)
Line 3	(235.5, 221.5)	(262.5, 221.5)	(289.5, 221.5)	(316.5, 220.5)
Line 4	(235.5, 248.5)	(262.5, 248.5)	(290, 248.5)	(316.5, 247.5)
Line 5	(236.5, 275.5)	(262.5, 275.5)	(290.5, 275.5)	(317, 275)
Line 6	(236.5, 302.5)	(263.5, 302.5)	(290.5, 302.5)	(317.5, 302.5)
(u,v)	Column 5	Column 6	Column 7	Column 8
Line 1	(344, 166)	(371, 166)	(397.5, 165.5)	(425, 165.5)
Line 2	(343, 193)	(370, 194)	(397.5, 192.5)	(425.5, 192.5)
Line 3	(343.5, 220.5)	(370.5, 220.5)	(398.5, 220.5)	(425.5, 219.5)
Line 4	(344, 247.5)	(371, 247.5)	(398.5, 247.5)	(425.5, 247.5)
Line 5	(344.5, 247.5)	(371.5, 274.5)	(398.5, 274.5)	(425.5, 274.5)
Line 6	(344, 301.5)	(371.5, 301.5)	(398.5, 301.5)	(398.5, 301.5)

4. Thresholding Segmentation

This design uses machine vision technology, the different colors of the workpiece intelligent sorting rules. When the camera calibration is completed, after the pixel coordinates to the scenecoordinate correspondence to obtain the target, will be on the video image is processed by computer.

First, the camera image acquired in the target workpiece is detected. Target detection is extracted from the image portion of interest, *i.e.*, the image segmentation; the image processing to the image analysis is a key step. At present, a lot of image segmentation methods, such as thresholding method, edge detection and region extraction method. In this paper, we use the thresholding method and edge detection method to detect the target workpiece.

Thresholding as the most common color distinction is a region-based image segmentation technique. Briefly, image segmentation threshold is transformed to a color image to grayscale, the first grayscale range of an image to determine the gray value threshold, then the image of the gray value of each pixel with this threshold value, and are classified according to the size of the result of the comparison, based on the threshold values of all the pixels in the image is divided into two areas.

A pair of the original image $f(x,y)$ is taken to define a single threshold value T of the divided

$$\text{image is}: g(x,y) = \begin{cases} 1 & f(x,y) > T \\ 0 & f(x,y) \le T \end{cases} \tag{7}$$

In the actual case, the difference between the image of the target workpiece and background brightness is small, so the threshold needs to be adjusted manually. **Figure 5** shows the use of maximum between-class variance method combined with manual adjustment method in Matlab to finalize the process threshold.

(a)

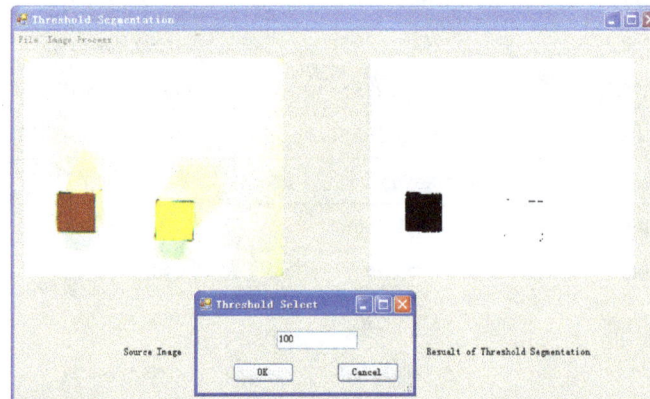

(b)

Figure 5. Threshold segmentation results. (a) When the threshold T to 80 threshold segmentation results; (b) When the threshold T to 100 threshold segmentation results.

From the segmentation results can be seen, when the threshold of T is 100, it's the best segmentation, can accurate segmentation of the target range of workpiece, which proves Ostu method combined with manual adjustment is the most suitable way for this system.

5. Determine the Centroid

In the sorting operation of the workpiece, the workpiece needs to be identified for positioning, the general centroid coordinates describing the location information of the workpiece. But the calculation method for multi-target centroid such as starting algorithms geometric center of the target group, the calculation is more complicated; through rows of pixels accumulating and averaging roughly centroid location, can not meet the precise positioning. According to the characteristics identified herein workpiece shape rules proposed rule graphics centroid location method [7].

When the function is applied to shape analysis, suppose there are binary function:

$$m_{pq} = \int_{-\infty}^{+\infty} \int_{-\infty}^{+\infty} x^p x^q f(x,y) \, dxdy \quad p, q \in N_0 = \{0,1,\cdots\} \tag{9}$$

Take $(p+q)$ as the moment of order. The moment for $M \times N$ digital image $f(i,j)$, $(p+q)$ is defined as:

$$m_{pq} = \sum_{i=0}^{M-1} \sum_{j=0}^{N-1} i^p i^q f(i,j) \tag{10}$$

0-order moment is the area of the sum of the image gray, m_{00} values represent object binary image, if used to standardize an order moments m_{10} and m_{01}, we get the coordinates of the object's center of gravity (i,j).

$$\begin{cases} \bar{i} = m_{10}/m_{00} = \sum_{i=0}^{M-1}\sum_{j=0}^{N-1} if(i,j) \Big/ \sum_{i=0}^{M-1}\sum_{j=0}^{N-1} f(i,j) \\ \bar{j} = m_{01}/m_{00} = \sum_{i=0}^{M-1}\sum_{j=0}^{N-1} jf(i,j) \Big/ \sum_{i=0}^{M-1}\sum_{j=0}^{N-1} f(i,j) \end{cases} \tag{11}$$

For binary images, grayscale only 0 and 1, the focus is the centroid. Finally, the first moment m_{10} and m_{01} is divided by zero-order moments to get the coordinates of the center of mass of the object:

$$(\bar{i}, \bar{j}) = (M_{10}/M_{00}, M_{01}/M_{00}) \tag{12}$$

After tests it is proved that the algorithm is simple and effective and is suitable for any graphics.

In this paper, the workpiece has been recognized as a rectangular, triangular, circular or a regular shape of the workpiece. The program is written in C#, in an ideal environment, workpiece recognition results are shown in **Figure 6**.

The rule workpiece centroid recognition results of **Figure 6** demonstrates that, it is simple and efficient to adopt rule centroid locolization algorithm and is suitable for any graphic. Besides, programming is simple, recognition is speedy and recognition results are accurate.

6. Conclusions

This paper describes the main algorithm in the industrial sorting machine vision technology used in the system, using thresholding to detect the target work area and describe the target workpiece position information combined centroid location method. The recognition algorithm is programmed in MATLAB environment to conduct a simulation algorithm and it is developed using C# rule artifacts recognition software. According to the results obtained, to identify the robot reaches the top of the target number of pulses required to send to the motion control card, by motion control card for controlling the motor and the drive to achieve sorting purposes.

The results show that the practical application of technical methods discussed in this article is simple and effective. The sorting system identifies the type of workpiece correct rate of 100%, error of the target location is less than 5 mm, and further validates the feasibility of the system for robot.

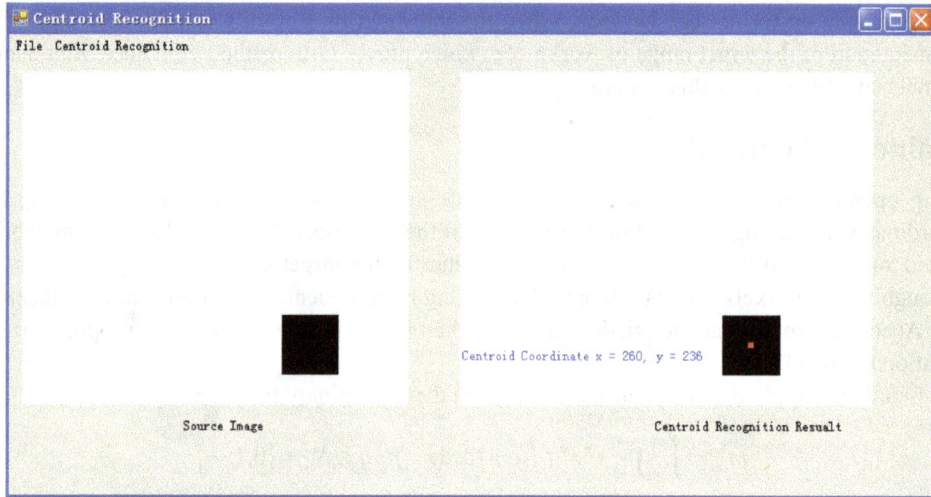

(a)

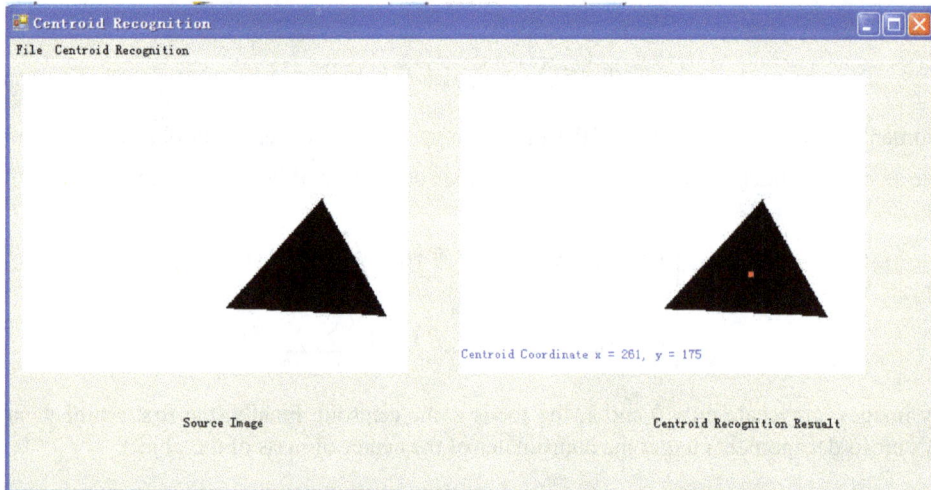

(b)

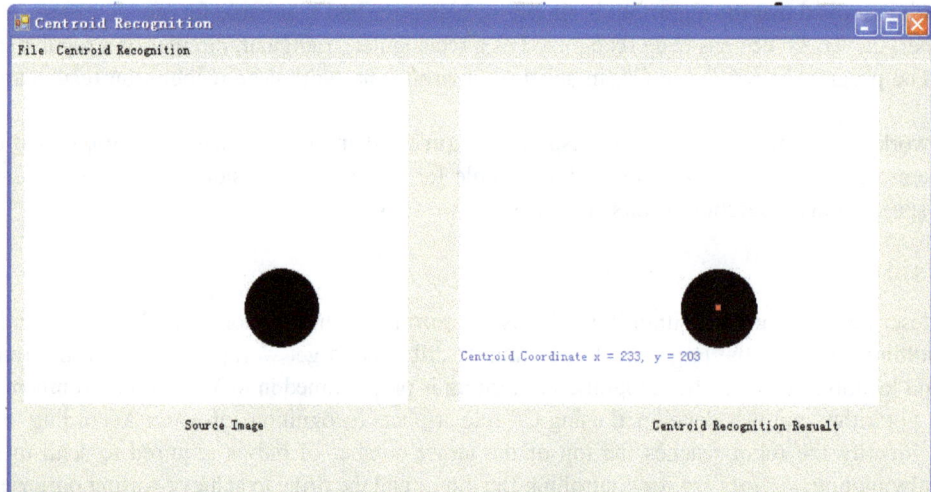

(c)

Figure 6. Rule workpiece centroid recognition results. (a) Rectangle centroid recognition results; (c) Triangle centroid recognition results; (c) Circular centroid recognition results.

References

[1]   Zhuang, K.L., Wang, J.Z. and Zhou, J. (2011) Application of Machine Vision Technology in Terms of Angle Detection. *Equipment Manufacturing Technology*, **4**, 9-11.

[2]   El Masry, G., Cubero, S. and Molto, E. (2012) In-Line Sourting of Irregular Potatoes by Using Automated Computer-Based Machine Vision System. *Journal of Food Engineering*, **112**, 60-68. http://dx.doi.org/10.1016/j.jfoodeng.2012.03.027

[3]   Liu, Z.Y., Zhao, B. and Zou, F.S. (2012) Machine Vision Technology in the Sorting of the Workpiece. *Computer Applications and Software*, **29**, 87-91.

[4]   Quan, H.H. and Zhang, L.P. (2007) Research on Machine Vision-Based Assembly Line Parts Recognition. *Modular Machine Tool & Automatic Manufacturing Technique*, **12**, 58-60.

[5]   Forsyth, D.A. and Ponce, J. (2002) Computer Vision: A Modern Approach. Prentice Hall, New Jersey.

[6]   Shapiro, R. (1976) Direct Linear Transformation Method for Three-Dimensional Cinematography. *Research Quarterly*, **49**, 197-205.

[7]   Ou, X.Y., Hou, X.Z. and Zheng, K. (2012) Machine Vision Based Automatic Teaching System of Six DOF Robot. *Manufacturing Automation*, **34**, 1-4.

Kinect-Based Humanoid Robotic Manipulator for Human Upper Limbs Movements Tracking

Mohammed Z. Al-Faiz[1], Ahmed F. Shanta[2]

[1]College of Information Engineering, Al-Nahrain University, Baghdad, Iraq
[2]Department of Computer Engineering, Al-Nahrain University, Baghdad, Iraq
Email: mzalfaiz@ieee.org, AhmedSlauf@gmail.com

Abstract

This paper presents a humanoid robotic arms controlled by tracking the human skeleton movement in real-time using Kinect upper limbbody tracking. Using Kinect tracking algorithm, the positions of upper limb arms of the body to the wrist in 3D space can be estimated by processing depth images from the Kinect. An extraction of 3D co-ordinates of the user's both arm in real-time then Arduino microcontroller is transferring the data between both of computer and the humanoid robotic arm. This method provides a way to send movement task to the humanoid robotic manipulator instead of sending the end position motion like gesture-based approaches and this method has been tested in detect, tracking and following the movement of human skeleton gesture. Designing complete prototype of a humanoid robotic arms with 4DOF three joints in shoulder and one elbow joint to the wrist that look like the Human arm Structure, Appearance and Action that represent human arm movement performed by the humanoid robotic arm. The error was and response time result generated is small (less than 4.6% and 105 ms).

Keywords

Humanoid Robotic Manipulator, Human Arm, Kinect

1. Introduction

In these days, the Robots can be helpful as possible and be able to assist humans in their everyday activities, it is proposed that they should have a human-like structure and behavior. In order to achieve such a tremendous target, Human Computer Interaction (HCI) and Human Robot Interaction (HRI) have to work closely together so as to allow the users to be able to interact with the robot's behavior in the easiest way possible, both to program the robot's functionalities as well as to control the triggering of these functionalities. When the robot is in a dangerous environment, robot human controlling may be necessary [1]. The idea of robots being able to have the abilities of a human brain is not entirely accepted. This is due to the possibility that, if robots have proper intelligence, they might turn against the human race and take over. Scientists take this into account and tend to program robots in such a way that the main target is to assist humans and make everyday life easier. Some human-robot interfaces like joysticks, dials and robot replicas, have been commonly used, but these contacting mechanical devices require unnatural arm motions to complete an operation task [2] [3]. Another way to communicate complex motions to a robot, which is more natural, is to track the operator arm motion which is used to complete the required task using contacting electromagnetic tracking sensors, inertial sensors and gloves instrumented with angle sensors. However, these contacting devices may hinder natural human-limb motions [4]-[6].

This paper presents a method of humanoid robotic arms controlled by tracking human upper body limp movement using Kinect-based 3D coordination arms to the wrist. Kinect arms tracking is used to acquire 3D skeleton position, and then it sends the data to the humanoid robotic manipulator by an Arduino interface to enable the robotic manipulator to copy the operator's arms motion in real-time. This natural way to communicate with the robot allows the operator to focus on the task instead of thinking in terms of limited separate commands that the human-robot interface can understand like gesture-based approaches [1]. Using the Kinect tracking device, **Figure 1** avoids the problem that physical sensors, cables and other contacting interfaces may hinder natural motions and that there may be marker occlusion and identification when using marker-based approaches [1].

2. Human Arm Tracking and Positioning System

Human arm detection and tracking is carried out by continuously processing depth images of an operator who is performing the arms motion to extract 3D coordination and sent it to the humanoid robotic manipulation. The RGB images and depth images are captured by the Kinect which is fixed in the front of the operator [1]. The Kinect has infrared projection that collaborate with infrared camera for depth detection and one RGB vision cameras.

3. Inverse Kinematics Model

Consider the human arm shown in **Figure 2** are modeled as a total 4-DOF, which the shoulder joint is represented by three intersecting revolute joint and elbow joint is represented by one revolute joints. From the D-H parameters it can get the four transformation matrices which transfer the movement from joint to another and they are denoted by $\left(T_{i-1}^{i}\right)$ where $(i = 1, \cdots, 4)$ **Table 1** gives the D-H parameters for human arm with

$d_1 = 25$ cm, $d_2 = 25$ cm.

The concept of inverse kinematics refers to the method of calculating the appropriate movements that a body needs to move from a point A to B. In this paper, the robotic arm is required to the observed posture of the user. As a human arms is made up of joints and bones, the joints have been treated as "start" and "end" points whereas the bones have been treated as vectors.

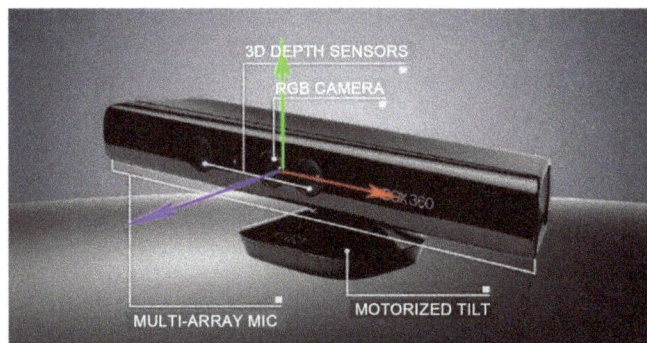

Figure 1. The Kinect device by Microsoft with the Kinect or camera reference frame.

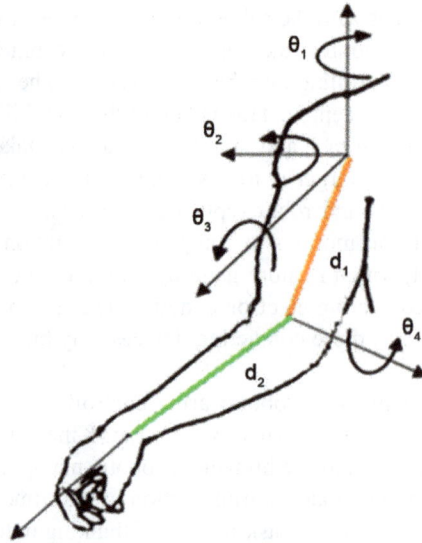

Figure 2. Human arm 4 DOF model.

Table 1. Numeric value for D-H parameters of human arm.

Frame	$\alpha_i\,(\text{rad})$	$a_i\,(\text{cm})$	$\theta_i\,(\text{rad})$	$d_i\,(\text{cm})$
1	$\dfrac{\pi}{2}$	0	θ_1	0
2	$-\dfrac{\pi}{2}$	0	θ_2	0
3	$\dfrac{\pi}{2}$	0	θ_3	d_1
4	$-\dfrac{\pi}{2}$	0	θ_4	0
5	$\dfrac{\pi}{2}$	0	θ_5	d_2

On the left side of **Figure 3** is illustrated where the Kinect is tracking a user's both arms. Points A, B and C represent the tracked shoulder, elbow and wrist joints respectively. As it can be seen on the right side of **Figure 3**, the joints A and B have been treated as "start" and "end" points respectively which have now formed a vector $\rho(AC)$. Using simple trigonometric methods, the angles $\theta_1, \theta_2, \theta_3$ and θ_4 are calculated. The equations to obtain the magnitudes of $\rho(AC)$, angles equations given below:

Step 1: Finding vector AB, BC and AC.

$$AB = \sqrt{(x_2 - x_1)^2 + (y_2 - y_1)^2} \tag{1}$$

Step 2: Finding $\theta_1, \theta_2, \theta_3$ and θ_4 angles between the vectors as following:

$$\theta_1 = \sin^{-1}\left(\frac{AB_x}{\sqrt{AB_x^2 + AB_y^2 + AB_z^2}}\right) \tag{2}$$

$$\theta_2 = \sin^{-1}\left(\frac{AB_y}{\sqrt{AB_x^2 + AB_y^2 + AB_z^2}}\right) \tag{3}$$

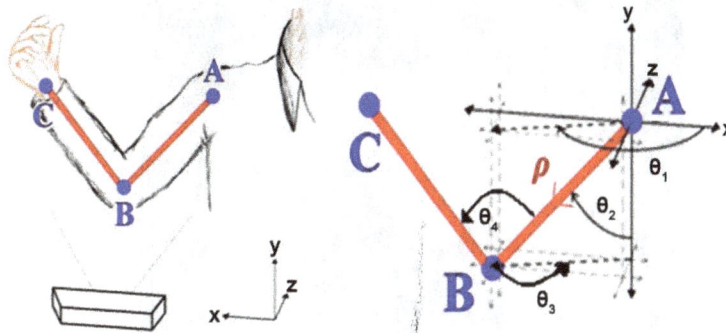

Figure 3. Geometric inverse kinematics.

$$\theta_3 = \sin^{-1}\left(\frac{BC_x}{\sqrt{BC_x^2 + BC_y^2 + BC_z^2}}\right) \tag{4}$$

$$\theta_4 = \cos^{-1}\left(AC^2 - AB^2 - BC^2\right) \tag{5}$$

Those functions has been implemented where the X, Y and Z coordinates of the joints A, B and C are passed as arguments and the vector AB, BC and AC values are returned. The function is then passed the vectors and calculate requires angles. This function is run on every frame (image) received, *i.e.* every 0.03 s as the system runs at 30 FPS. For every frame, the angles values are then sent to the Arduino using serial communication so as for the robot to move accordingly.

4. Designing Robot Manipulation System

The humanoid robotic arm was designed to have four degrees of freedom from shoulder up to and including the forearm for each arms. Each degree of freedom has been represented an angle of movement that can be performed by the human arm. A servo motor has been used to implement each angle of movement ($\theta_1 \rightarrow \theta_4$). As the shoulder joint feature 3D motion and elbow joint feature 1D motion, pan and tilt mechanisms [7] have been implemented so as to allow servo motors at each joint to perform the required motion. The arm's bones have been summed to two parts, the upper arm (shoulder to elbow) and the lower arm (elbow to wrist). As it can be seen in the complete humanoid robotic arm design **Figure 4**. The upper and lower arm parts have been represented and printing by using 3D printer. So as to represent the human body, a rectangular chest shape (center pillar) has been used to elevate the arm to a height of 460 mm from its base. The chest had been design so that Servo A can fitted exactly in the hole inside the chest indicating the first joint, the shoulder. All servo motors have been fitted to the plastic parts by using bolts and nuts to allow future modifications while maintaining consistency through the structure. Finally overall system was implement as shown in **Figure 5**.

5. Experiments

To evaluate the Kinect and humanoid robotic arms operation algorithm described in this paper, we use Processing language to develop a humanoid robotic interface system and this system is used to operate the a four-axis humanoid robot. This experimental system includes:

1) Use the human arm tracking and following system to get the arm skeleton 3D positions arm points, and then calculate the required Invers kinematic angles of the arms.

2) Humanoid robotic manipulation system based on the joint angles which are calculated through inverse kinematic and then they will be transmitted to the real robot to control it.

In the experiment, the operator stand in front of the Kinect to control the robotic arm. After the angles been calculated by invers kinematics algorithm the data are send by USB cable to the robot through an Arduino interface microcontroller and measure values can be taken from potentiometer sensor build inside the servo motors.

Arduino communication with pc without any delaying to ensure that data is correct functionality. Some jittering and excessive oscillations in the structure of humanoid robotic arm encourages oscillations due to the arm

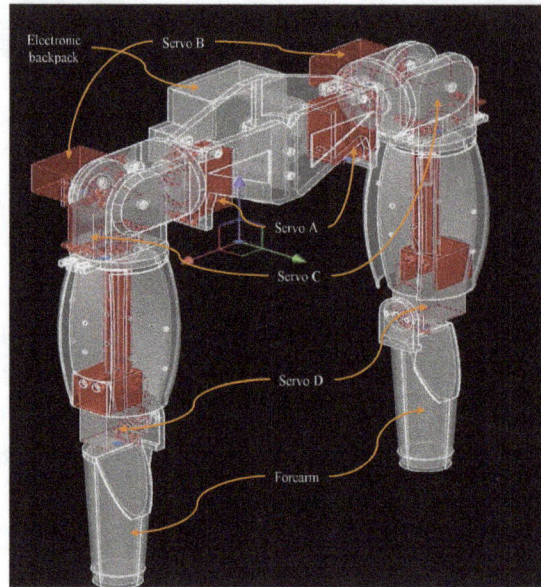

Figure 4. Humanoid roboticmanipulators.

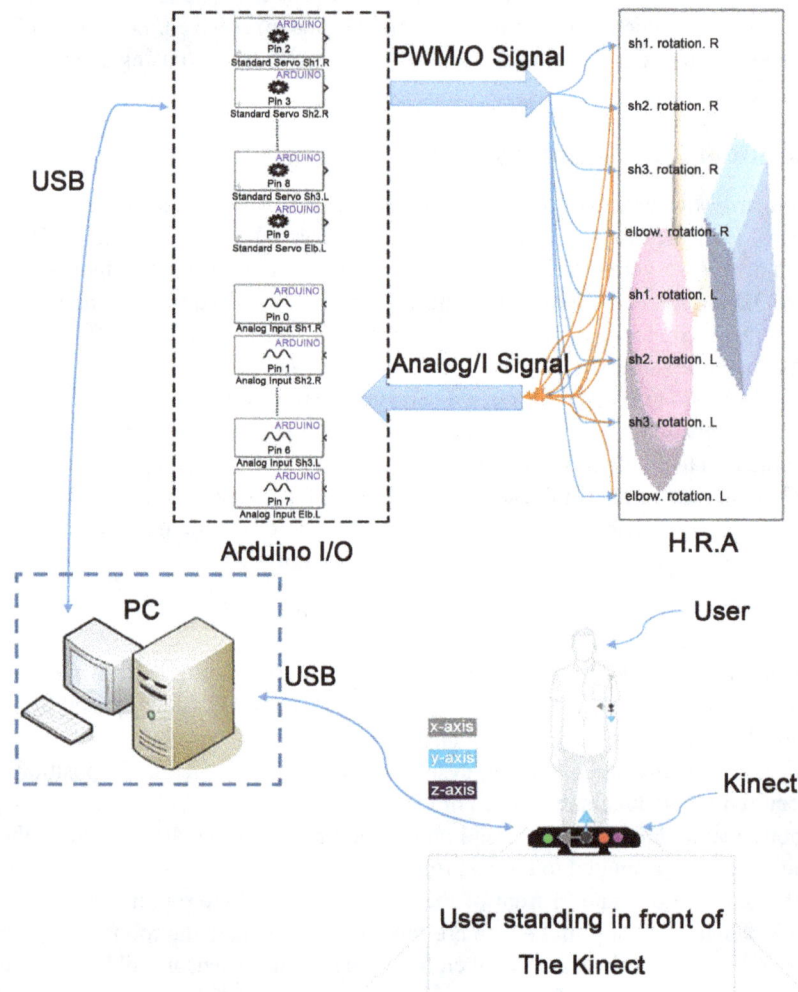

Figure 5. System architecture.

being suspended only from servo A without any additional support. Jittering is present due to the hypersensitivity of the skeletal tracking algorithms. Controlling and error tolerance servos feature an embedded PID controller proved to work efficiently and effectively minimizes rotational errors and overshoots. By the structure consisting of only four servos and embedded PID controllers on each individual servo, the control of the robotic arm is straightforward and robust. The use of a single medium used between the Windows PC and the servos (*i.e.* the Arduino), allows low latency times between the time of user input and the time of output performed by the robotic arm.

The quality of output performed by the humanoid robotic arm is justified by its similarity to the movement of that of the human arm. Fluidity and human like smoothness is required so as for the developed structure to be considered humanoid. **Figure 6** shows the entire system data flowchart and how information is passing through every component. Even though the actuators consist of pre-calibrated built in PID controllers, the output movement can be characterized to have a fast settling time but yet a critically damped approach to the given settling point.

The fluctuations visible within the structure refer to a poor construction link between servo A and servo B. This is due to the fact that the pan and tilt mechanism (containing servo B and the rest of the arm), connects to servo A by screwing on the servo's shaft. As the diameter of the servo shaft is small (5 mm), it fails to keep the arm in a sturdy posture but still allowing full control of the arm's movements. The rest of the arm construction comprises of solid connections between the servos themselves as well as tight fittings between the servos and the chassis.

6. Result

Next ten cases were used to test the developed system controller in different posture **Figure 7** from 1-10 positions, the result in **Table 2**, the table contains in the first column the posture number according to **Figure 7**, the second column contains the desired angles (calculated angles by IK Equation (1)-(5)) in degree, the third column contains measure, actual angles that the robotic arms move to in degree. The respond time was calculated as the time required to competed desire posture position. The error percentage was calculated as Equation (6). **Figure 8** shows the position of the robot's end-effector and the operator's arm during operation experiments for posture number 7 in **Figure 7** and **Table 2**, the dashed red line represents the end-effector's path. The solid blue line represents the path of the operator's arm, the values in the Seventh position in **Table 2** was taken for the worse result represent by a point in **Figure 8** for each angle.

$$e_g\% = \sqrt{\Delta\theta_1^2 + \Delta\theta_2^2 + \Delta\theta_3^2 + \Delta\theta_4^2} \qquad (6)$$

Table 2. Result of arm movement.

Position	Desired Angles $[\theta_1,\theta_2,\theta_3,\theta_4]$	Measure Angles $[\theta_1,\theta_2,\theta_3,\theta_4]$	Error Percentage
1	[87,139,102,38]	[88,144,100,38]	5.47%
2	[85,171,173,10]	[86,171,173,11]	1.41%
3	[165,101,93,11]	[160,100,97,10]	6.55%
4	[93,86,80,14]	[90,85,80,14]	3.16%
5	[20,109,51,82]	[20,110,54,82]	3.16%
6	[79,151,70,84]	[76,145,69,84]	6.78%
7	[59,97,84,81]	[52,94,86,81]	7.87%
8	[13,102,94,17]	[16,101,94,17]	3.74%
9	[95,170,87,88]	[95,173,85,88]	3.60 %
10	[165,102,90,12]	[163,105,90,13]	3.74%

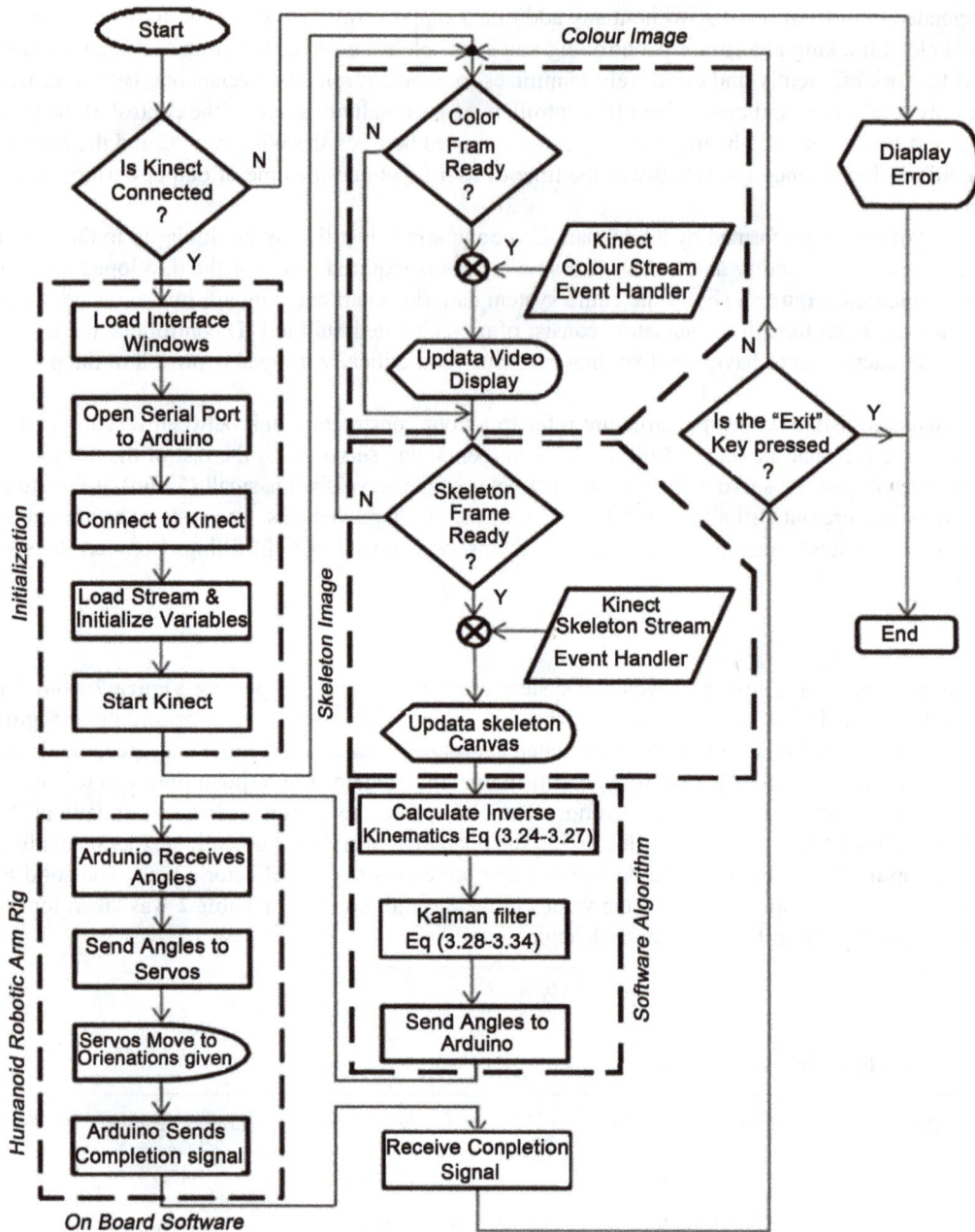

Figure 6. Data system flowchart.

7. Conclusions

The aim of designing such a structure was to replicate the movement of the human arm (from the shoulder down to the wrist joint). Such movement is capable and proven to be capable by the developed humanoid robotic arm through the extensive testing carried out. The possible scenarios were chosen to investigate the accuracy and validity of the skeletal tracking algorithms performed by the Microsoft Kinect sensor.

This work proves that 3D humanoid robot control does not necessarily require the user to utilize any specialized devices attached to their bodies so as to perform 3D tracking of their movements. In addition, this work allows comprehension of the function of the Microsoft Kinect sensor and analyses the extent to which this may be used for general Human Robot Interaction purposes.

The human skeletal tracking algorithms appear to be operating normally but during some of the testing

Figure 7. Cases study.

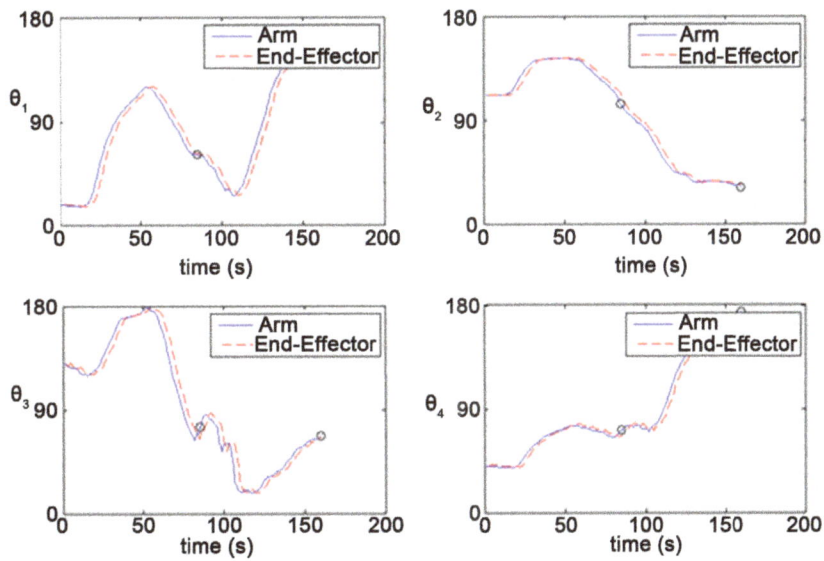

Figure 8. Analysis of the experiment for position 7.

undertaken, weaknesses have appeared. When the user moves their arm in front any other part of the body (e.g. other arm, main body, head or legs), the accuracy of the tracked joints drops significantly, causing the system to perform erratically indicating system instability in these cases. The algorithm seems to be affected when tracked joints get close together or overlap. Such an issue can be considered as one of the main weaknesses of the system. Even though such events are only experienced in extreme cases, they are still possible to occur and cause instabilities in the system.

To complete the complex tasks, multi-Kinect will be used to work together, more DOF, full size humanoid and wireless communication in future work.

References

[1]   Du, G.L., Zhang, P., Mai, J.H. and Li, Z.L. (2012) Markerless Kinect-Based Hand Tracking for Robot Teleoperation. *International Journal of Advanced Robotic Systems*, **9**, 36.

[2]   Yussof, H., Capi, G., Nasu, Y., Yamano, M. and Ohka, M. (2011) A CORBA-Based Control Architecture for Real-Time Teleoperation Tasks in a Developmental Humanoid Robot. *International Journal of Advanced Robotic Systems*, **8**, 29-48.

[3]   Mitsantisuk, C., Katsura, S. and Ohishi, K. (2010) Force Control of Human-Robot Interaction Using Twin Direct-Drive Motor System Based on Modal Space Design. *IEEE Transactions on Industrial Electronics*, **57**, 1338-1392. http://dx.doi.org/10.1109/TIE.2009.2030218

[4]   Hirche, S. and Buss, M. (2012) Human-Oriented Control for Haptic Teleoperation. *Proceedings of the IEEE*, **100**, 623-647. http://dx.doi.org/10.1109/JPROC.2011.2175150

[5]   Villaverde, A.F., Raimundez, C. and Barreiro, A. (2012) Passive Internet-Based Crane Teleoperation with Haptic Aids. *International Journal of Control Automation and Systems*, **10**, 78-87.

[6]   Wang, Z., Giannopoulos, E., Slater, M., Peer, A. and Buss, M. (2011) Handshake: Realistic Human-Robot Interaction in Haptic Enhanced Virtual Reality. *Presence-Teleoperators and Virtual Environments*, **20**, 371-392. http://dx.doi.org/10.1162/PRES_a_00061

[7]   Xu, D. and Acosta, A. (2005) An Analysis of the Inverse Kinematics for a 5-DOF Manipulator. *International Journal of Automation and Computing*, **2**, 114-124. http://dx.doi.org/10.1007/s11633-005-0114-1

3

Robust Sliding Mode Control for Nonlinear Discrete-Time Delayed Systems Based on Neural Network

Vishal Goyal, Vinay Kumar Deolia, Tripti Nath Sharma

Department of Electronics and Communication Engineering, G. L. A. University, Mathura, India
Email: vishal.glaitm@gmail.com, vinayk.deolia@gmail.com, tns11@yahoo.co.in

Abstract

This paper presents a robust sliding mode controller for a class of unknown nonlinear discrete-time systems in the presence of fixed time delay. A neural-network approximation and the Lyapunov-Krasovskii functional theory into the sliding-mode technique is used and a neural-network based sliding mode control scheme is proposed. Because of the novality of Chebyshev Neural Networks (CNNs), that it requires much less computation time as compare to multi layer neural network (MLNN), is preferred to approximate the unknown system functions. By means of linear matrix inequalities, a sufficient condition is derived to ensure the asymptotic stability such that the sliding mode dynamics is restricted to the defined sliding surface. The proposed sliding mode control technique guarantees the system state trajectory to the designed sliding surface. Finally, simulation results illustrate the main characteristics and performance of the proposed approach.

Keywords

Discrete-Time Nonlinear Systems, Lyapunov-Krasovskii Functional, Linear Matrix Inequality (LMI), Sliding Mode Control (SMC), Chebyshev Neural Networks (CNNs)

1. Introduction

Time delay is undesirable parameter which is often encountered in various engineering systems, such as mechanical systems, chemical systems, and so on. The time delay degrades the system performance and leads to instability of the system. As a design tool for robust motion control system, SMC has been well designed for a wide range of nonlinear systems in both continuous time and discrete time. SMC is robust to parametric uncertainties and insensitive to unknown disturbance. SMC has been studied in seventies by the name of variable structure control (VSC) [1] [2]. VSC, is characterized by discontinuous feedback control law which switches the system in a predefined subspace [3]. Its implementation by a digital system requires sampling interval which leads to chattering. Over the past few decades, considerable attention has been reported to the stability analysis of continuous time delay systems by using different approaches [4]-[10]. In [4], robust controller has been designed for continuous time delay system using Ricacati equation approach. Considerable, attention has been given using LMI approach for stabilization of continuous time delay systems [5]-[8]. An adaptive control approach has been proposed for the control of time delay system [9] [10]. A discrete SMC (DSMC) is important when we implement robust control digitally with slow sampling rate. It is important to note that DSMC cannot be obtained from its continuous counterpart by simple conversion. Since modeling inaccuracy and external conditions lead to uncertainties, disturbances and nonlinearities in systems. Hence, the stability analysis of uncertain discrete-time delay systems have been studied over past few years with different control approaches have been well documented in [11] and reference therein. Moreover, in the above papers, the unknown nonlinearities have not been investigated. In [12], a robust control of uncertain nonlinear state delayed system, which gives a conservative condition of control, is presented. In recent years, many papers have reported the problem of SMC for state delay uncertain systems [13]-[17]. Most of these papers for uncertain time delay systems involve norm-bound nonlinearities which are treated as external disturbances. Adaptive multilayer neural control schemes for the control of complex nonlinear systems have shown great results over past few years. Now, it is an established fact that unknown nonlinear functions can be approximated from neural network. Neural network appears a powerful tool for nonlinear control problems [18]-[20]. In [21], the SMC have been used for control of uncertain state-delay system with unknown nonlinearity. In this work, Chebyshev Neural Network is used to estimate the unknown nonlinearity and linear matrix inequalities (LMI) conditions were derived to ensure the asymptotic stability on the defined sliding surface. Adaptive SMC for a class of discrete nonlinear systems was proposed. The proposed controller uses switching function with adaptive term to reduce the problem of chattering. Artificial neural network (ANN) was used for approximation of modeling errors. The nonlinearity is strictly positive and bounded away from zero [22]. In [23], a new SMC has been used to control the unknown nonlinear discrete-time systems. The chattering is reduced as compared to normal discrete-time sliding mode control using time varying gain.

This paper proposes a discrete-time sliding mode controller for a class of state delay nonlinear discrete systems. The unknown nonlinear functions in system dynamics is approximated using Chebyshev Neural Networks (CNNs). New weight update laws are derived to make this scheme adaptive. The stability of state delay system is taken care by carefully selecting Lyapunov-Krasovskii functional candidate. Thus conservative, sufficient conditions were derived which was represented by an appropriate set of LMIs.

The paper organization is as follows. Section 2 presents the CNN structure. Problem formulation and preliminaries are elaborated in Section 3. Section 4 presents controller design is stated in detail. The stability analysis is presented in Section 5. The effectiveness of proposed scheme is validates through simulation results in Section 6. The note ends with concluding remarks in Section 7.

Notations: $\|.\|$ denotes Euclidean norm, $\|.\|_F$ implies Frobenius norm. The $\text{tr}(.)$ stands for trace of matrix.

2. CNN Structure

An ANN is a simple interconnected group of nonlinear elements, which has the capability to represents nonlinear functions. The representation accuracy depends on the ANN complexity, *i.e.*, the number of elements and the way in which they are interconnected [24].

There is different ANN configuration available, like feed forward network such as multilayer perceptron (MLP), radial basis function (RBF) networks, Chebyshev neural network (CNN) etc. The MLP network has certain disadvantage that it requires a large amount of computation for learning. The RBF network can effectively learn from discontinuities and local variations. The problem with this network is choosing an appropriate set of RBF centers for effective learning. A single-layer functional link artificial neural network (FLANN) in which the need of hidden layer is eliminated by expanding the input pattern using Chebyshev polynomials. The main advantage of this network is that it requires much less computation as compared to a multilayer perceptron (MLP).

CNN is a functional link network (FLN) based on Chebyshev polynomials. CNN architecture has two main parts, namely, numerical transformation and learning [25]. In numerical transformation we use finite set of Chebyshev polynomials as a functional expansion (FE) of input pattern. The learning part is a functional-link neural network based on Chebyshev polynomials. The Chebyshev polynomials can be obtained by a recursive formula

$$T_{i+1}(x) = 2xT_i(x) - T_{i-1}(x), \quad T_0(x) = 1 \tag{1}$$

where, $T_1(x)$ are Chebyshev polynomials, i is the order of polynomials chosen and here x is a scalar quantity. The different choices of $T_1(x)$ are $x \& 2x$.

The output of single layer neural network is given by

$$\hat{g}(x) = \hat{w}^{\mathrm{T}}\phi \tag{2}$$

where, w are the weights and ϕ is the suitable basis function of neural network. Based on the approximation property of CNN [27]-[30], there exist ideal weights w, so that the function $g(x)$ to be approximated can be represented as

$$g(x) = w^{\mathrm{T}}\phi + \varepsilon \tag{3}$$

where, ε is the CNN functional reconstruction error vector and $\|\varepsilon\| \leq \varepsilon_N$ is bounded.

3. Problem Formulation

Consider the following discrete-time state delay system as in [26] (**Figure 1**)

$$x(k+1) = Ax(k) + A_d x(k-h) + g(x(k))u(k) \tag{4}$$

where, $x(k) \in R^n$ and $u(k) \in R^m$ denote the state and input vectors respectively. A and A_d are real constant matrices with appropriate dimensions. $g(x(k))$ is a unknown nonlinear function of a given system in Equation (4), and h is a positive number representing delay.

For the system given in (4) the sliding mode controller is obtained as

$$u(k) = -\frac{1}{\hat{g}(x(k))}\left[Ax(k) + A_d x(k-h)\right] \tag{5}$$

where $\hat{g}(x(k))$ is the approximated value of the nonlinear function.

The objective of this work is to guarantee the stability of sliding mode controller in Equation (5) of the nonlinear system Equation (4), so that the system stays on the sliding surface.

4. Controller Design

The first step in the design of discrete-time SMC control algorithm would be the design of sliding surface. The linear sliding surface is defined as:

$$s(k) = Cx(k) \tag{6}$$

where $C \in R^{m \times n}$ is a real matrix of appropriate dimensions.

For a system to be asymptotically stable, the sliding surface is defined as follows.

$$s(k) = 0 \tag{7}$$

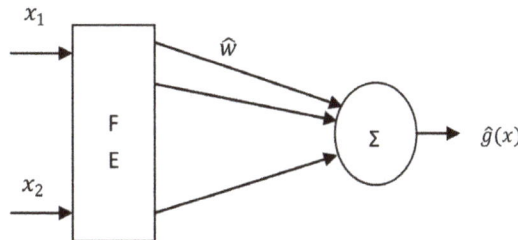

Figure 1. Chebyshev neural network.

The second step is to design a control law which can guarantee the sliding mode reaching condition of the given linear sliding surface. The obtained control law is given in Equation (5) will force the trajectory of the system to move towards the sliding surface monotonically and causes zigzag motion around the sliding surface.

5. Stability Analysis

The following assumptions are needed for the stability analysis of the given unknown nonlinear system [26].

Assumption 1: The state delay h is a constant time delay that is basically induced by the network transmission. For constant time delay the lower and upper bounds are assumed to be identical.

Assumption 2: The nonlinear function $g(x(k))$ in the system is unknown and bounded.

Assumption 3: (Bounded Ideal NN Weights): The ideal NN weights w are bounded so that $\|w\| \le w_M$, with w_M a known bound. The symbol $\|\cdot\|_F$ denotes the Frobenius norm, *i.e.* given a matrix A, the Frobenius norm is given by,

$$\|A\|_F^2 = \mathrm{tr}\left(A^\mathrm{T} A\right)$$

Assumption 4: Let $\tilde{g}(x(k)) = G\hat{g}(x(k))$, where $G = G^\mathrm{T}$ is a $n \times n$ symmetric matrix, and $\tilde{g}(x(k))$ and $\hat{g}x(k)$ are the n-column vectors.

Theorem 1:

Given the system in Equation (4) and Assumptions 1 - 4, sliding mode control law Equation (5), the estimated NN weights are given by

$$\hat{w}(k+1) = \left[\left(\hat{w}(k)\right) + \left(x^\mathrm{T}(k)(Q-P-Z)x(k)\right)^{1/2} + \left(\hat{M}^\mathrm{T}(x(k))Z\hat{M}(x(k))\right)^{1/2}\right] \tag{8}$$

with the condition are

$$4\left(\left(\hat{w}(k)\right)\left(\hat{M}^\mathrm{T}(x(k))Z\hat{M}(x(k))\right)^{1/2}\left(x^\mathrm{T}(k)(Q-P-Z)x(k)\right)^{1/2}\right)^{1/2} > 0 \tag{9}$$

$$2\left(x^\mathrm{T}(k)(Q-P-Z)x(k)\right)^{1/2}\left(\hat{M}^\mathrm{T}(x(k))Z\hat{M}(x(k))\right)^{1/2} > 0 \tag{10}$$

Suppose there exist an $n \times n$ positive-definite matrix P, an $n \times n$ nonnegative-definite matrix Q, an $n \times n$ nonnegative-definite matrix z and $n \times n$ symmetric matrix G such that   following LMI holds,

$$H1) = \begin{bmatrix} A^\mathrm{T}G^\mathrm{T}PGA - A^\mathrm{T}G^\mathrm{T}zGA - A^\mathrm{T}G^\mathrm{T}z - zGA & A^\mathrm{T}G^\mathrm{T}PGA_d - A^\mathrm{T}G^\mathrm{T}zGA_d - GzA_d \\ * & A_d^\mathrm{T}G^\mathrm{T}PGA_d - A_d^\mathrm{T}G^\mathrm{T}zGA_d - Q \end{bmatrix} < 0 \tag{11}$$

Thus by properly selecting the control gain and the design parameters, the state trajectory is reaching on the designed sliding surface.

Proof: Choose Lyapunov-Krasovskii functional candidate,

$$V(k) = V_1(k) + V_2(k) + V_3(k) + V_4(k) \tag{12}$$

where

$$V_1(k) = x^\mathrm{T}(k)Px(k) \tag{13}$$

$$V_2(k) = \sum_{i=k-h}^{k-1} x^\mathrm{T}(i)Qx(i) \tag{14}$$

$$V_3(k) = \mathrm{tr}\left(\tilde{w}^\mathrm{T}(k)\tilde{w}(k)\right) \tag{15}$$

$$V_4(k) = \eta^\mathrm{T}(k)Z\eta(k) \tag{16}$$

And

$$\eta(k) = x(k+1) - x(k) \tag{17}$$

Substituting Equations (13)-(16) in Equation (12)

$$V(k) = x^{\mathrm{T}}(k)Px(k) + \sum_{i=k-h}^{k-1} x^{\mathrm{T}}(i)Qx(i) + \mathrm{tr}\left(\tilde{w}^{\mathrm{T}}(k)\tilde{w}(k)\right) + \eta^{\mathrm{T}}(k)Z\eta(k) \tag{18}$$

and

$$V(k+1) = x^{\mathrm{T}}(k+1)Px(k+1) + \sum_{i=k-h+1}^{k} x^{\mathrm{T}}(i)Qx(i) + \mathrm{tr}\left(\tilde{w}^{\mathrm{T}}(k+1)\tilde{w}(k+1)\right) + \hat{M}^{\mathrm{T}}(x(k))Z\hat{M}(x(k)) \tag{19}$$

where

$$\eta(k+1) = \hat{M}(x(k)) \tag{20}$$

Since P, Z is a positive-definite and Q is a nonnegative-definite, $V(k)$ is then positive-definite. Therefore,

$$\Delta V(k) = V(k+1) - V(k) \tag{21}$$

Substituting Equation (18) and (19) in Equation (21),

$$\begin{aligned}
\Delta V(k) = {}& x^{\mathrm{T}}(k+1)Px(k+1) + \sum_{i=k-h+1}^{k} x^{\mathrm{T}}(i)Qx(i) + \mathrm{tr}\left(\tilde{w}^{\mathrm{T}}(k+1)\tilde{w}(k+1)\right) + \hat{M}^{\mathrm{T}}(x(k))Z\hat{M}(x(k)) \\
& - x^{\mathrm{T}}(k)Px(k) - \sum_{i=k-h}^{k-1} x^{\mathrm{T}}(i)Qx(i) - \mathrm{tr}\left(\tilde{w}^{\mathrm{T}}(k)\tilde{w}(k)\right) - \eta^{\mathrm{T}}(k)Z\eta(k).
\end{aligned} \tag{22}$$

Substituting Equation (5) in Equation (21) and using Assumption 3 in Equation (22)

$$\begin{aligned}
\Delta V(k) = {}& \left[Ax(k) + A_d x(k-h) + \tilde{g}(x(k))u(k) + \hat{g}(x(k))u(k) \right] \\
& \times P\left[Ax(k) + A_d x(k-h) + \tilde{g}(x(k))\ u(k) + \hat{g}(x(k))u(k) \right] \\
& + x^{\mathrm{T}}(k)Qx(k) + \hat{M}^{\mathrm{T}}(x(k))Z\hat{M}(x(k)) - x^{\mathrm{T}}(k)Px(k) - x^{\mathrm{T}}(k-h)Qx(k-h) \\
& - \left[x(k+1) - x(k) \right]^{\mathrm{T}} Z\left[x(k+1) - x(k) \right] + \left\| \tilde{w}^{\mathrm{T}}(k+1) \right\|^2 - \left\| \tilde{w}^{\mathrm{T}}(k) \right\|^2.
\end{aligned} \tag{23}$$

After some mathematical manipulations in Equation (23),

$$\begin{aligned}
\Delta V(k) = {}& x^{\mathrm{T}}(k)Qx(k) + \hat{M}^{\mathrm{T}}(x(k))Z\hat{M}(x(k)) - x^{\mathrm{T}}(k)Px(k) - x^{\mathrm{T}}(k-h)Qx(k-h) - x^{\mathrm{T}}(k)Zx(k) \\
& + \left\| \tilde{w}^{\mathrm{T}}(k+1) \right\|^2 - \left\| \tilde{w}^{\mathrm{T}}(k) \right\|^2 + Z(k).
\end{aligned} \tag{24}$$

where

$$\begin{aligned}
Z(k) = {}& x^{\mathrm{T}}(k)A^{\mathrm{T}}P\tilde{g}(x(k))u(k) + x^{\mathrm{T}}(k-h)A_d^{\mathrm{T}}P\tilde{g}(x(k))u(k) + u^{\mathrm{T}}(k)\hat{g}^{\mathrm{T}}(x(k))P\tilde{g}(x(k)) \\
& + u^{\mathrm{T}}(k)\tilde{g}^{\mathrm{T}}(x(k))PAx(k) + u^{\mathrm{T}}(k)\tilde{g}^{\mathrm{T}}(x(k))PA_d^{\mathrm{T}}x(k-h) + u^{\mathrm{T}}(k)\tilde{g}^{\mathrm{T}}(x(k))P\hat{g}(x(k))u(k) \\
& + u^{\mathrm{T}}(k)\tilde{g}^{\mathrm{T}}(x(k))P\tilde{g}(x(k))u(k) - x^{\mathrm{T}}(k)A^{\mathrm{T}}Z\tilde{g}(x(k))u(k) - x^{\mathrm{T}}(k-h)A_d^{\mathrm{T}}Z\tilde{g}(x(k))u(k) \\
& - u^{\mathrm{T}}(k)\hat{g}^{\mathrm{T}}(x(k))Z\tilde{g}(x(k)) - u^{\mathrm{T}}(k)\tilde{g}^{\mathrm{T}}(x(k))ZAx(k) - u^{\mathrm{T}}(k)\tilde{g}^{\mathrm{T}}(x(k))ZA_d^{\mathrm{T}}x(k-h) \\
& - u^{\mathrm{T}}(k)\hat{g}^{\mathrm{T}}(x(k))Z\hat{g}(x(k))u(k) - u^{\mathrm{T}}(k)\tilde{g}^{\mathrm{T}}(x(k))Z\tilde{g}(x(k))u(k) + u^{\mathrm{T}}(k)\tilde{g}^{\mathrm{T}}(x(k))Zx(k) \\
& + x^{\mathrm{T}}(k)Z\tilde{g}(x(k))u(k).
\end{aligned} \tag{25}$$

Collecting the terms together and substitute control law Equation (5) in Equation (24) yields

$$\begin{aligned}
\Delta V(k) = {}& -\left\| \hat{w}(k+1) \right\|^2 + \left[\left(\hat{w}(k)\right) + \left(x^{\mathrm{T}}(k)(Q-P-Z)x(k) \right)^{1/2} + \left(\hat{M}^{\mathrm{T}}(x(k))Z\hat{M}(x(k)) \right)^{1/2} \right]^2 \\
& + S(k) + Z(k)
\end{aligned} \tag{26}$$

where

$$S(k) = -2\big(\hat{w}(k)\big)\big(\hat{M}^{\mathrm{T}}\big(x(k)\big)Z\hat{M}\big(x(k)\big)\big)^{1/2} - 2\big(\hat{w}(k)\big)\big(x^{\mathrm{T}}(k)(Q-P-Z)x(k)\big)^{1/2}$$
$$-2\big(x^{\mathrm{T}}(k)(Q-P-Z)x(k)\big)^{1/2}\big(\hat{M}^{\mathrm{T}}\big(x(k)\big)Z\hat{M}\big(x(k)\big)\big)^{1/2} - x^{\mathrm{T}}(k-h)Qx(k-h). \tag{27}$$

with tuning law in Equation (8) , Equation (26) will be,

$$\Delta V(k) = S(k) + Z(k) \tag{28}$$

substituting Equations (25), (27) in Equation (28)

$$\Delta V(k) = -2\big(\hat{w}(k)\big)\big(\hat{M}^{\mathrm{T}}\big(x(k)\big)Z\hat{M}\big(x(k)\big)\big)^{1/2} - 2\big(\hat{w}(k)\big)\big(x^{\mathrm{T}}(k)(Q-P-Z)x(k)\big)^{1/2}$$
$$-2\big(x^{\mathrm{T}}(k)(Q-P-Z)x(k)\big)^{1/2}\big(\hat{M}^{\mathrm{T}}\big(x(k)\big)Z\hat{M}\big(x(k)\big)\big)^{1/2}$$
$$+ x^{\mathrm{T}}(k)\big[A^{\mathrm{T}}G^{\mathrm{T}}PGA - A^{\mathrm{T}}G^{\mathrm{T}}ZGA - A^{\mathrm{T}}G^{\mathrm{T}}Z - ZGA\big]x(k)$$
$$+ x^{\mathrm{T}}(k)\big[A^{\mathrm{T}}G^{\mathrm{T}}PGA_d - A^{\mathrm{T}}G^{\mathrm{T}}ZGA_d - ZGA_d\big]x(k-h)$$
$$+ x^{\mathrm{T}}(k-h)\big[A_d^{\mathrm{T}}G^{\mathrm{T}}PGA - A_d^{\mathrm{T}}G^{\mathrm{T}}ZGA - A_d^{\mathrm{T}}ZG^{\mathrm{T}}\big]x(k)$$
$$+ x^{\mathrm{T}}(k-h)\big[A_d^{\mathrm{T}}G^{\mathrm{T}}PGA_d - A_d^{\mathrm{T}}G^{\mathrm{T}}ZGA_d - Q\big]x(k-h). \tag{29}$$

Manipulating the nonquadratic terms using the following inequality $\sqrt{ab} \le \dfrac{a+b}{2}$ (which turns into equality if and only if $a = b$ we get,

$$\Delta V(k) \le -4\left(\big(\hat{w}(k)\big)\big(\hat{M}^{\mathrm{T}}\big(x(k)\big)Z\hat{M}\big(x(k)\big)\big)^{1/2}\big(x^{\mathrm{T}}(k)(Q-P-Z)x(k)\big)^{1/2}\right)^{1/2}$$
$$-2\big(x^{\mathrm{T}}(k)(Q-P-Z)x(k)\big)^{1/2}\big(\hat{M}^{\mathrm{T}}\big(x(k)\big)Z\hat{M}\big(x(k)\big)\big)^{1/2}$$
$$+ x^{\mathrm{T}}(k)\big[A^{\mathrm{T}}G^{\mathrm{T}}PGA - A^{\mathrm{T}}G^{\mathrm{T}}ZGA - A^{\mathrm{T}}G^{\mathrm{T}}Z - ZGA\big]x(k)$$
$$+ x^{\mathrm{T}}(k)\big[A^{\mathrm{T}}G^{\mathrm{T}}PGA_d - A^{\mathrm{T}}G^{\mathrm{T}}ZGA_d - ZGA_d\big]x(k-h)$$
$$+ x^{\mathrm{T}}(k-h)\big[A_d^{\mathrm{T}}G^{\mathrm{T}}PGA - A_d^{\mathrm{T}}G^{\mathrm{T}}ZGA - A_d^{\mathrm{T}}ZG^{\mathrm{T}}\big]x(k)$$
$$+ x^{\mathrm{T}}(k-h)\big[A_d^{\mathrm{T}}G^{\mathrm{T}}PGA_d - A_d^{\mathrm{T}}G^{\mathrm{T}}ZGA_d - Q\big]x(k-h). \tag{30}$$

where

$$a = -2\big(\hat{w}(k)\big)\big(\hat{M}^{\mathrm{T}}\big(x(k)\big)Z\hat{M}\big(x(k)\big)\big)^{1/2} \tag{31}$$

$$b = -2\big(\hat{w}(k)\big)\big(x^{\mathrm{T}}(k)(Q-P-Z)x(k)\big)^{1/2} \tag{32}$$

In Equation (30), $\Delta V(k)$ is guaranteed to remain negative as long as

$$\begin{bmatrix} A^{\mathrm{T}}G^{\mathrm{T}}PGA - A^{\mathrm{T}}G^{\mathrm{T}}zGA - A^{\mathrm{T}}G^{\mathrm{T}}z - zGA & A^{\mathrm{T}}G^{\mathrm{T}}PGA_d - A^{\mathrm{T}}G^{\mathrm{T}}zGA_d - GzA_d \\ * & A_d^{\mathrm{T}}G^{\mathrm{T}}PGA_d - A_d^{\mathrm{T}}G^{\mathrm{T}}zGA_d - Q \end{bmatrix} < 0$$

Since the first two terms in Equation (30) are satisfying the condition in Equation (9), (10), next four terms are satisfying the LMI in Equation (11). Therefore, we conclude that the system in Equation (4) is stable with control law Equation (5) and LMI in Equation (11).

6. Simulation Results

In this section, a numerical example is presented to validate the performance and effectiveness of the nonlinear discrete-time system proposed in Equation (4). Consider the set of parameters for the given system

$$A = \begin{bmatrix} 0.8 & 0.4 \\ -0.14 & 0.1 \end{bmatrix}, \quad A_d = \begin{bmatrix} 0.2 & 0 \\ 0 & 0.1 \end{bmatrix}$$

and

$$g\big(x(k)\big) = \begin{bmatrix} \dfrac{1.4x_1^2(k)}{1+x_1^2(k)} \\[2ex] \dfrac{x_1(k)}{1+x_1^2(k)+x_2^2(k)} \end{bmatrix}$$

The fixed time delay is assumed to be $h = 2$. The proposed system has initial condition of states x_1 and x_2 are chosen as $\begin{bmatrix} 0.01 & 0,02 \end{bmatrix}^T$. The LMI in H1) are solved by using Matlab LMI Toolbox and the values of P, Q, z and G are obtained as

$$P = \begin{bmatrix} 108.8411 & 0 \\ 0 & 108.8411 \end{bmatrix}, \quad Q = 10^{-11} \begin{bmatrix} 0.1040 & 0.0338 \\ 0.0338 & -0.0186 \end{bmatrix}$$

$$G = 10^{-10} \begin{bmatrix} -0.1198 & 0.0001 \\ 0.0001 & 0.0247 \end{bmatrix}, \quad z = 10^{-11} \begin{bmatrix} 0.3485 & -0.0103 \\ -0.0103 & 0.4639 \end{bmatrix}$$

The trajectories of the system states x_1 and x_2 are shown in **Figure 2** and **Figure 3**. It is observed in **Figure 2** and **Figure 3** that the states of the sliding motion approach to zero quickly. **Figure 4** demonstrates that the controller robustly stabilizes the system by sliding mode technique with fixed time delay. The simulation results confirm the stability of the system and show the effectiveness of proposed scheme.

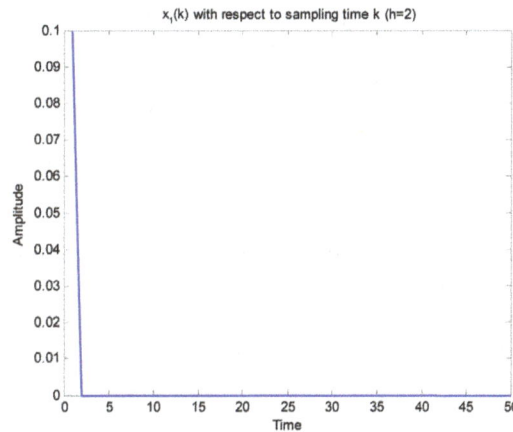

Figure 2. $x_1(k)$ with respect to sample time k $(h = 2)$.

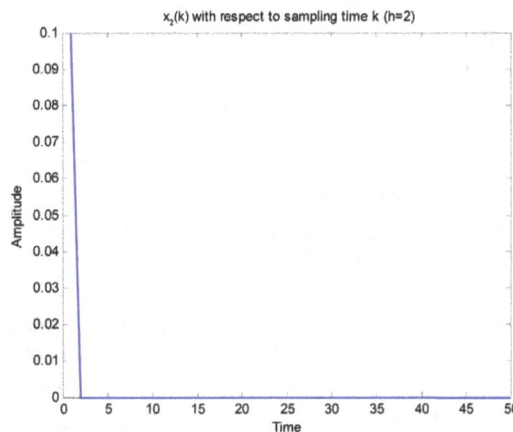

Figure 3. $x_2(k)$ with respect to sample time k $(h = 2)$.

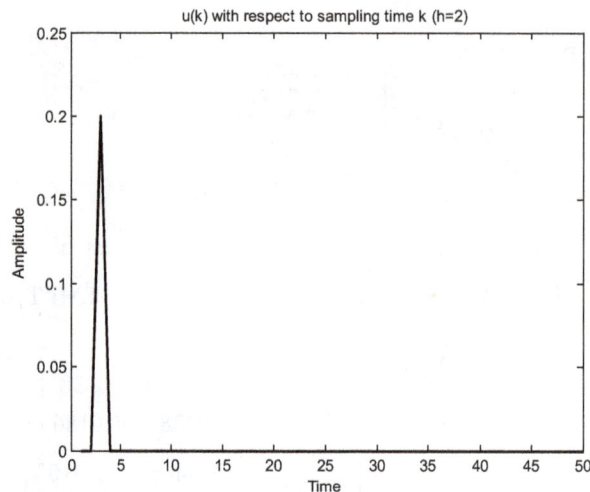

Figure 4. $u(k)$ with respect to sample time k $(h = 2)$.

7. Conclusion

In this paper, a sliding mode control for a class of unknown nonlinear discrete-time system is proposed ,which results in small chattering motion in both control signal and system output. A Chebyshev Neural Network is used to approximate the unknown system dynamics. A new learning algorithm for neural network approximation is proposed. This neural network based sliding mode control approach guarantees the system state trajectory to the defined sliding surface. An LMI based sufficient condition for the asymptotic stability of the sliding mode dynamics is derived by means of a Lyapunov-Krasovskii approach. Simulation results are validating the effectiveness of proposed scheme.

References

[1]   Utkin, V.I. (1977) Variable Structure Systems with Sliding Modes. *IEEE Transactions on Automatic Control*, **AC-22**, 212-222. http://dx.doi.org/10.1109/TAC.1977.1101446

[2]   Utkin, V.I. (1978) Sliding Modes and Their Applications in Variable Structure Systems. Nauka, Moscow.

[3]   Sarpturk, S.Z., Istefanopolos, Y. and Kaynak, O. (1987) On the Stability of Discrete-Time Sliding Mode Control Systems. *IEEE Transactions on Automatic Control*, **32**, 930-932. http://dx.doi.org/10.1109/TAC.1987.1104468

[4]   Jeung, E.T., Oh, D.C., Kim, J.H. and Park, H.B. (1996) Robust Controller Design for Uncertain Systems with Time Delays, LMI Approach. *Automatica*, **32**, 1229-1231. http://dx.doi.org/10.1016/0005-1098(96)00055-6

[5]   Yue, D. (2004) Robust Stabilization of Uncertain Systems with Unknown Input Delay. *Automatica*, **40**, 331-336. http://dx.doi.org/10.1016/j.automatica.2003.10.005

[6]   Basin, M.V., Perez, J., Acosta, P. and Fridman, L. (2006) Optimal Filtering for Nonlinear Polynomial Systems over Linear Observations with Delay. *International Journal of Innovative Computing Information and Control*, **2**, 863-874.

[7]   Boukas, E.K. and Al-Muthairi, N.F. (2006) Delay-Dependent Stabilization of Singular Linear Systems with Delays. *International Journal of Innovative Computing Information and Control*, **2**, 283-291.

[8]   Chen, M., Lam, J. and Xu, S. (2006) Memory State Feedback Guaranteed Cost Control for Neutral Delay Systems. *International Journal of Innovative Computing Information and Control*, **2**, 293-303

[9]   Ge, S.S., Hong, F. and Lee, T.H. (2004) Adaptive Neural Control of Nonlinear Time-Delay Systems with Unknown Virtual Control Coefficients. *IEEE Transactions on Systems, Man, and Cybernetics, Part B: Cybernetics*, **34**, 449-516. http://dx.doi.org/10.1109/TSMCB.2003.817055

[10]   Ge, S.S., Hong, F. and Lee, T.H. (2003) Adaptive Neural Network Control of Nonlinear Systems with Unknown Time Delays. *IEEE Transactions on Automatic Control*, **48**, 2004-2010. http://dx.doi.org/10.1109/TAC.2003.819287

[11]   Xia, Y. and Jia, Y. (2003) Robust Sliding-Mode Control for Uncertain Time-Delay Systems: An LMI Approach. *IEEE Transactions on Automatic Control*, **48**, 1086-1092. http://dx.doi.org/10.1109/TAC.2003.812815

[12]   Wang, Z., Huang, B. and Unbehauen, H. (1999) Robust Reliable Control for a Class of Uncertain Nonlinear State-Delayed System. *Automatica*, **35**, 955-963. http://dx.doi.org/10.1016/S0005-1098(98)00233-7

[13] Shyu, K. and Yan, J. (1993) Robust Stability of Uncertain Time-Delay Systems and Its Stabilization by Variable Structure Control. *International Journal of Control*, **57**, 237-246. http://dx.doi.org/10.1080/00207179308934385

[14] Roh, Y.H. and Oh, J.H. (1999) Robust Stabilization of Uncertain Input-Delay Systems by Sliding Mode Control with Delay Compensation. *Automatica*, **35**, 1861-1865. http://dx.doi.org/10.1016/S0005-1098(99)00106-5

[15] Basin, M.V., Gonzalez, J.R. and Fridman, L. (2003) Optimal and Robust Sliding Mode Control for Linear Systems with Time Delays in Control Input. *International Journal of Pure and Applied Mathematics*, **5**, 395-420.

[16] Xia, Y., Liu, G.P., Shi, P., Chen, J. and Rees, D. (2008) Robust Delay-Dependent Sliding Mode Control for Uncertain Time-Delay Systems. *International Journal of Robust and Nonlinear Control*, **18**, 1142-1161. http://dx.doi.org/10.1002/rnc.1272

[17] Yan, M. and Shi, Y. (2008) Robust Discrete-Time Sliding Mode Control for Uncertain Systems with Time—Varying State Delay. *IET Control Theory & Applications*, **2**, 662-674. http://dx.doi.org/10.1049/iet-cta:20070460

[18] Zhang, T., Ge, S.S. and Hang, C.C. (1999) Design and Performance Analysis of a Direct Adaptive Controller for Nonlinear Systems. *Automatica*, **35**, 1809-1817. http://dx.doi.org/10.1016/S0005-1098(99)00098-9

[19] Chen, S. and Billings, S.A. (1992) Neural Networks for Nonlinear Dynamic System Modelling and Identification. *International Journal of Control*, **56**, 319-346. http://dx.doi.org/10.1080/00207179208934317

[20] Jagannathan, S. and Lewis, F.L. (1996) Identification of Nonlinear Dynamical Systems Using Multilayered Neural Networks. *Automatica*, **36**, 1707-1712. http://dx.doi.org/10.1016/S0005-1098(96)80007-0

[21] Niu, Y., Lam, J., Wang, X. and Ho, D.W.C. (2003) Sliding-Mode Control for Nonlinear State-Delayed Systems Using Neural-Network Approximation. *IEEE Proceedings—Control Theory and Applications*, **150**, 233-239.

[22] Munoz, D. and Sbarbaro, D. (2000) An Adaptive Sliding-Mode Controller for Discrete Nonlinear Systems. *IEEE Transaction on Industrial Electronics*, **47**, 574-581. http://dx.doi.org/10.1109/41.847898

[23] de Jesús Rubio, J. and Yu, W. (2006) Discrete-Time Sliding-Mode Control Based on Neural-Networks. Springer-Verlag, Berlin, 956-961.

[24] Patra, J.C. and Kot, A.C. (2002) Nonlinear Dynamic System Identification Using Chebyshev Functional Link Artificial Neural Networks. *IEEE Transaction on Systems, Man and Cyberntics*, **32**, 505-511.

[25] Purwar, S., Kar, I.N. and Jha, A.N. (2007) Nonlinear System Identification Using Neural Networks. *IETE Journal of Research*, **53**, 35-42. http://dx.doi.org/10.1080/03772063.2007.10876119

[26] Deolia, V.K., Purwar, S. and Sharma, T.N. (2012) Stabilization of Unknown Nonlinear Discrete-Time Delay Systems Based on Neural Network. *Intelligent Control and Automation*, **3**, 337-345.

[27] Namatame, A. and Ueda, N. (1992) Pattern Classification with Chebyshev Neural Network. *International Journal Neural Network*, **3**, 23-31.

[28] Lee, T.T. and Jeng, J.T. (1998) The Chebyshev Polynomial Based Unified Model Neural Networks for Functions Approximations. *IEEE Transactions on Systems, Man & Cybernetics, Part B*, **28**, 925-935. http://dx.doi.org/10.1109/3477.735405

[29] Purwar, S., Kar, I.N. and Jha, A.N. (2008) Adaptive Output Feedback Tracking Control of Robot Manipulators Using Position Measurements Only. *Expert Systems with Applications*, **34**, 2789-2798. http://dx.doi.org/10.1016/j.eswa.2007.05.030

[30] Purwar, S., Kar, I.N. and Jha, A.N. (2005) On-Line System Identification of Complex Systems Using Chebyshev Neural Networks. *Applied Soft Computing*, **7**, 364-372. http://dx.doi.org/10.1016/j.asoc.2005.08.001

4

DXF File Identification with C# for CNC Engraving Machine System

Huibin Yang, Juan Yan

College of Mechanical Engineering, Shanghai University of Engineering Science, Shanghai, China
Email: webin@sues.edu.cn

Abstract

This paper researches the main technology of open CNC engraving machine, the DXF identification technology. Agraphic information extraction method is proposed. By this method, the graphic information in DXF file can be identified and transformed into bottom motion controller's code. So the engraving machine can achieve trajectory tracking. Then the open CNC engraving machine system is developed with C#. At last, the method is validated on a three axes motion experiment platform. The result shows that this method can efficiently identify the graphic information including line, circle, arc etc. in DXF file and the CNC engraving machine can be controlled well.

Keywords

DXF, CNC Engraving Machine, GALIL, C#

1. Introduction

With the development of pattern recognition techniques, modern CNC engraving machine needn't be programmed manually. By importing graphics file, the corresponding shape will be engraved by the machine immediately. The operating process of the machine is simplified enormously, and the rich programming knowledge is no longer need for operators. Among them, DXF identification is a key technology of CNC engraving machine. By reading and recognition of the DXF file, the machining track can be directly generated, so the motion control of the CNC engraving machine can be achieved.

2. Research Status

Researchers have done a lot of researches on how to contact CAD software to NC code. Omirou and Barouni-proposed a series of machine codes, with which the advanced programming ability is integrated into the control of modern CNC milling machine system [1]. Kovacic and Brezocnik proposed the concept of which using the genetic algorithm to program the CNC machine based on the CAD model under manufacturing environment [2]. But some problems are still existed in this kind of CNC programming (such as the artificial participation degree is higher and the efficiency is lower).

The research direction of Chinese researchers mainly includes two aspects. One is the theoretical study of DXF file and NC machining, the other is the application of DXF file reading. ZhaiRui and Zhang Liang proposed a program structure, which is used to read data information of DXF file and do some preprocess based on the cross platform open source library DXF Lib by the analysis of DXF file structure characteristic [3]. Huang Jieqiong and Yuan Qun wrote the interface program to read the stored parts graphic information in DXF file by use of the object-oriented secondary development tools, Object ARX and C++, in the research of stamping parts machining. The stamping parts geometric model is automatically created by the automatic generation algorithm of closed contour [4].

3. DXF File and Graphic Information Extraction

3.1. DXF File

DXF (Drawing Exchange File) is a representation of all information labeled data contained in the AutoCAD graphics file, and the ASCII or binary file format of AutoCAD file. It can be used as input/output interface and graphics file exchange between AutoCAD and other graphics applications [5].

A complete DXF file is composed of six segments called SECTION. These segments were HEADER, CLASSES, TABLES, BLOCKS, ENTITIES and file ending character (group code is 0, group value is EOF). The DXF file structure and meaning of each segment is shown in **Figure 1**.

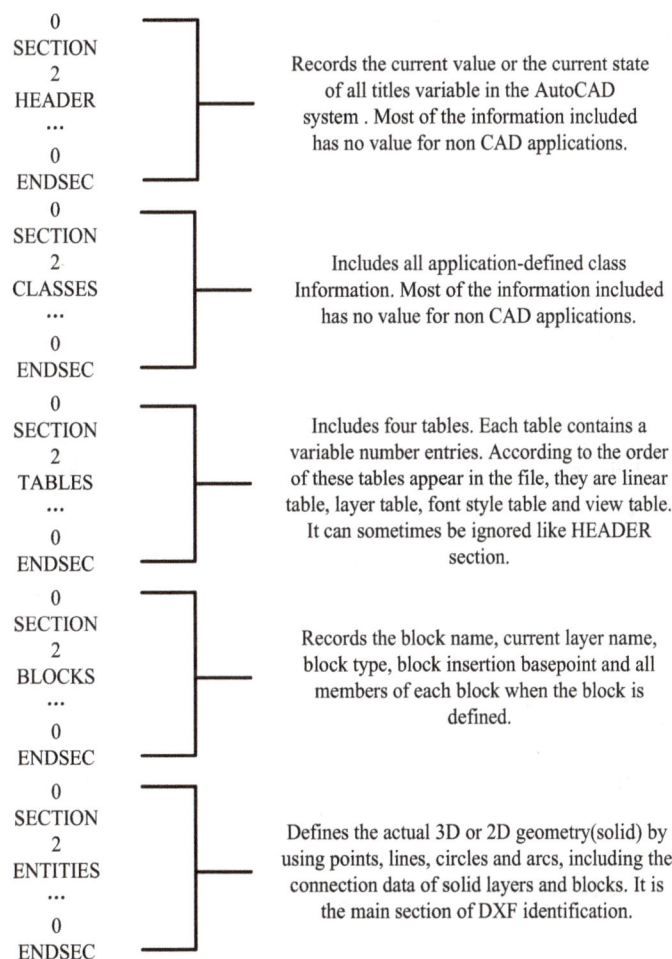

Figure 1. DXF file structure.

3.2. Graphic Information Extraction Method

In order to extract useful information of the graphic, many parts in the file can be ignored. The corresponding geometric description can be completed as long as the sections of TABLES, BLOCK, ENTITIES are obtained. Each graphic element in the DXF file are stored with a fixed format, so it is convenient for data exchange, and also called its readability. The characteristics of each individual graphic element in DXF file is described by the parameter (group) consisted by paired group code and group value. Therefore, according to the target of open CNC engraving machine, it is enough to describe the target geometry contour by reading the ENTITIES section in DXF files only. The particular identification process is: First search the DXF file until the "ENTITLES" is found, then build a graphic element object. Then search the graphic element type (LINE, CIRCLE, ARC), and search the corresponding value followed by the group code. For example, if the program has found the ENTITLES section and confirm the first graphic element is LINE (The program found "LINE" after "ENTITLES"). Then it will search the group code which represents the parameters of the line. The number at the next line after the group code is the value of the parameter.(e.g. The number at the next line after "10" represent the X value of start point of this line, and "20" for Y value of start point, "11" for X value of end point, "21" for Y value of end point, etc.). **Table 1** shows an example of an ENTITIES section.

By getting these parameters and values, system then "sees" the graph and "knows" the specific parameters of the graph which is drew by AutoCAD. **Figure 2** is the flow diagram of extraction of graphic information.

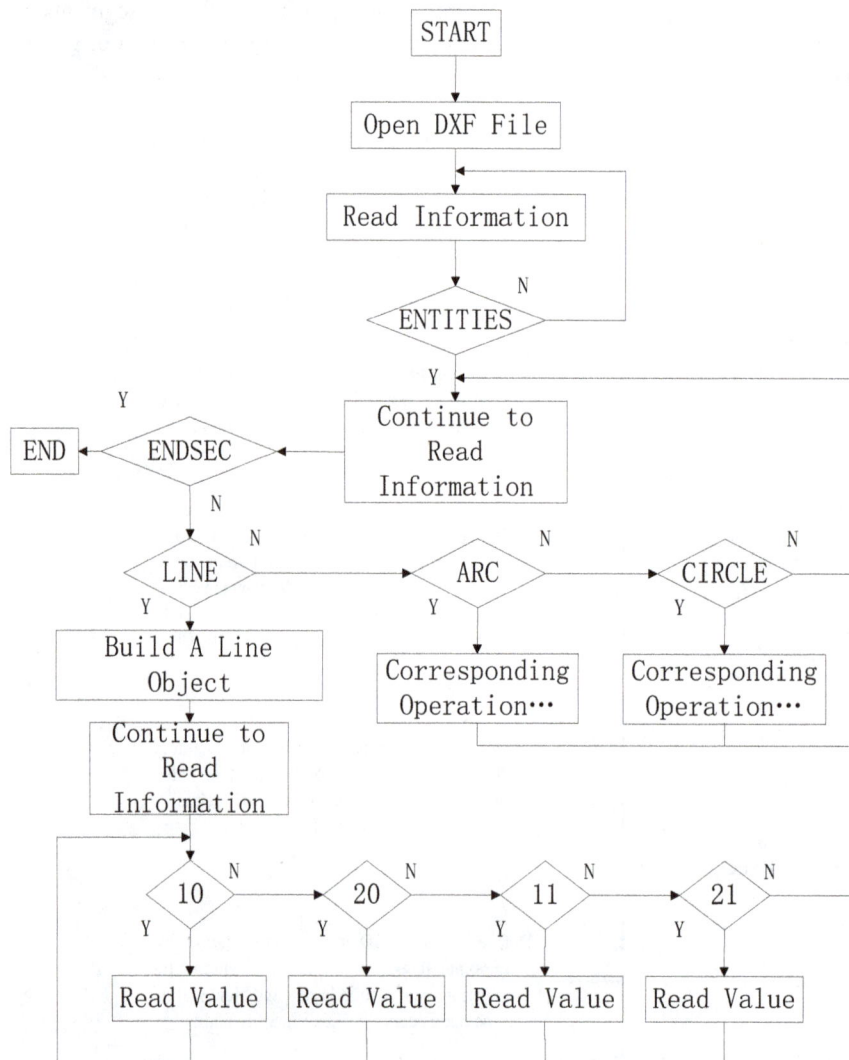

Figure 2. Flow diagram of extraction of graphic information.

Table 1. An example of an ENTITIES section.

Parts of DXF File	Explanation
ENTITIES	Section name
0	Group code
LINE	Graphic element type
…	
10	
50.0	The X value of start point
20	
100.0	The Y value of start point
30	
0.0	The Z value of start point
11	
350.0	The X value of end point
21	
500.0	The Y value of end point
31	
0.0	The Z value of end point
0	
ENDSEC	Section end symbol

3.3. C# Realization of Graphic Information Extraction

In order to store the graph data, the convenient method is to store numeric variables by using array, and it is also very convenient for call and assignment operation. First define a 2D array:

$s[i, j]$ ($i \leq 100$, $j \leq 20$), define a 100 lines and 20 lows array at initialization, in which, every line i stores a graphic element, every element j in a line stands for the value after the group code. The format and meaning are shown in **Table 2**.

Then, the graphic element storage state is $s[i, 0]$, $s[i, 1]$, …, $s[i, 15]$ ($I = 0, 1, 2, …$).

The advantage of this design is: For each graphic element, all the geometric elements associated with the trajectory can be stored in an array variable space which has a fixed serial number. It is convenient and not easy to make mistakes in the calculation or logical judgment. But to any entire graphics trajectory, the number of lines or curves is not consistent, so it is important to apply for enough variable memory space to adapt to different requirements of graphics trajectory.

Part of the C# program of reading arc graphic element in DXF are as follows：

```
do
  {
    Line = mysr. ReadLine ();
    if (Line == "ENTITIES")
    {
          ……
        if (Line == "10")
          {
             Line = mysr.ReadLine();
string m;
             m = Line;
double n;
             n = Convert.ToDouble(m);
             s[i, j] = n;
             j++;
  }
  ……
} while (Line! = null)
```

Table 2. Data storage format table.

Array Variable Position	Data Meaning
[i, 0]	Property mark: 1 is line, 2 is circle, 3 is arc
[i, 1]	The X axis coordinate value of start point
[i, 2]	The Y axis coordinate value of start point
[i, 3]	The Z axis coordinate value of start point
[i, 4]	The X axis coordinate value of end point
[i, 5]	The Y axis coordinate value of end point
[i, 6]	The Z axis coordinate value of end point
[i, 7]	The X axis coordinate value of the centre of a circle or an arc
[i, 8]	The Y axis coordinate value of the centre of a circle or an arc
[i, 9]	The Z axis coordinate value of the centre of a circle or an arc
[i, 10]	The radius value of a circle or an arc
[i, 11]	Arc start angle
[i, 12]	Arc end angle
[i, 13]	The identification number of subsequent data sorting process
[i, 14]	For subsequent operations
[i, 15]	For subsequent operations

4. Graphics Trajectory Generation

To open CNC engraving machine, the key point is how to convert the graphic element information in DXF file into motion controller code, so as to control the machine's motion according to the machining trajectory.

4.1. DXF Analysis Principle

The so-called DXF analysis is the standardization of each graphic element which has been read in order to according with the standard instructions of motion controller. Considering the basic type of graphic element is line, circle or arc, the standardization requirements of different graphic element type are different. The specific principles are as follows:

1) LINE

Line has only start and end point coordinate. According to **Table 2**, the actual useful memory space is s[i, 0], s[i, 1],…, s[i, 6], other parts are all zero.

2) ARC

As the format of arc in the DXF is include the center coordinates value, radius, start angle and end angle. So the center coordinates value, radius, start angle and end angle can be recognized and stored in s[i, 7], s[i, 8], …, s[i, 12], according to **Table 2**. But for the GALIL DMC2143 motion controller which is used in the open CNC engraving machine, the arc instruction requires start and end point coordinate and rotation angle of the arc. So, the analysis of arc includes two aspects: a) Calculate the start and end point coordinate. b) Calculate the rotation angle and store in s [i, 15].

3) CIRCLE

Because the rotation angle of circle is 360°, it can be set as a fixed value. For the sake of convenience, the starting position of circle is set to the left or right quadrantal points.

4.2. DXF Analysis Method

According to 3.1, the difficulty of graphic element analysis is arc. Although the information in DXF file can confirm geometry feature, for the track sequencing, the start and end point coordinates are needed; and for the motion controller programs, it also need to change the format for direct connection. By four elements of center, radius, start angle and end angle as well as simple trigonometric function calculation, the start and end point position as well as the rotation angle of the arc can be determined. For example, if center of the arc is o(x_0, y_0), radius is r, start angle is θ ($0 < \theta < 90°$) and end angle is δ ($0 < \theta < 90°$), according to the parametric equation of the circle, the start point a(x_1, y_1), end point b(x_2, y_2), and rotation angle ε can be calculated using Equation (1) to Equation (3):

$$\begin{cases} x_1 = x_0 + r \cdot \cos\theta \\ y_1 = y_0 + r \cdot \sin\theta \end{cases} \tag{1}$$

$$\begin{cases} x_2 = x_0 + r \cdot \cos\delta \\ y_2 = y_0 + r \cdot \sin\delta \end{cases} \tag{2}$$

$$\varepsilon = \delta - \theta \tag{3}$$

Parts of the C# program of arc analysis are as follows:

```
for (h = 0; h < shu; h++)
{ if(s[i,0] == 3)
    { ...
      if (s[h, 11] > 180 && s[h, 11] < 270)
      {
        s[h, 1] = s[h, 7] - Math.Cos(s[h, 11] - 180) * s[h, 10];
        s[h, 2] = s[h, 8] - Math.Sin(s[h, 11] - 180) * s[h, 10];
      ......
```

5. Development of Open CNC Engraving Machine System

The hardware of the open CNC engraving machine system includes a motion controller and an upper computer (PC). The real-time control of the CNC engraving machine body is done by the motion controller. The main task of the motion controller is servo motor control and IO logic control. The PC runs The DXF analysis algorithm, Human-Machine Interface (HMI) and sends the motion control instructions got from the DXF analysis algorithm to the motion controller, so the engraving machine can be controlled.

The software of the system includes PC program and motion controller program.

5.1. PC Program

The PC program includes HMI and DXF analysis program running in the background. DXF analysis program are mainly programmed based on DXF analysis principles and methods on 3. The HMI developed by C# is shown in **Figure 3**.

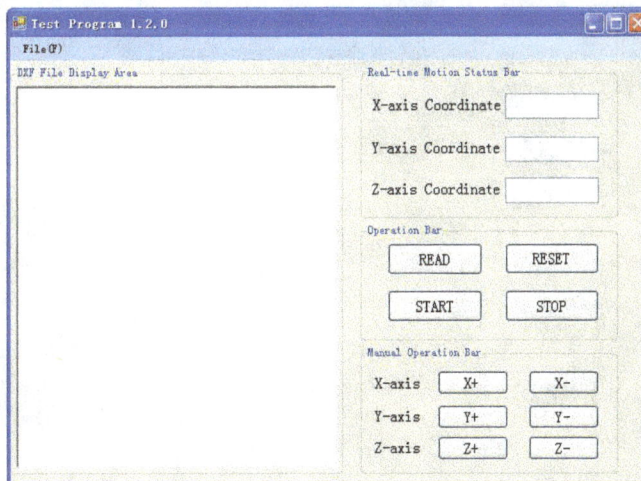

Figure 3. Human-machine interface.

5.2. Program Design of Motion Controller

In this design, the subprograms of linear and circular interpolation are programmed in GALIL motion controller. According to the results of DXF analysis in PC, call different subprogram in proper order and assign variable, the continuous tracking trajectory can be realized. The linear interpolation program of GALIL motion controller is as follows:

```
#LINEAR          //the subprogram name of linear interpolation
MT 2, 2          // specify the type of motor
VMAB             // specify the motion plane
VS 5000          //specify the vector speed of 5000 counts/sec
VA 100000        // specify the vector acceleration of 100000 counts
VD 100000        // specify the vector deceleration of 100000 counts
VP X, Y          // specifies the coordinates of the end points (X,Y are variables which is identified from the
DXF file.)
VE               // specify the end of the coordinated motion
BGS              // motion begin
EN               //end of the subprogram
```

6. Test Running Result

By C#, the authors first finished the DXF file identification as well as the extraction and storage of graphic element information. The graphic element ordering operations were also achieved. At last, the graphics trajectories were generated by calling the bottom GALIL software instructions and achieved motion tacking. The test was carried out on a three axes motion experiment platform which is shown in **Figure 4**, the carving cutter was replaced with pen. Pen was fixed on the experiment platform.

The test used a trajectory graph drawn by AutoCAD which is shown in **Figure 5(a)**. With the author's method, the graph was identified and the results were stored in a 2D array [i, j] which is defined in **Table 2**. The result is shown in **Table 3**. **Table 3** indicates that the identify result is consistent with the CAD graph. **Figure 5(b)** is the

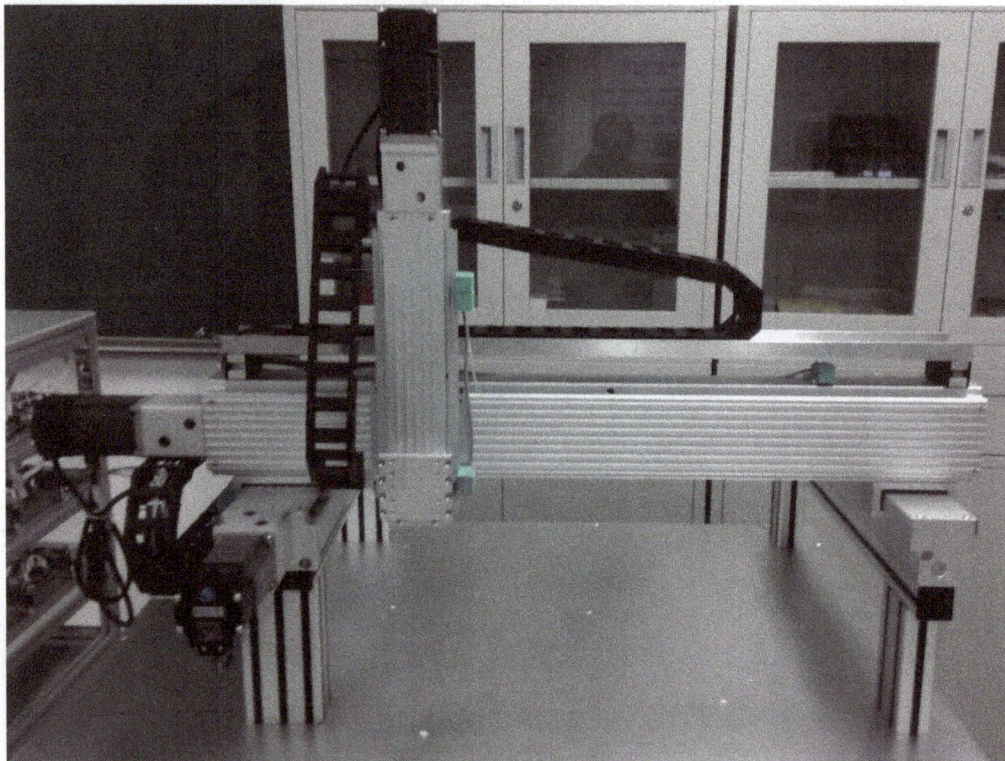

Figure 4. Experiment platform.

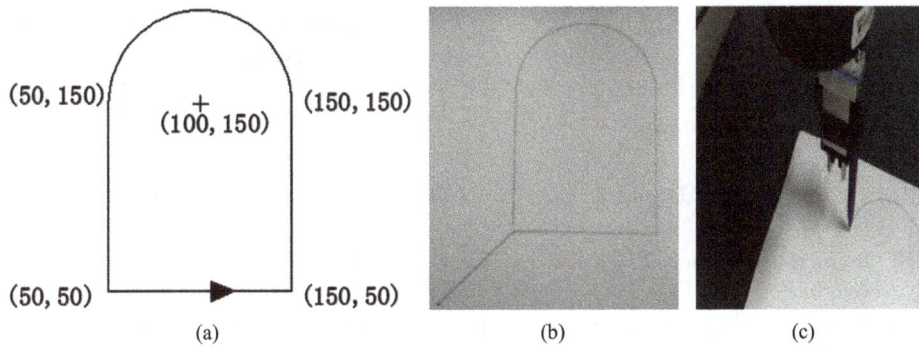

Figure 5. Test result.

Table 3. The identify result of CAD graph.

Array Variable Position	Result Data	Result Data Meaning
[0, 0]	1	The first graphic element is line.
[0, 1]	50	The X axis coordinate value of the start point of the first graphic element is 50.
[0, 2]	50	The Y axis coordinate value ofthe start point of the first graphic element is 50.
[0, 3]	0	The Z axis coordinate value ofthe start point of the first graphic element is 0.
[0, 4]	150	The X axis coordinate value ofthe end point of the first graphic element is150.
[0, 5]	50	The Y axis coordinate value ofthe end point of the first graphic element is 50.
[0, 6]	0	The Z axis coordinate value ofthe end point of the first graphic element is 0.
[1, 0]	1	The second graphic element is line.
[1, 1]	150	The X axis coordinate value of the start point of the second graphic element is 150.
[1, 2]	50	The Y axis coordinate value of the start point of the second graphic element is 50.
[1, 3]	0	The Z axis coordinate value of the start point of the second graphic element is 0.
[1, 4]	150	The X axis coordinate value of the end point of the second graphic element is 150.
[1, 5]	150	The Y axis coordinate value of the end point of the second graphic element is 150.
[1, 6]	0	The Z axis coordinate value of the end point of the second graphic element is 0.
[2, 0]	3	The third graphic element is arc.
[2, 7]	100	The X axis coordinate value of the centre of the third graphic element is 100.
[2, 8]	150	The Y axis coordinate value of the centre of the third graphic element is 150.
[2, 9]	0	The Z axis coordinate value of the centre of the third graphic element is 0.
[2, 10]	50	The radius value of the third graphic element is 50.
[2, 11]	0	The start angle of the third graphic element is 0.
[2, 12]	180	Theend angle of the third graphic element is 180.
[3, 0]	1	The fourth graphic element is line.
[3, 1]	50	The X axis coordinate value of the start point of the fourth graphic element is 50.
[3, 2]	150	The Y axis coordinate value of the start point of the fourth graphic element is 150.
[3, 3]	0	The Z axis coordinate value of the start point of the fourth graphic element is 0.
[3, 4]	50	The X axis coordinate value of the end point of the fourth graphic element is 50.
[3, 5]	50	The Y axis coordinate value of the end point of the fourth graphic element is 50.
[3, 6]	0	The Z axis coordinate value of the end point of the fourth graphic element is 0.

trajectory graph which is drawn by the three axes motion experiment platform according to the identify result. **Figure 5(c)** shows the motion process. The final result shows that the developed open CNC engraving machine system can accurately complete the identification of DXF file, and the walk path is consistent with the CAD file.

References

[1]   Omirou Sotiris, L. and Barouni Antigoni, K. (2005) Integration of New Programming Capabilities into a CNC Milling System. *Robotics and Computer-Integrated Manufacturing*, **21**, 518-527. http://dx.doi.org/10.1016/j.rcim.2004.10.002

[2]   Kovacic, M., Brezocnik, M., Pahole, I., Balic, J. and Kecelj, B. (2005) Evolutionary Programming of CNC Machines. *Journal of Materials Processing Technology*, **164-165**, 1379-1387. http://dx.doi.org/10.1016/j.jmatprotec.2005.02.047

[3]   Zhai, R. and Zhang, L. (2011) Reading Frame Design Based on the DXF File Format. *Fujian Computer*, **4**, 107-109.

[4]   Huang, J.Q. and Yuan, Q. (2012) Automatic Input and Identification for Stamping Graph Based on AutoCAD. *Machinery Design & Manufacture*, **2**, 82-84.

[5]   Bai, X.C. and Chen, Y.M. (2010) Automatic Programming of Bridge Cutting Machine Based on the DXF File. *Equipment Manufacturing Technology*, **2**, 110-112.

5

A Novel Technique for Detection of Time Delay Switch Attack on Load Frequency Control

Arman Sargolzaei[1], Kang K. Yen[1], Mohamed N. Abdelghani[2], Abolfazl Mehbodniya[3], Saman Sargolzaei[4]

[1]Department of Electrical and Computer Engineering, Florida International University, Miami, FL, USA
[2]Department of Mathematics and Statistics, University of Alberta, Edmonton, Canada
[3]Graduate School of Engineering-Sendai, Tohoku University, Sendai, Japan
[4]Department of Electrical and Computer Engineering, Wentworth Institute of Technology, Boston, MA, USA
Email: arman.sargolzaei@rancsgroup.com

Abstract

In this paper, we focus on the estimation of time delays caused by adversaries in the sensing loop (SL). Based on the literature review, time delay switch (TDS) attacks could make any control system, in particular a power control system, unstable. Therefore, future smart grids will have to use advanced methods to provide better situational awareness of power grid states keeping smart grids reliable and safe from TDS attacks. Here, we introduce a simple method for preventing time delay switch attack on networked control systems. The method relies on an estimator that will estimate and track time delays introduced by an adversary. Knowing the maximum tolerable time delay of the plant's optimal controller for which the plant remains stable, a time-delay detector issues an alarm signal when the estimated time delay is larger than the minimum one and directs the system to alarm state. In an alarm state, the plant operates under the control of an emergency controller that is local to the plant and remains in this mode until the networked control system state is restored. This method is an inexpensive and simple way to guarantee that an industrial control system remains stable and secure.

Keywords

Time Delay Switch Attack, Load Frequency Control, Detection and Estimation, Emergency Controller

1. Introduction

Modern power grids rely on telecommunication technologies for control and monitoring, in a way to improve efficiency and reliability distribution. However, their reliance on computers and multi-purpose networks makes them vulnerable to cyber-attacks [1]-[3]. Recently, the investigation of attack methods on industrial control systems and their countermeasures have been the focus of many academic, industries, and governments. Here, we will investigate one type of attacks on a power control system recently proposed by Sargolzaei *et al.* [4]-[6] that is time delay switch attack and we provide a simple detection/estimation technique for time delays caused by an adversary. State estimation is one of the pivotal features of future power systems [7]. According to [8], improving the resiliency and reliability of smart distribution networks could be obtained via deploying renewable resources and demand side management programs. As it has been mentioned in [9], demand response can ameliorate the load curve by motivating smart customers utilizing smart communication infrastructure and smart meters.

The controller design for systems with time delay is one of the interests of many researchers [10]-[12] but they have not studied that time delay can be injected to the system as an attack. Time delays injected by a hacker in a control system would, in general, destabilize the system or cause inefficiency in performance of the system. This is a new attack in the context of power systems (e.g., load frequency control (LFC)) on networked control systems (NCS) [13] and is named time-delay-switch attack or TDS for short [4]. To avoid possible damage of TDS attacks, systems and controllers must be redesigned in a way to detect and correct for variable time delays.

Time delays exist in power systems, in the sensing and control loops. The traditional controllers of power systems are designed based on current information being available and ignoring time delays even if they present. However, power grids technologies are continuously being improved by introducing new telecommunication technologies for monitoring to improve efficiency, reliability and sustainability of supply and distribution. For example, the introduction of a wide area measurement system (WAMS) provides synchronized near real-time measurements in phase measurement units (PMUs). WAMS which are used for stability analysis of power systems can also be used for designing more robust controllers. Nevertheless, time delays are present in PMUs measurements as a result of natural transmission lines [14].

Several studies have considered the problem of stability of power systems with time delays [10] [11]. The impact of time delays on the power system controllers was discussed in [15]-[19]. In [20], authors studied the effects of delays on the small signal stability of power systems. In [15] and [16], methods to eliminate oscillations that resulted from time-delayed feedback control were proposed. Authors in [21] presented a wide-area control system for damping generator oscillations. Using phasor measurements, but with delays, a controller was proposed in [14] and small signal stability of the power system was considered. A feedback controller designed for power systems with delayed states was proposed in [22]. This controller deals with the combined effects of the instantaneous as well as delayed states using the quadratic Lyapunov function for systems with delays. Additional studies on power systems with delay can be found in [2] [3] [10]-[12] [23]-[25] and the references therein. However, most of these studies considered either the construction of controllers that were robust to time delays or controllers that used offline estimate of time delays. As far as we know, there are no control methods that perform online estimation of dynamic time delays and real time control of power systems. Furthermore, few studies [26] [27] considered control of power systems with time delays introduced by adversary.

In this paper, we will describe a simple yet effective method to address a TDS attack on the observed states of a controlled system. Our method utilizes a time delay estimator, a communication protocol to alarm for time delay switch attack, a buffer to store the history of controller commands and an optimal controller to stabilize or track a reference signal and a local to the plant emergency controller to stabilize the plant if large time delays are detected. For now, we will only deal with LTI systems in state feedback.

2. Method

All control methods developed in the past compensate for time delays either rely on controller robust to a maximum time delay, off-line estimates of time delays or approximation of time delayed signals [26] [27]. Here, we develop an alternative time delay estimator which is simpler and will help with systems under time delay attack. In this paper, we propose a general method for control of systems under TDS attack. We have developed this method for continuous linear time-invariant systems. We have implemented our models and control strategies in MATLAB to demonstrate the performance with simulation.

The proposed method is shown in **Figure 1**. Its basic elements are: plant model, time delay detector, emergency controller and controller. The controller can either be a PID controller or an optimal feedback controller. The emergency controller's job is to stabilize the system in case of an attack on the communication lines of the distant controller. The time delay detector estimates the time delay in the communication channel from telemetered data. If the delay is larger than a maximum allowed time delay it sends alarm signals to the controller and emergency controller. If an alarm signal is received by the emergency controller it will begin to operate in a way to stabilize the plant to a desired state while the distant controller stops to operate until the delay is corrected.

Suppose the system we are dealing with is linear time invariant (LTI) or can be approximated in a region of interest by a LTI system,

$$\dot{x}(t) = Ax(t) + Bu(t), \tag{1}$$

where x and u are state and control vectors, respectively. Matrices A and B are constant matrices with suitable dimensions.

Then, the solution is given by

$$x(t) = e^{At}x_0 + \int_0^t e^{A(t-s)}Bu(s)\,ds \tag{2}$$

with time delay τ, either a time-delay switch attack or a natural delay, the solution of Equation (2) becomes

$$x(t-\tau) = e^{A(t-\tau)}x_0 + \int_0^{t-\tau} e^{A(t-\tau-s)}Bu(s)\,ds. \tag{3}$$

Let us write the solution $x(t)$ at the time t in terms of the solution with the time delay τ; we have

$$
\begin{aligned}
x(t) &= e^{At}x_0 + e^{A\tau}\int_0^{t-\tau} e^{A(t-s)}e^{-A\tau}Bu(s)\,ds + \int_{t-\tau}^t e^{A(t-s)}Bu(s)\,ds \\
&= e^{At}x_0 + e^{A\tau}\left[x(t-\tau) - e^{A(t-\tau)}x_0\right] + \int_{t-\tau}^t e^{A(t-s)}Bu(s)\,ds,
\end{aligned} \tag{4}
$$

In general, the time delay τ is an unknown variable. Let's assume that τ is slowly varying, compared to the changes in u and x, and $\hat{\tau}$ is our estimate of the time delay τ. Then, $\varepsilon = \hat{\tau} - \tau$ is the estimation error in the time delay. The predicted state $\hat{x}(t)$ of the system based on the estimate of time delay $\hat{\tau}$ is given by

$$\hat{x}(t) = e^{At}x_0 + e^{A\hat{\tau}}\left[\hat{x}(t-\hat{\tau}) - e^{A(t-\hat{\tau})}x_0\right] + \int_{t-\hat{\tau}}^t e^{A(t-s)}Bu(s)\,ds \tag{5}$$

where $\hat{x}(t-\hat{\tau})$ is the estimate of the delayed state given the estimate of the delay $\hat{\tau}$ (*i.e.* a simulated signal).

It should be noted that, $x(t-\tau)$ is what we actually measures and deliver to the plant model. So, at every instance of time, the variables $\hat{x}(t)$, $\hat{x}(t-\hat{\tau})$, $u(t)$, A, B and $x(t-\tau)$ are known to the controller and the plant model. On the other hand, the current state $x(t)$ and the time delay τ are unknown. It's essential that the plant model estimates state $x(t)$ accurately. Because of the delay, an accurate enough estimation of $x(t)$ requires a good estimate of the delay τ. We will show how to estimate the delay τ.

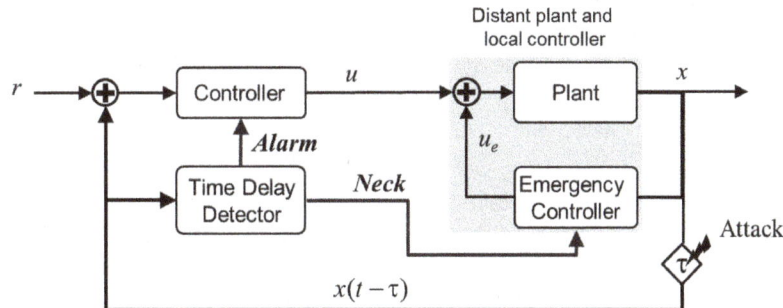

Figure 1. Block diagram of the proposed control system.

The estimation error in states can be described by $e_m(t) = x(t) - \hat{x}(t)$ and with delay it is given by

$$e_m(t; \tau, \hat{\tau}) = x(t - \tau) - \hat{x}(t - \hat{\tau}). \tag{6}$$

Then the idea is to estimate $\hat{\tau}$ overtime as fast as possible to minimize the error $e_m(t; \tau, \hat{\tau})$. Using the gradient descent method, we set

$$\frac{d\hat{\tau}}{dt} = -\eta \frac{\partial v}{\partial \hat{\tau}}, \tag{7}$$

where η is the learning parameter and select $v = \frac{1}{2}e_m^2$. With some manipulation, we have

$$
\begin{aligned}
\frac{d\hat{\tau}}{dt} &= -\eta e_m \frac{\partial e_m}{\partial \hat{\tau}} = -\eta e_m \frac{\partial \left[x(t - \tau) - \hat{x}(t - \hat{\tau}) \right]}{\partial \hat{\tau}} \\
&= \eta e_m \frac{\partial \hat{x}(t - \hat{\tau})}{\partial \hat{\tau}} = \eta e_m \frac{\partial}{\partial \hat{\tau}} \left[e^{A(t - \hat{\tau})} x_0 + \int_0^{t - \hat{\tau}} e^{A(t - \hat{\tau} - s)} Bu(s) \, ds \right]. \\
&= \eta e_m \frac{\partial}{\partial \hat{\tau}} \left[\int_0^{t - \hat{\tau}} e^{A(t - \hat{\tau} - s)} Bu(s) \, ds \right] - \eta e_m A e^{A(t - \hat{\tau})} x_0 \\
&= -\eta e_m \left[Bu(t - \hat{\tau}) - e^{A(t - \hat{\tau})} Bu(0) - A e^{A(t - \hat{\tau})} x_0 \right]
\end{aligned}
\tag{8}
$$

Assuming that, $u(0) = 0$, which is reasonable for initial time, we arrive at

$$\frac{d\hat{\tau}}{dt} = -\eta e_m Bu(t - \hat{\tau}) - A e^{A(t - \hat{\tau})} x_0, \qquad 0 \leq \hat{\tau} \leq t \tag{9}$$

Equation (9) is the one we will use to estimate the time delay τ. However, there are practical issues that we need to consider. Computing machines have finite memory and temporal resolution. Therefore, we are unable to implement Equation (9) without discrete approximation and boundedness assumptions. To guarantee stability of calculations and limit our memory usage we have to add the following condition, $\tau < \tau_{\max}$. This condition will allow us to construct a finite buffer to store the history of $u(t)$ from t to $t - \tau_{\max}$.

After designing the time delay estimator, we turn our attention to the controller and emergency controller. The controller can either be a PID controller or an optimal controller depending on the requirements of the application. Equation (10) is the PID controller and Equation (11) is the optimal controller,

$$u(t) = K_P e(t) + K_D \frac{de}{dt}(t) + K_I \int_0^t e(s) \, ds, \tag{10}$$

$$u(t) = Ke(t). \tag{11}$$

where the error is $e(t) = r(t) - x(t)$. Either controller, *i.e.*, the PID or the optimal controller, can be designed in a way to be robust to some maximum time-delay τ_{stable}. The emergency controller sits close to the plant and operates only in case of emergency, for example, an attack on the communication line between the plant and controller. The emergency controller could either be a PID controller or optimal controller with the goal to stabilize the system to a particular reference trajectory r_E.

Suppose there is a time delay attack on the system with delay τ; the time delay estimator will estimate the delay $\hat{\tau}$. The cyber-attack detector will use the time delay estimate to perform the following function

$$D = \begin{cases} 1 & \hat{\tau} > c\tau_{\text{stable}} \\ 0 & \text{otherwise} \end{cases}, \tag{12}$$

where c is a constant between 0 and 1. In case $D = 1$ an alarm signal is sent to the controller to shut it down and a negative acknowledgement is sent to the emergency controller to stabilize the plant. The control strategy is shown in **Figure 1**.

We have implemented this method in MATLAB and verified its performance using the load frequency control (LFC) of a two-area distributed power system. In the next section we present and discuss the simulation results.

3. Result

We focus on the LFC system where the controller's function is to regulate the states of a networked power plant. The multi interconnect LFC dynamic system description can be found in [4]. Here we focus on two-area power system with the attack model as

$$\begin{cases} \dot{X}(t) = AX(t) + BU(t) + D\Delta P_l \\ X(0) = X_0 \end{cases}.$$

(13)

The optimal feedback controller is given by

$$U(t) = -K\hat{X}(t)$$

(14)

and the new state after the attack can be modeled by

$$\hat{X}(t) = X(t - \tau),$$

(15)

where $\tau = [t_{d1}, t_{d2}, \cdots, t_{dN}]^T$ are different/random time-delays and are positive values. When $t_{d1}, t_{d2}, \cdots, t_{dN}$ are all zero, the system is in its normal operation. An adversary can get access to the communication link and inject a delay attack on the line to direct the system to abnormal operations.

In (13), $X(t) = [x_1(t) \quad x_2(t)]^T$ denotes the states in the first and second power areas. Then the state vector in the ith power area is defined as

$$x_i(t) = \left[\Delta f^i(t) \quad \Delta P_g^i(t) \quad \Delta P_{tu}^i(t) \quad \Delta P_{pf}^i(t) \quad \Lambda^i(t)\right]^T \qquad i = 1, 2,$$

(16)

where $\Delta f^i(t)$, $\Delta P_g^i(t)$, $\Delta P_{tu}^i(t)$, $\Delta P_{pf}^i(t)$ and $\Lambda^i(t)$ are frequency deviation, power deviation of the generator, position value of the turbine, tie-line power flow, and control error on the ith power area, respectively [19]. The control error of the ith power area is expressed as

$$\Lambda^i(t) = \int_0^t \beta_i \Delta f^i(s)\,\mathrm{d}s,$$

(17)

where β_i denotes the frequency bias factor.

In the dynamic model of the two-area LFC (13):

$$A = \begin{bmatrix} A_{11} & A_{12} \\ A_{21} & A_{22} \end{bmatrix},$$

(18)

$$B = \mathrm{diag}\left\{\begin{bmatrix} B_1^T & B_2^T \end{bmatrix}^T\right\},$$

(19)

$$D = \mathrm{diag}\left\{\begin{bmatrix} D_1^T & D_2^T \end{bmatrix}^T\right\},$$

(20)

and A_{ii}, A_{ij}, B_i and D_i are represented by

$$A_{ii} = \begin{bmatrix} -\dfrac{\mu_i}{J_i} & \dfrac{1}{J_i} & 0 & -\dfrac{1}{J_i} & 0 \\[2mm] 0 & -\dfrac{1}{T_{tu\,i}} & \dfrac{1}{T_{tu\,i}} & 0 & 0 \\[2mm] -\dfrac{1}{\omega_i T_{g\,i}} & 0 & -\dfrac{1}{T_{g\,i}} & 0 & 0 \\[2mm] \sum_{\substack{i \neq j \\ j=1}}^{N} 2\pi T_{ij} & 0 & 0 & 0 & 0 \\[2mm] \beta_i & 0 & 0 & 0 & 1 \end{bmatrix}$$

(21)

$$B_i = \begin{bmatrix} 0 & 0 & \dfrac{1}{T_{g\,i}} & 0 & 0 \end{bmatrix}^T \tag{22}$$

$$A_{ij} = \begin{bmatrix} 0 & 0 & 0 & 0 & 0 \\ 0 & 0 & 0 & 0 & 0 \\ 0 & 0 & 0 & 0 & 0 \\ -2\pi T_{ij} & 0 & 0 & 0 & 0 \\ 0 & 0 & 0 & 0 & 0 \end{bmatrix} \tag{23}$$

and

$$D_i = \begin{bmatrix} -\dfrac{1}{J_i} & 0 & 0 & 0 & 0 \end{bmatrix}^T \tag{24}$$

where i and j can only be values one and two, J_i, ω_i, μ_i, $T_{g\,i}$ and $T_{tu\,i}$ are the generator moment of inertia, the speed-droop coefficient, generator damping coefficient, the governor time constant, the turbine time constant in the ith power area and T_{ij} is the stiffness constant between the ith and the jth power areas, respectively.

The analysis starts with the design of an optimal controller for the LFC in the normal operation (*i.e.*, with no attack). Consider the system model described by (13) with the performance index described by

$$J = \frac{1}{2}\int_0^{t_f} \left\{ X^T(t)QX(t) + U^T(t)RU(t) \right\} dt, \tag{25}$$

where matrix $Q \in R^{n\times n}$ is positive semi-definite and $R \in R^{m\times m}$ is positive definite. Then, the optimal control problem is to obtain optimal control $U^*(t)$ that minimizes the performance index (25), subject to the dynamic of the system with no time-delay in its states.

Simulation studies have been conducted to evaluate the effects of TDS attacks on the dynamics of the system and detect TDS attacks to direct the controller to emergency state. By solving the Riccati matrix equation we obtain the close loop control law in the form of state feedback. For the simulations we have used discrete linear-quadratic regulator design from continuous cost function called "lqrd" function in MATLAB 2013a. **Table 1** shows parameter values used in this process. We also set both ΔP_l^1 and ΔP_l^2 to zero.

In our simulation, the total simulation time is 50 second and the sampling time is 0.01 s as its common in industrial applications. To show accuracy of our proposed TDS attack detector/tracker, we didn't send negative acknowledgment to the local controller for the first simulation to only track the injected time delay to the system. The simulation has been done for three different scenarios: 1) single TDS attack on one power area, 2) Simultaneous TDS attack on both power areas, 3) Complex varied TDS attack on both power areas.

1) Single TDS attack to one power area: Here, we considered that adversary attacks the third state of the first power area at time of 2 second for a delay value of 3 seconds and the time delay is increased to 4.5 seconds at time of 7 seconds. **Figure 2** shows that the detector accurately tracked and monitored the TDS attack or any

Table 1. Parameter values for a two-area power system controller design.

Parameter	Value	Parameter	Value
J_1	10	ω_1	0.05
μ_1	1.5	T_{g1}	0.12 s
T_{tu1}	0.2 s	T_{tu2}	0.45 s
T_{12}	0.198 pu/rad	T_{21}	0.198 pu/rad
J_2	12	ω_2	0.05
μ_2	1	T_{g2}	0.18 s
R	100 I	Q_f	0
Q	100 I	t_f	∞
β_1	21.5	β_2	21

natural delay on the system.

2) Simultaneous TDS attack to both power areas: In this scenario, we attack the third state of both power areas. TDS attacks 1 and 2 have been injected to the first and the second power areas respectfully. TDS attack 1 is started at time 2 seconds for value 3 seconds and increased at time 8 seconds to the value of 4.5 seconds. Also TDS attack 2 is started simultaneously with TDS attack 1 for the value of 1.5 seconds and increased to the value of 6 seconds. **Figure 3** shows the result.

3) Complex varied TDS attack to both power areas: In the last scenario we injected the TDS attack at different time with different time-delay values. We assume that an adversary injects TDS attack to feedback lines of both power areas. In our simulation, an attacker starts TDS attack to the second power area (third state) at time 1 second for $t_{d3} = 5$ s and then increases it to $t_{d3} = 10$ s at 20 second. Furthermore, adversary starts to attack the first power area (third state), at 1 second with $t_{d8} = 3$ s and increased it to $t_{d8} = 4.5$ s at 30 second. **Figure 4** and **Figure 5** show TDS attack tracking real time. **Figure 4** shows TDS attack on the third state of the first power area and **Figure 5** shows the attack on the same state of the second power area.

For the second part of our simulation we enabled the emergency controller to show the effect of our proposed technique for overcoming the TDS attack. We assume that an adversary injects TDS attack to the feedback lines of both power areas. In the simulation, a TDS attack was applied to the second power area (the third state) at time of 1 second for $t_{d3} = 5$ s and then increases it to $t_{d3} = 10$ s at 20 second. Furthermore, adversary starts to attack the first power area (the third state), at 1 second with $t_{d8} = 3$ s and increased it to $t_{d8} = 4.5$ s at 30 second. We also set the $\tau_{stable} = 0.4$ s based on stability analysis of LFC system in [4]. We send the negative acknowledgment to emergency controller in the case that detected value of time-delay is more than the maximum allowed time-delay. **Figure 6** and **Figure 7** show the third state of the first and the second power areas

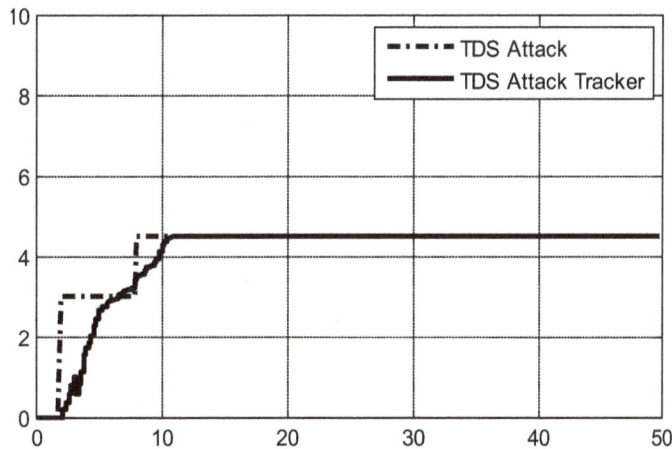

Figure 2. TDS attack detection and tracking for third state of first power area.

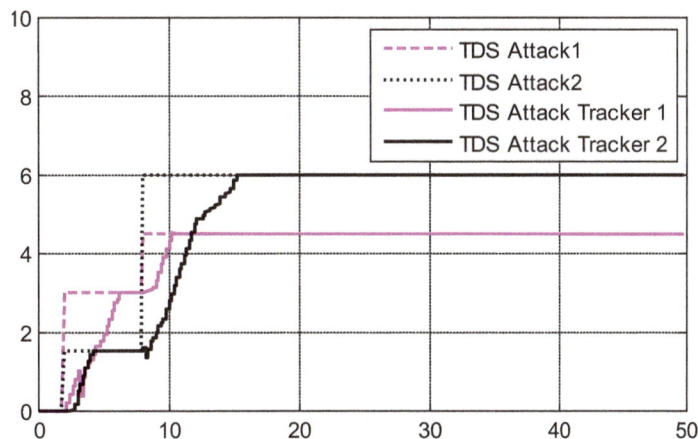

Figure 3. TDS attack detection and tracking for simultaneous TDS attack on both power areas.

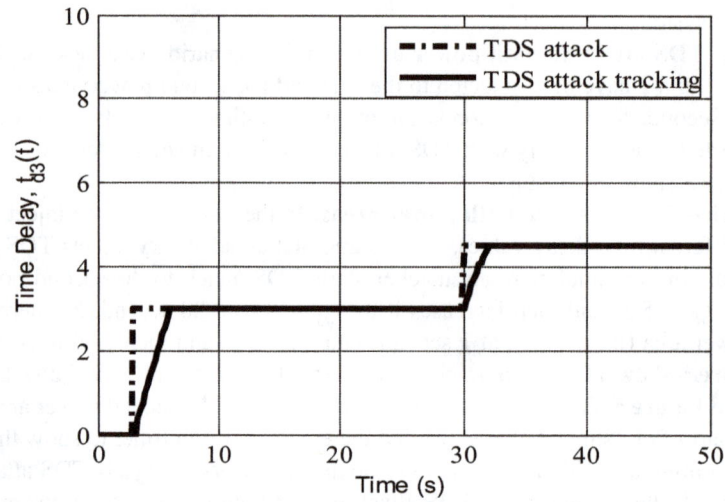

Figure 4. TDS attack detection and tracking for third state of first power area.

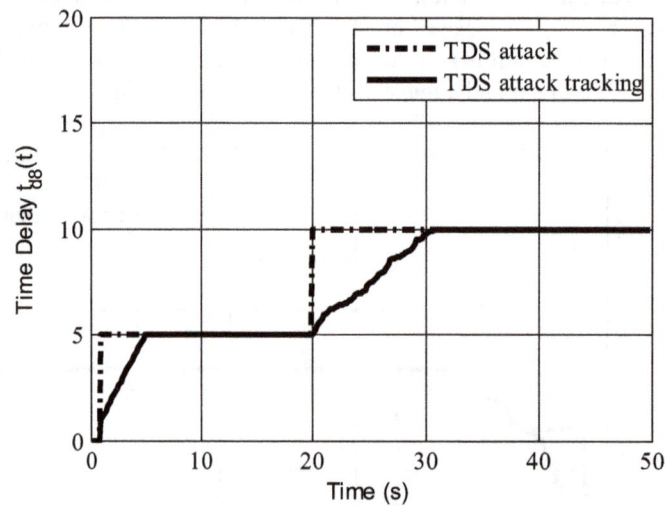

Figure 5. TDS attack detection and tracking for third state of second power area.

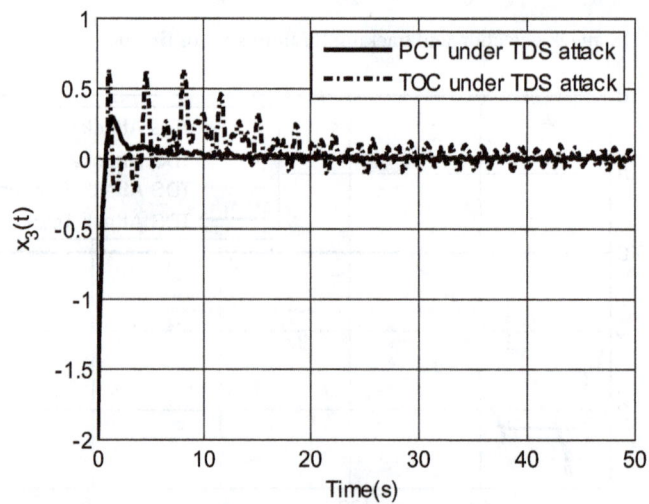

Figure 6. Value position of the turbine system under TDS attack for the first power-area control system.

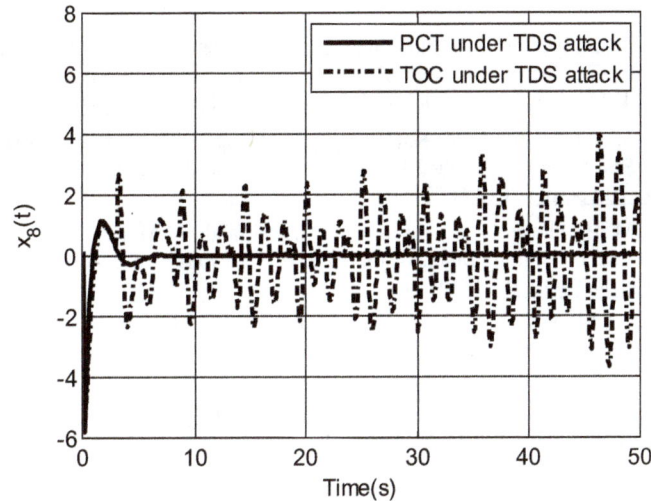

Figure 7. Value position of the turbine system under TDS attack for the second power-area control system.

under attack with traditional optimal controller (TOC) and the proposed control technique (PCT) respectively. As it's clear on the result, the simulated attack makes the system unstable. With the proposed technique, we could overcome TDS attack on the simulated system.

4. Conclusion

In summary, we have demonstrated a simple method for estimating and detecting time delay switch attack on a networked control system. The method relies on a time delay estimator that estimates and tracks time-delays introduced by an adversary. With knowledge of the maximum time delay of the control system, for which the plant remains stable and secure, the time-delay detector compares the estimated time delay to the maximum allowed time delay and issues an alarm signal when the estimated time delay is larger than this value. It also directs the system to an alarm state. In an alarm state, the plant is under the control of the emergency controller, which is local to the plant. The plant remains in this mode until the networked control system state is restored and time-delay switch is eliminated. We think that this method is simple and an inexpensive way to assure that an industrial control system remains stable and secure.

Acknowledgements

The authors would like to thank the Resilient, Autonomous Networked Control Systems (RANCS) group for support during publishing this paper.

References

[1] Ericsson, G.N. (2010) Cyber Security and Power System Communication—Essential Parts of a Smart Grid Infrastructure. *IEEE Transactions on Power Delivery*, **25**, 1501-1507. http://dx.doi.org/10.1109/TPWRD.2010.2046654

[2] Zhang, C.-K., Jiang, L., Wu, Q.H., He, Y. and Wu, M. (2013) Delay-Dependent Robust Load Frequency Control for Time Delay Power Systems. *IEEE Transactions on Power Systems*, **28**, 2192-2201.

[3] Yuan Y.L., Li Z.Y. and Ren, K. (2011) Modeling Load Redistribution Attacks in Power Systems. *IEEE Transactions on Smart Grid*, **2**, 382-390.

[4] Sargolzaei, A., Yen, K. and Abdelghani, M.N. (2014) Time-Delay Switch Attack on Load Frequency Control in Smart Grid. *Journal of Advanced Communication Technologies*, **5**, 55-64.

[5] Sargolzaei, A., Yen, K. and Abdelghani, M.N. (2014) Delayed Inputs Attack on Load Frequency Control in Smart Grid. *Innovative Smart Grid Technologies Conference (ISGT)*, 2014 *IEEE PES*, Washington DC, 19-22 February 2014, 1-5. http://dx.doi.org/10.1109/isgt.2014.6816508

[6] Sargolzaei, A., Yen, K.K. and Abdelghani, M.N. (2014) Control of Nonlinear Heartbeat Models under Time-Delay-

Switched Feedback Using Emotional Learning Control. *International Journal on Recent Trends in Engineering & Technology*, **10**, 2.

[7]   Amini, M.H., Sarwat, A.I., Iyengar, S.S. and Guvenc, I. (2014) Determination of the Minimum-Variance Unbiased Estimator for DC Power-Flow Estimation. *40th IEEE Industrial Electronics Conference* (*IECON* 2014), Dallas, 29 October-1 November 2014, 114-118. http://dx.doi.org/10.1109/IECON.2014.7048486

[8]   Amini, M.H., Nabi, B. and Haghifam, M.-R. (2013) Load Management Using Multi-Agent Systems in Smart Distribution Network. *IEEE PES General Meeting* 2013, Vancouver, 21-25 July 2013, 1-5. http://dx.doi.org/10.1109/pesmg.2013.6672180

[9]   Kamyab, F., Amini, M. H., Sheykhha, S., Hasanpour, M. and Jalali, M.M. (2015) Demand Response Program in Smart Grid Using Supply Function Bidding Mechanism. *IEEE Transactions on Smart Grid*, 25 May 2015, 1949-3053. http://dx.doi.org/10.1109/TSG.2015.2430364

[10] Li, Y., Tong, S.C. and Li, Y.M. (2012) Observer-Based Adaptive Fuzzy Backstepping Control for Strict-Feedback Stochastic Nonlinear Systems with Time Delays. *International Journal of Innovative Computing, Information and Control*, **8**, 8103-8114.

[11] Benzaouia, A., Ouladsine, M., Naamane, A. and Ananou, B. (2012) Fault Detection for Uncertain Delayed Switching Discrete-Time Systems. *International Journal of Innovative Computing, Information and Control*, **8**, 8049-8062.

[12] Sargolzaei, A., Yen, K., Noei, S. and Ramezanpour, H. (2013) Assessment of He's Homotopy Perturbation Method for Optimal Control of Linear Time-Delay Systems. *Applied Mathematical Sciences*, **7**, 349-361.

[13] Benéıtez-Péerez, H., Benéıtez-Péerez, A. and Ortega-Arjona, J. (2012) Networked Control Systems Design Considering Scheduling Restrictions and Local Faults. *International Journal of Innovative Computing, Information and Control*, **8**, 8515-8526.

[14] Dotta, D., Silva, A.S. and Decker, I.C. (2009) Wide-Area Measurements-Based Two-Level Control Design Considering Signal Transmission Delay. *IEEE Transactions on Power Systems*, **24**, 208-216. http://dx.doi.org/10.1109/TPWRS.2008.2004733

[15] Kamwa, I., Grondin, R. and Hebert, Y. (2001) Wide-Area Measurement Based Stabilizing Control of Large Power Systems—A Decentralized/Hierarchical Approach. *IEEE Transactions on Power Systems*, **16**, 136-153. http://dx.doi.org/10.1109/59.910791

[16] Wu, H.X., Tsakalis, K.S. and Heydt, G.T. (2004) Evaluation of Time Delay Effects to Wide-Area Power System Stabilizer Design. *IEEE Transactions on Power Systems*, **19**, 1935-1941. http://dx.doi.org/10.1109/TPWRS.2004.836272

[17] Jiang, Q., Zou, Z. and Cao, Y. (2005) Wide-Area TCSC Controller Design in Consideration of Feedback Signals' Time Delays. *IEEE Power Engineering Society General Meeting*, **2**, 1676-1680.

[18] Saad, M.S., Hassouneh, M.A., Abed, E.H. and Edris, A.A. (2005) Delaying Instability and Voltage Collapse in Power Systems Using SVCs with Washout filter-Aided Feedback. *American Control Conference*, **6**, 4357-4362. http://dx.doi.org/10.1109/acc.2005.1470665

[19] Chaudhuri, B., Majumder, R. and Pal, B. (2005) Wide-Area Measurement Based Stabilizing Control of Power System Considering Signal Transmission Delay. *IEEE Power Engineering Society General Meeting*, **2**, 1447-1450. http://dx.doi.org/10.1109/pes.2005.1489106

[20] Milano, F. and Anghel, M. (2012) Impact of Time Delays on Power System Stability. *IEEE Transactions on Circuits and Systems I: Regular Papers*, **59**, 889-900. http://dx.doi.org/10.1109/TCSI.2011.2169744

[21] Ray, S. and Venayagamoorthy, G.K. (2008) Real-Time Implementation of a Measurement-Based Adaptive Wide-Area Control System Considering Communication Delays. *IET Generation, Transmission & Distribution*, **2**, 62-70. http://dx.doi.org/10.1049/iet-gtd:20070027

[22] Alrifai, M.T., Zribi, M., Rayan, M. and Mahmoud, M.S. (2013) On the Control of Time Delay Power Systems. *International Journal of Innovative Computing, Information and Control*, **9**, 769-792.

[23] Yang, Q.Y., An, D. and Yu, W. (2013) On Time Desynchronization Attack against IEEE 1588 Protocol in Power Grid Systems. 2013 *IEEE Energytech*, Cleveland, 21-23 May 2013, 1-5.

[24] Schenato, L. (2008) Optimal Estimation in Networked Control Systems Subject to Random Delay and Packet Drop. *IEEE Transactions on Automatic Control*, **53**, 1311-1317. http://dx.doi.org/10.1109/TAC.2008.921012

[25] Mahmoud, M.S. (2000) Robust Control and Filtering for Time-Delay Systems. Marcel Dekker Inc., New York.

[26] Tan, Y.H. (2004) Time-Varying Time-Delay Estimation for Nonlinear Systems Using Neural Networks. *International Journal of Applied Mathematics and Computer Science*, **14**, 63-68.

[27] Li, C.M. and Xiao, J. (2006) Adaptive Delay Estimation and Control of Networked Control Systems. *International Symposium on Communications and Information Technologies*, 2006, *ISCIT'06*, Bangkok, 18-20 October 2006, 707-710. http://dx.doi.org/10.1109/iscit.2006.339832

6

Simple, Flexible, and Interoperable SCADA System Based on Agent Technology

Hosny Abbas[1*], Samir Shaheen[2], Mohammed Amin[1]

[1]Department of Electrical Engineering, Assiut University, Assiut, Egypt
[2]Department of Computer Engineering, Cairo University, Giza, Egypt
Email: *hosnyabbas@aun.edu.eg, mhamin@aun.edu.eg, sshaheen@eng.cu.edu.eg

Abstract

SCADA (Supervisory Control and Data Acquisition) is concerned with gathering process information from industrial control processes found in utilities such as power grids, water networks, transportation, manufacturing, etc., to provide the human operators with the required real-time access to industrial processes to be monitored and controlled either locally (on-site)or remotely (*i.e.*, through Internet). Conventional solutions such as custom SCADA packages, custom communication protocols, and centralized architectures are no longer appropriate for engineering this type of systems because of their highly distribution and their uncertain continuously changing working environments. Multi-agent systems (MAS) appeared as a new architectural style for engineering complex and highly dynamic applications such as SCADA systems. In this paper, we propose an approach for simply developing flexible and interoperable SCADA systems based on the integration of MAS and OPC process protocol. The proposed SCADA system has the following advantages: 1) simple (easier to be implemented); 2) flexible (able to adapt to its environment dynamic changes); and 3) interoperable (relative to the underlying control systems, which belongs to diverse of vendors). The applicability of the proposed approach is demonstrated by a real case study example carried out in a paper mill.

Keywords

SCADA, Real-Time Monitoring, Process Control, Agent Technology, Multi-Agent Systems, Open Process Control (OPC)

1. Introduction

A SCADA system is responsible for gathering information and real-time data from variety of plants and provid-

*Corresponding author.

ing this data to operators located at anywhere at any time. Furthermore, SCADA systems can be considered as critical information systems; their criticality comes from the fact that they are currently vital components of most nations' critical infrastructures. They control pipelines, water and transportation systems, utilities, refineries, chemical plants, and a wide variety of manufacturing operations. Failure of controlled systems can lead to direct loss of life due to equipment failure or indirect losses due to failure of critical infrastructure controlled by SCADA. SCADA systems have evolved in parallel with the growth and sophistication of modern computing technology. That means that SCADA is technology-dependent and it is not a standalone science but it is the result of integrating variety of applied sciences such as communication, computers, software engineering, networking, security, etc. Therefore and according to the evolution of technology, SCADA developers architect SCADA by selecting the appropriate technologies and mechanisms for handling SCADA challenges such as complexity resulted from the continuous increasing of the size of SCADA systems. SCADA systems should be scalable because their components and the amount of exchanged data increase with rapid rate. Further, a SCADA system should be flexible and able to adapt to internal or external changes. SCADA is no longer restricted to concern real-time monitoring and control inside a factory, but it currently has new concerns. It has been moved from the local sense to the global one. For example, in advanced countries, a global SCADA system is used for supervising and real-time monitoring of the power grid (thousands of electric power generation and distribution stations).

Designing, monitoring and controlling of modern industrial systems is getting more challenging as a consequence of the steady growth of their size, complexity, level of uncertainty, unpredictable behavior, and interactions. Unfortunately, the conventional SCADA systems are not capable of providing information management and high-level intelligent approaches. That is because achieving those functionalities requires comprehensive information management support and coordination among system devices, and the control of many different types of task, such as data transportation, data display, data retrieval, information interpretation, control signals and commands, documentation sorting and database searching…,etc. These tasks operate at different timescales and are widely distributed over the global system and its subsystems. Without sophisticated software architectures and hardware structures, it is impossible to handle these tasks efficiently, safely and reliably, with the possibility of online reconfiguration and flexibly embedding applications [1] [2].

It becomes obvious that computing systems, especially those related to modern industrial applications such as SCADA systems, are becoming increasingly interconnected and more difficult to maintain. Due to the increase in the size, complexity and the number of components, it is no longer practical to anticipate and model all possible interactions and conditions that the system may experience at design time. Similarly, the systems are becoming too large and too complex for system managers to maintain them at run-time [3].

The agent-based approach seems to be the promising solution. The rapid development of the field of agent-based systems offers a new and exciting paradigm for the development of sophisticated programs in dynamic and open environments [4]. The agent-based approach is considered as a new software engineering architectural style for the development of complex, decentralized and open software applications. What distinguishes the agent-based approach from other traditional approaches is its unique ability to handle simultaneously many challenges of the current software applications specially those applications which are highly distributed and their working environments are highly dynamic and uncertain. MAS provide a suitable paradigm for decentralized systems in which autonomous individuals engage in flexible high-level interactions.

By an agent-based system, we mean the one in which the key abstraction used is that of an agent. We therefore expect an agent-based system to be both designed and implemented in terms of agents. An agent-based system may contain any non-zero number of agents. The multi-agent case where a system is designed and implemented as several interacting agents, is both more general and significantly more complex than the single-agent case. Originally, MAS emerged as a scientific area, from the previous research efforts in Distributed Artificial Intelligence (DAI) started in the early eighties. MAS are now seen as a major trend in R & D, mainly related to artificial intelligence and distributed computing techniques. This research has attracted attention in many application domains where difficult and inherently distributed problems have to be tackled [5]. A MAS is defined as a set of interacting autonomous agents in a common environment in order to solve a common, coherent task. These agents try to achieve individual objectives which are sometimes conflicting. There are many definitions of the meaning of agent we found in their literature but the widely accepted definition of an agent is that which has been stated by Wooldridge and Jennings [6] who defined an agent as:

"... *a hardware or* (*more usually*) *software-based computer system that enjoys the following properties: –autonomy: agents operate without the direct intervention of humans or others, and have some kind of control over their actions and internal state; –social ability: agents interact with other agents (and possibly humans) via some kind of agent-communication language; –reactivity: agents perceive their environment, (which may be the physical world, a user via a graphical user interface, a collection of other agents, the Internet, or perhaps all of these combined), and respond in a timely fashion to changes that occur in it; –pro-activeness: agents do not simply act in response to their environment, they are able to exhibit goal-directed behavior by taking the initiative*".

A MAS is autonomous, means that there is no external entity which controls this system. This property is enforced because agents inside the system are autonomous. Inside a MAS, data (knowledge) are distributed inside all its agents. Moreover, the control is decentralized (there is no supervisor). MAS allow the design and implementation of software systems by using the same ideas and concepts that are the very founding of human societies and habits. These systems often rely on the delegation of goals and tasks among autonomous software agents, which can interact and collaborate with others to achieve common goals [7]. It provides an approach to solve a software problem by decomposing the system into a number of autonomous entities embedded in an environment in order to achieve the functional and quality requirements of the system [8]. The agent-based computing has been hailed as "the next significant breakthrough in software development" [9], and "the new revolution in software" [10].

Currently, agents are the focus of intense interest on the part of many sub-fields of computer science and artificial intelligence. Agents are being used in an increasingly wide variety of applications, ranging from comparatively small systems such as email filters to large, open, complex, mission critical systems such as air traffic control. At first sight, it may appear that such extremely different types of system can have little in common. And yet this is not the case: in both, the key abstraction used is that of an agent [11]. MAS are claimed to be especially suited to the development of software systems that are decentralized, can deal flexibly with dynamic conditions, and are open to system components that come and go. That is why they are used in domains such as manufacturing control, automated vehicles, and e-commerce markets.

There are many reviews and surveys tackled with the applications of agent and multi-agent systems in variety of application domains. For instance, in [11] the authors described the suitability of intelligent and autonomous agents to model and develop applications for certain types of software system which are inherently more difficult to correctly design and implement than others. And they subdivide these systems into three classes: open systems, complex systems, and ubiquitous computing systems. Moreover, they classified these systems or applications according to the application domain such as industrial applications (*i.e.* manufacturing, process control…, etc.), commercial applications (*i.e.* information management, e-commerce…, etc.), medical applications (*i.e.* patient monitoring, health care…, etc.), and entertainment (*i.e.* games, interactive cinema…, etc.). The authors tackled many types of agent-based applications and described generally their functionality. They tackled the design and development methods and challenges of MAS from a general viewpoint independent from a specific application domain.

In [12] the author presented a summary of the state-of-the-art of the Distributed Artificial Intelligence (DAI) applied to Intelligent Manufacturing and presented main applications along with different technologies applied in these areas. Also, he presented areas of agent technology applications and the agent development tools. Moreover, he described briefly and generally the agents and multi-agent architecture types and design approaches.

In [13] the authors reviewed multi-agent systems power engineering applications and stated that the flexibility offered by an open architecture of agents with good social ability easily led to the design of a fault-tolerant system. Also, they reviewed the application of MAS as a modeling and simulation approach and its application in grid computing and web services composition, etc. Moreover they presented a bibliographical analysis of agent research aiming to provide an indication of the active areas of agent research with respect to power systems and related applications. They concentrated on the functionalities and the different power systems aspects which can be designed and developed by MAS.

Shehory [14] stated that one aspect of multi-agent systems that had been only partially studied was their role in software engineering and especially their merit as a software architecture style. In his report he provided analysis guidelines which supported designers in their assessment of the suitability of MAS as a solution to computational problem they addressed. Moreover he discussed the architectural properties that should be considered when analyzing such systems and he supported his work with case studies of several MAS.

Still there is a debate about the identity of multi-agent systems: is it a radically new way for systems engineering? Is it a new software engineering modeling style? Weyns *et al.* [8] stated that the trend in agent-oriented software engineering was to consider multi-agent systems as a radically new way of engineering software, and this position isolated agent-oriented software engineering from mainstream software engineering and could be one important reason why MAS were not widely adopted in industry yet. What we understand from this is that the agent-based modeling should be considered similar to the object-oriented modeling both of them are software engineering modeling styles. In other words, MAS are considered now as a novel general-purpose paradigm for software development.

Moreover, Agents and multi-agent systems constitute one of the most prominent and attractive technologies in computer science at the beginning of this new century. Agents and multi-agent systems technologies, methods, and theories are currently contributing to many diverse domains. Jennings and Wooldridge [11] stated that intelligent agents were a new paradigm for developing software applications. Agent technology is seen as a fundamentally important new tool for building a wide array of systems (*i.e.* open systems, complex systems, and ubiquitous computing systems). A number of software tools exist that allow a user to implement software systems as agents, and as societies of cooperating agents, by tools we mean agent platforms. An agent platform provides a basis for the implementation of MAS, and the means to manage agent execution and message passing. For the sake of interoperability, it is intended that the agent platform architecture should be implemented by using the Foundation of Intelligent and Physical Agents [15] specifications of agent platforms abstract architecture. FIPA is an IEEE Computer Society standards organization that promotes agent-based technology and the interoperability of its standards with other technologies. FIPA, the standards organization for agents and MAS was officially accepted by the IEEE as its eleventh standards committee on June 8, 2005. The specifications define an abstract agent platform, a number of services that must or may be provided by such a platform and a standard communications language.

From the above discussion it is clear that the agent technology can offer a feasible approach for handling the challenges and requirements of modern SCADA systems [16]. This paper proposes a simple but flexible and interoperable SCADA system based on the agent-based approach. The proposed system-to-be can be used for providing not only a local access but also a remote access through the Internet to local control processes. The remaining of this paper is organized as follows: Section 2 provides a short background of the concerned problem; Section 3 presents the proposed practical approach; Section 4 presents the real case study application of the proposed practical approach; Section 5 concerns the deployment and testing of the proposed SCADA system; and finally, Section 6 concludes the paper and highlights the future intentions.

2. Background

Traditionally, SCADA developers adopted the web-based approach to realize a remote real-time monitoring of production control processes, a survey of some web-based SCADA systems can be found in [17]. But the problem of the web-based applications which use the web browser as a remote client and a web server as the onsite server through which the remote operators have access to control processes, is an example of centralized control architecture which has disadvantages such as:

1) Single point of failure problem;
2) These systems are not flexible because they cannot adapt to changes such as increasing the number of system components;
3) These systems are not scalable because increasing system components can result in web server overload and crash.

Figure 1 demonstrates why the web-based SCADA systems are not flexible, not scalable, and are not able to adapt to working environments dynamic changes. As shown in **Figure 1(a)** a remote operator uses the web browser to connect to an onsite web server to access a control process. In case the system changes by increasing the number control processes and the number of remote operators as shown in **Figure 1(b)** the web server which has limited resources such as the computation power, memory and the network bandwidth, is vulnerable to be overloaded and crash.

Many of SCADA developers considered the web-based SCADA systems as a solution to some of modern SCADA challenges and shortcomings [17] [18]. Although they succeeded to develop web-based SCADA applications that had a number of important quality attributes such as efficiency, interoperability, and real-time

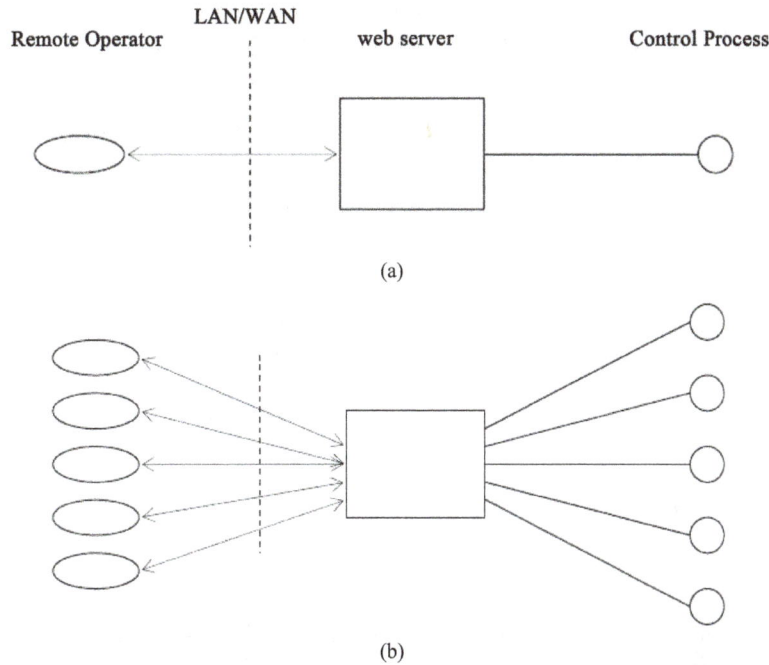

(a)

(b)

Figure 1. Traditional centralized web-based SCADA architecture.

monitoring, they failed to obtain other important attributes such as scalability and adaptivity. The reason is that the Hypertext Transfer Protocol (HTTP) uses a client-server architecture with which it is difficult to achieve higher scalability degree because of the dominant adopted centralized control regime. The author of [19] emphasized that the future infrastructures will contain a huge amount of data is generated by real world devices and needs to be integrated and processed within a specific context and communicated on demand and on time. As a result, traditional approaches aiming at the efficient data inclusion in enterprise services need to be changed. The main challenges facing current and future SCADA systems are generally related to quality attributes. The extent to which the system possesses a desired combination of quality attributes such as scalability, usability, performance, reliability, and security indicates the success of the design and the overall quality of the software system. New design and development approaches should be adopted for building future complex and critical SCADA systems.

The agent-based approach considers a SCADA system as a multi-agent system. A multi-agent system provides a powerful computational technology, for which dynamic aspects are based on interactions between autonomous agents, rather than centralized control. To do their required functions, agents have the ability to communicate with each other, with their environment, and with human operators. Agents are particularly adapted to complex systems modeling where environments are unpredictable [20]. **Figure 2** provides an abstract architecture of the agent-based real-time monitoring. As shown, the system is flexible and can easily adapt to changes such as increasing the number of control processes which can easily assigned to new agents, and also increasing the number of remote operators by assigning a remote agent to each operator. The agent-based approach adopts a decentralized control and behavior, which had proven to be the appropriate approach for handling system complexity and unpredictable work environments.

The agent-based approach can be considered as the promising solution for the design and development of future SCADA because it has the ability to handle future SCADA challenges such as complexity, scalability, and flexibility, etc. Although these quality attributes are not unique to MAS, but combining them in a single system is unique to MAS. This combination results in the suitability of MAS for solving problems where information, location and control are highly distributed, heterogeneous, autonomous components comprise the system, the environment is open and dynamically changing, and uncertainty is present [14] [21].

The agent-based approach reinforces decentralization which is the effective way for obtaining higher degree of scalability. Today, agents are being applied in a wide range of industrial applications, for example, process control, manufacturing, air traffic control, etc. These applications have been relatively successful, suggesting

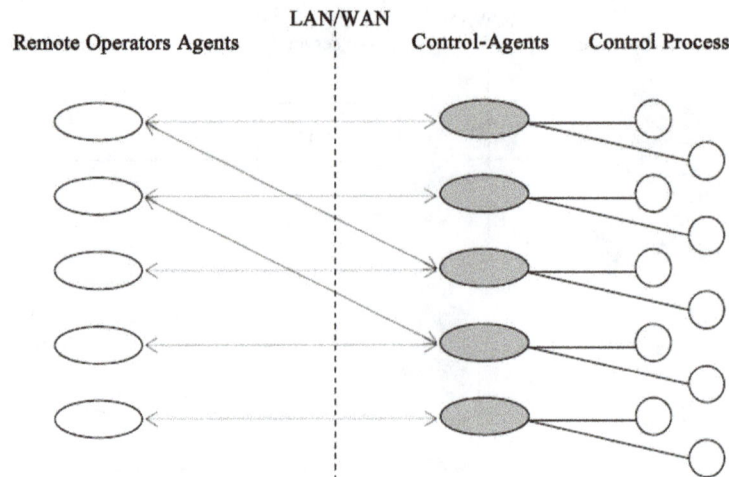

Figure 2. The decentralized agent-based SCADA.

that the multi-agent approach is a promising method for the implementation of industrial automation systems and that encourages us to adopt MAS for the design and development of SCADA systems. Metzger *et al.* [22] surveyed the applications of the agent technology in the industrial process control and concluded that the agent technology is particularly popular in the manufacturing domain, while the applications in other domains of industrial control are scarce. They related their conclusions to the lack of the technology support on the part of control instrumentation vendors. In manufacturing automation, the process consists of discrete and countable components and actions. The natural approach is to assign the software agents to each of the components and each of the actions performed. On the other hand, the process automation deals with the continuous physical phenomena, such as chemical reactions. When a process automation system is designed, the phenomena are represented as mathematical models, for which control algorithms are chosen in order to keep the process parameters within a desired range. Therefore, in a single continuous control loop, there is not much place for any additional computational techniques, including the agent technology.

3. The Proposed Approach

The proposed approach is based on the integration of MAS and OPC process protocol [23], realizing this integration enables us to achieve two goals. First, it will be possible to transfer the process data from the process domain to the information domain (MAS), where executive management can use this data for decision making. Second, it will be possible to take the benefit of control devices interoperability provided by the OPC process protocol. Interoperability is assured through the creation and maintenance of non-proprietary open standards specifications. OPC initially meant *Ole for Process Control*, but after it becomes familiar for achieving control systems interoperability it was redirected to mean *Open Process Control*. OPC is open connectivity in industrial automation and the enterprise systems that support the industry. The first OPC standard specification resulted from the collaboration of a number of leading worldwide automation suppliers working in cooperation with Microsoft. Originally based on Microsoft's OLE COM/DCOM technologies, the specification defined a standard set of objects, interfaces and methods for use in process control and manufacturing applications to achieve interoperability. There are now hundreds of OPC Data Access (OPC DA) servers and clients.

   Figure 3 shows the basic architecture of the OPC protocol usage. The OPC server is connected to the PLC (programmable Logic Controller) which is responsible of directly controlling a certain control process. By this way the OPC server can be considered as the driver of the PLC, in the same sense as the printer driver. The communication between the OPC server and the PLC is vendor-specific and depends on the custom technologies used by the vendors for manufacturing their control systems. On the other hand, the communication between the OPC server and software applications adopts the COM/DCOM technologies created by Microsoft. That is considered as a limitation because it means that the OPC protocol is bonded only to Microsoft Windows operating systems. In this research and as shown in figure a Siemens S7-400 PLC is used as a real control system but the same system and approach can be used with other control systems manufactured by other vendors such as ABB,

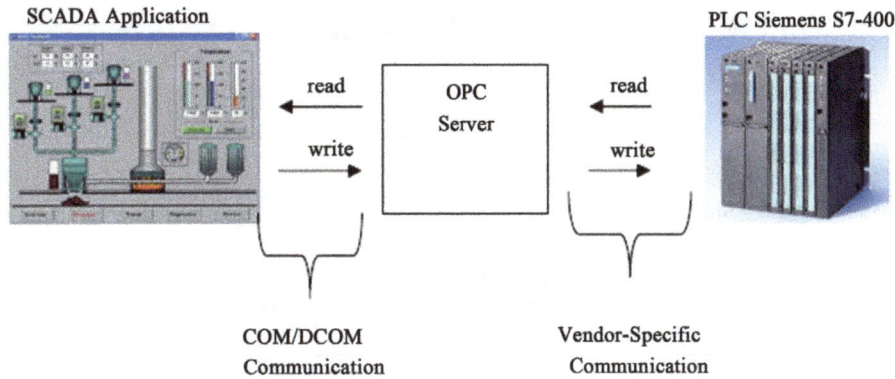

Figure 3. Basic OPC protocol architecture.

Honeywell, etc. That is possible because of the vendors interoperability provided by the standard OPC protocol. This feature represents the second goal (interoperability) of the aimed two goals mention in the beginning of this section.

Most MAS platforms are implemented in Java programming language [24], therefore to establish a connection between a Java agent and an OPC server, a Java-COM bridge or adapter is required as shown in **Figure 4**. For the purpose of this thesis the *JEasyOPC* Java-OPC adapter [25] [26] was selected to interface a Java MAS agent with OPC servers and of course the other bridges can be used in a similar way. *JEasyOPC* is a Java OPC client that is now greatly enhanced. It uses a JNI layer coded in Delphi. The current version supports both OPC DA 2.0 and OPC DA 3.0. *JEasyOPC* is free source and can be downloaded easily from the Internet. **Appendix A** provides some information about realizing the Java-OPC interactions through JEasyOPC.

Figure 5 demonstrates the layered nature of the system-to-be. The system contains four layers (or worlds) with three interfaces in between. The layers are (from bottom to up): the physical process layer, control systems layer, OPC communication layer, and the agents' layer which represents the application world. The figure also demonstrates the appropriate interface between any two layers. For example, the interface between the physical process and its control system is the field interface, which is constructed using analog/digital signals transferred through wires, or it can be a digital field bus. Further, the interface between the control system and the OPC driver layer is vendor specific. Finally, the interface between the OPC layer and the agents' world layer a COM/ DCOM interface. In this paper, we are interested in the later (top) layer and interface (Shown in Dashed line in **Figure 5**).

To connect non-agent Java applications to OPC servers, it is required to download the *JEasyOPC* library, which includes a sample Eclipse [27] java project that contains basic sample examples programmed to connect to a default OPC server (Matrikon. OPC. Simulation). These examples can be modified as required, for instance the Makitron OPC server can be replaced with a Siemens one (*i.e.* OPC. Simatic Net). Furthermore, the examples don't provide a graphical user interface (GUI) and therefore the developer has to modify them to provide a suitable GUI. Before running Eclipse, a recent Java run-time environment (JRE) should be installed on the operating system. Using a Java development environment such as Eclipse frees the developer from caring about modifying related system variables such as CLASPATH and PATH as it does these issues automatically. From the other hand, to connect a Java agent (under Eclipse) to an OPC server, it is required first to install a MAS platform such as Jade (Java Agent Development Environment) [28]. Jade is a software framework fully implemented in Java language. It simplifies the implementation of multi-agent systems through a middleware that claims to comply with the FIPA specifications and through a set of tools that supports the debugging and deployment phase. The agent platform can be distributed across machines with different operating systems and the configuration can be controlled via a remote GUI. The configuration can be even changed at run-time by creating new agents and moving agents from one machine to another one as and when required. The only system requirement is the Java Run-Time version 5 or later. JADE is distributed in open source. To run Jade under Eclipse, the developer should add Jade libraries to Eclipse Java build path (*project → prosperities → Java Build path → Libraries → add external Jars*), then through the Windows file system find *Jade.jar* file in the Jade home as shown in **Figure 6**. Now Eclipse is ready for creating a new java class that extends *jade. core. Agent* class and start programming the required agent let us call it OPC-Agent.

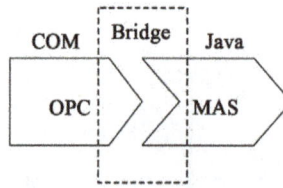

Figure 4. Integrating MAS with OPC protocol.

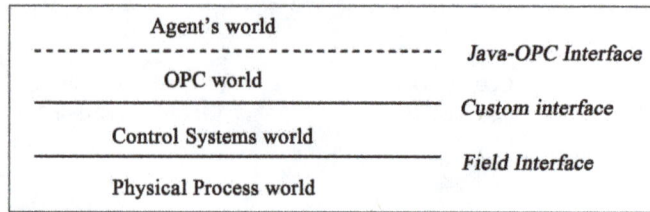

Figure 5. System layers and the interfaces in between.

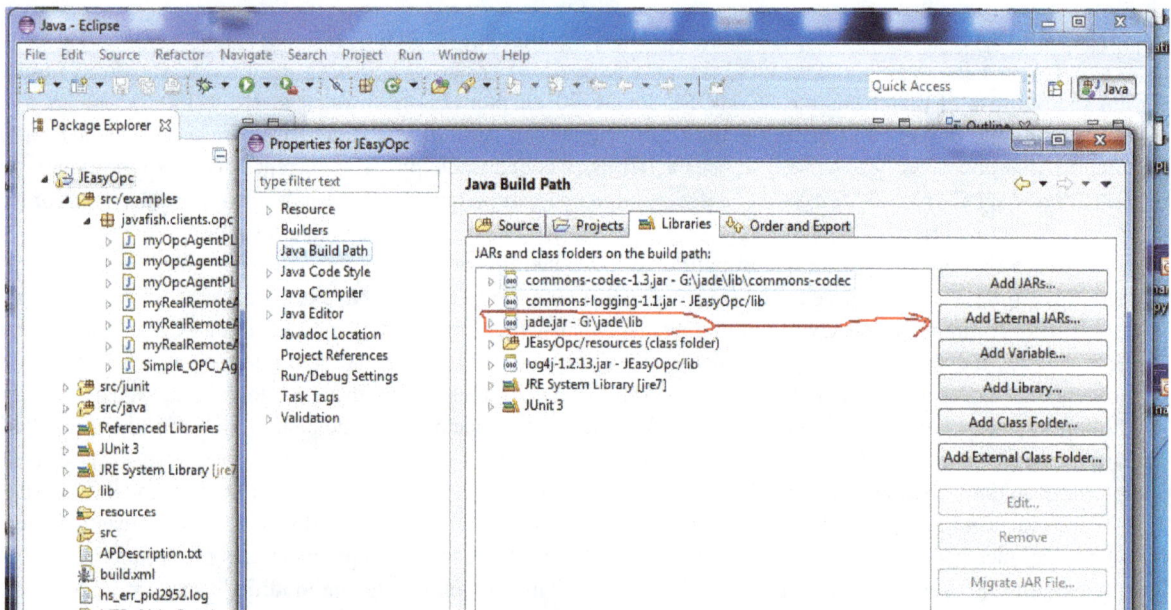

Figure 6. Running JADE from ECLIPSE.

4. Real Case Study Example

The case study is concerned with the real adoption of the agent-based approach for connecting locally and remotely to real control processes. The case study was executed in a paper mill, specifically in the finishing area within a paper mill. The paper mill finishing area comprises three separated stations as follows:

1) The Winder station, which takes a 6 meters width paper spool as its input and gives a smaller width paper rolls (*i.e.*, 6 × 1 meter) as output.

2) The wrapping station, which is used for paper rolls packaging and labeling.

3) The Salvage winder station for preparing sample rolls; it is a smaller Winder station.

Each control process is controlled by a Siemens S7-400 PLC; also the three stations are connecting a local LAN including the operator stations and the control systems. Further, each control process contains many process variables that should be monitored continuously, for instance, for the winder station the following variables required to be monitored in real-time: Machine Speed (m/min), Paper Tension (N/m), Drums Torques (N/m), and so on. **Figure 7** shows the winder station while it is running. The architecture of the experimental practical project is shown in **Figure 8**. As shown in the figure, all hosts are connected through a LAN with the

Figure 7. Qena paper winder station.

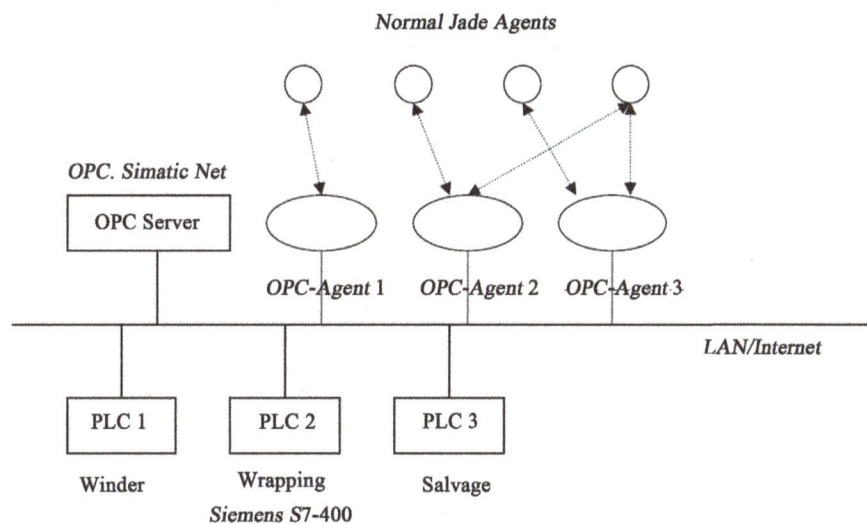

Figure 8. Architecture of the concrete example.

possibility to be connected to the Internet to enable real-time data accessing from outside the paper mill. Furthermore, the normal remote Jade agents are considered as the process real-time data consumers while each one can be hosted on a different host in the LAN or in the Internet. In this case study project, it is assumed that:

1) Each PLC of a control process is assigned to an OPC-Agent, which in turn registers its services to the JADE DF (Directory Facilitator) agent, represents the yellow page service in the JADE platform. Therefore, in the case study example there are three OPC-Agents, one for each real PLC system.

2) The remote operator agents initially don't know which OPC-Agent is responsible of which PLC system. Accordingly, after start running they contact the JADE DF agent to search for the required OPC-Agents, as illustrated in **Figure 9**.

3) After identifying the required OPC-Agents, the operator agents send an ACL message with per-formative (Request) to the OPC-Agent asking her to continuously send back the real-time process data of its assigned PLC. This request is done only once then the OPC-Agent will continuously send the required real-time process data through ACL messages with per-formative (Inform), this mechanism is called subscription behavior.

4) Each operator agent can access many OPC-Agents simultaneously, in other words each remote agent can be used to supervise and monitor more than one control process.

5) The project JADE platform was named as SCADA, so an agent with local name "H1" has a full name "H1@SCADA".

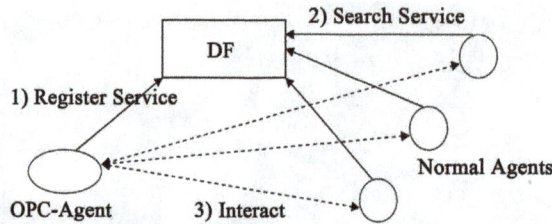

Figure 9. Exploiting JADE yellow page service.

6) The implemented agents were made simple with light GUI because it is just an illustrative experimental example. In real applications, the agents GUI may become complex and user friendly. **Figure 10** shows a simple remote operator agent continuously receives changed data from an OPC-agent situated on a host inside the mill. The remote agent designed to provide real-time monitoring, alarm service, and trend service, as shown in the figure. This feature represents our first goal (remote real-time monitoring) as we mentioned in the beginning of this section. This feature is very important for executive management stuff, which might be located far from the factory to be able to take the suitable quick decision in the suitable time.

It is possible to run an operator agent from any host in the LAN or even from the Internet. The point will be with the agents launching commands which defined in the Eclipse class arguments setting (*run as →run configuration → new configuration → arguments*), as shown in **Figure 11**. For example (considering *javafish. clients.opc* as the package name):

- For initially booting and running an OPC-Agent called H1 in the main container of the JADE platform called SCADA:
  -gui-name SCADA -local -agents H1: javafish.clients.opc.myOpcAgent
- For running an operator agent called R1 on the same host as the OPC-Agent H1:
  -container-agents R1:javafish.clients.opc.myRealRemotAgent1
- For running an operator agent called R2 from another host in the LAN Where the host with IP (192.168.100.31) is the one on which the Jade platform main container is situated:
  -container-host 192.168.100.31 R2:javafish.clients.opc.myRemoteAgent

For debugging purposes, the JADE platform provides a built-in sniffer tool, which is used to visualize messages as a low-level UML sequence diagram, was used as shown in **Figure 12**, which demonstrates how JADE agents interacts together through exchanging asynchronous messages. The sniffer tool is very powerful for tracing the system-to-be behaviors in run-time

Returning to the concerned case study, **Figure 13** presents the system sequence diagram. As shown in the figure the remote operator agent depends on the yellow page service provided by the JADE directory facilitator (DF) to find the control agent responsible of a required control process. The operator agent will wait for the DF until it provides it with the required control process. After getting a response from the DF the remote agent starts interacting with the control agent to get the process real-time data and presenting this data to the human operator in text or graphical format. As we have illustrated before the remote operator agent can be hosted locally onsite or remotely outside, *i.e.* using the Internet.

5. Deployment and Testing

The proposed approach solves the problem of accessing the OPC server (which is a COM/DCOM server) from the Internet because using DCOM communication is not recommended through the Internet because of its delay time and configuration security problems. The agent-based approach depends on the HTTP protocol, which is a firewall-friendly protocol. **Figure 14** shows the concrete project while it is running. The figure includes two running OPC-Agent in the left side:

- WinderOpcAgent 1;
- WrappingOpcAgent 1;

And three remote normal agents in the right side:

- WinderRemoteAgent 1;
- WinderRemoteAgent 2;
- WrappingRemoteAgent 2;

Figure 15 demonstrates how the OPC-Agents continuously send real-time data to the remote operator agents

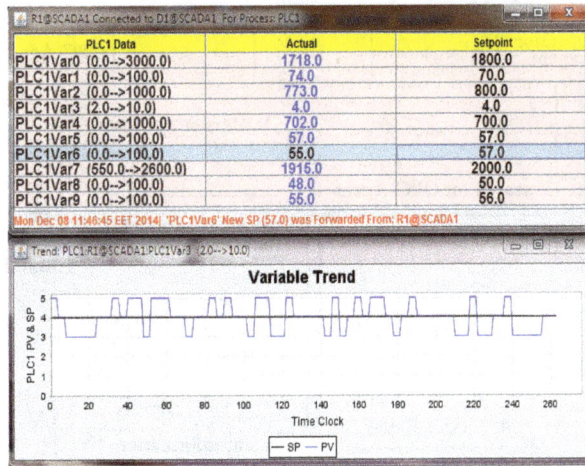

Figure 10. A remote operator agent connected to an onsite OPC-agent and provides real-time process monitoring, Alarm service, and Trend service.

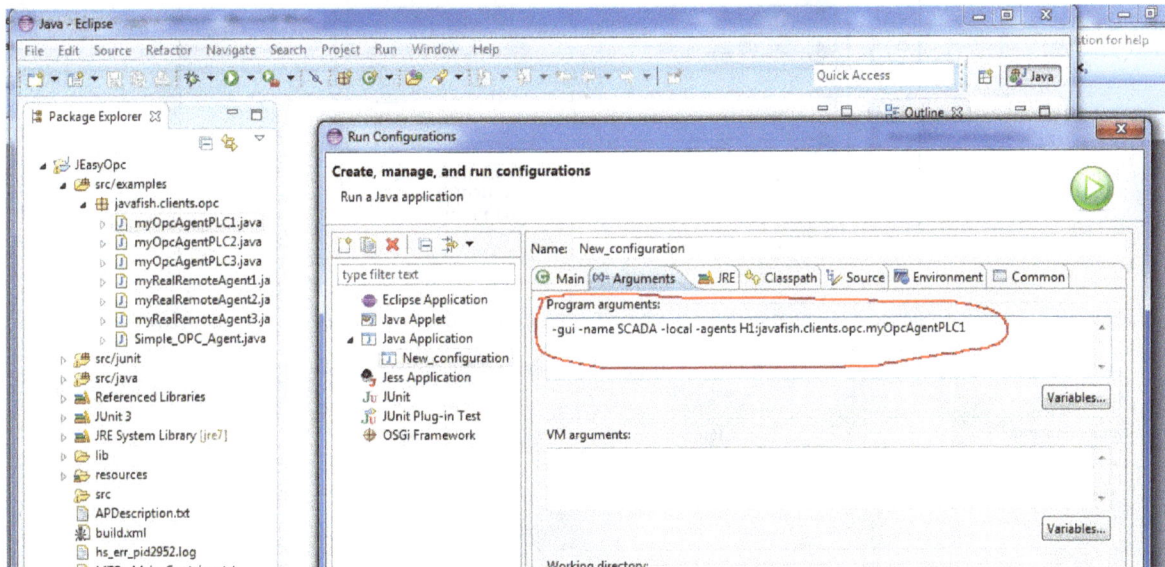

Figure 11. Configuring a JADE agent in Ellipse.

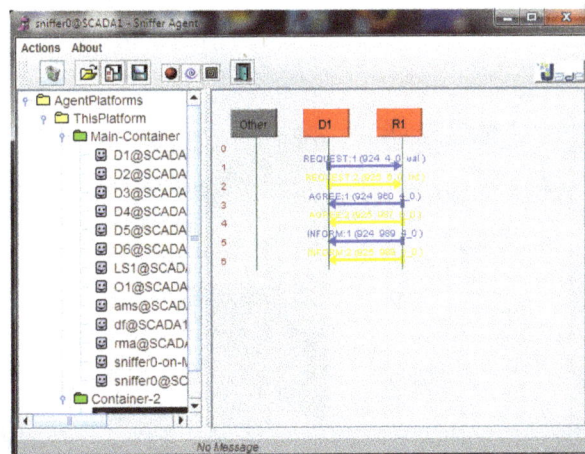

Figure 12. Jade sniffer tool shows message flow between two agents.

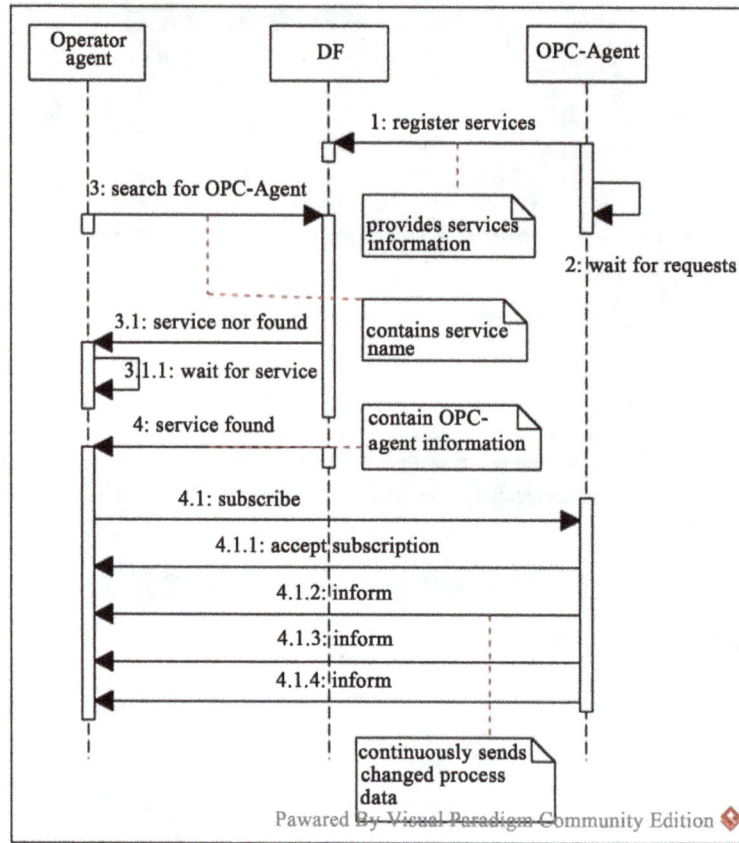

Figure 13. System sequence diagram demonstrated how a remote agent finds a control agent to interact with.

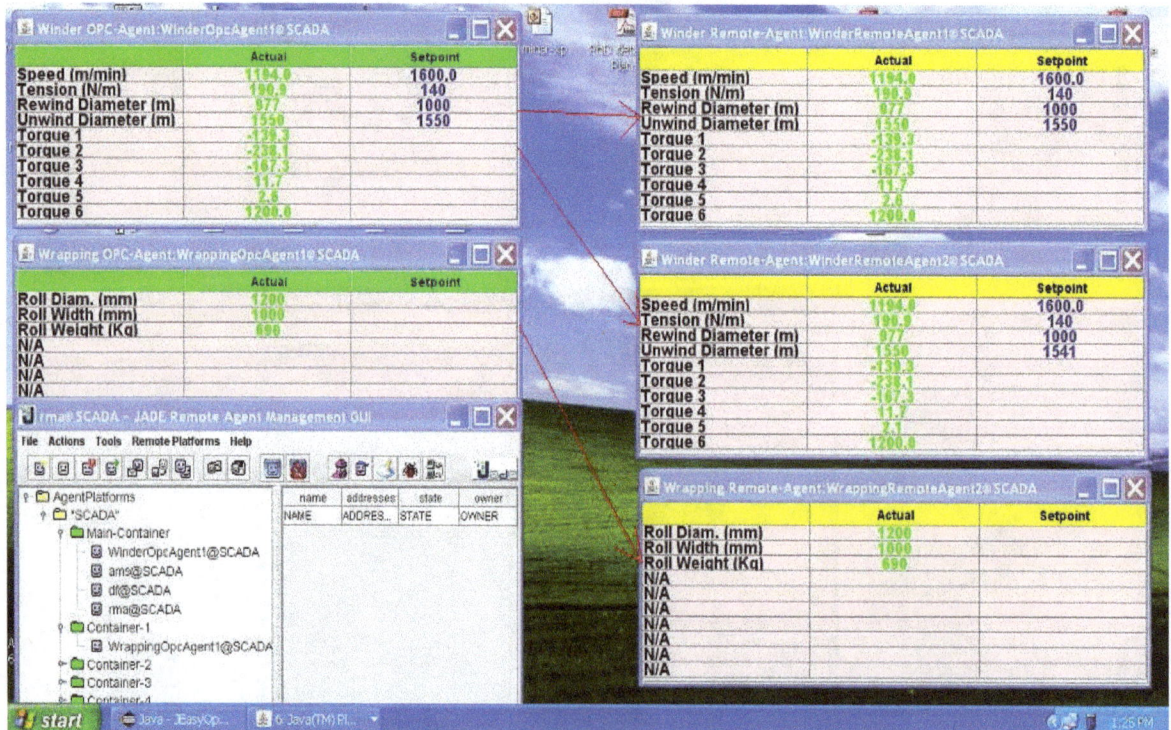

Figure 14. Running the concrete example: Two OPC-Agents (left side) and three normal Jade Agents (right side).

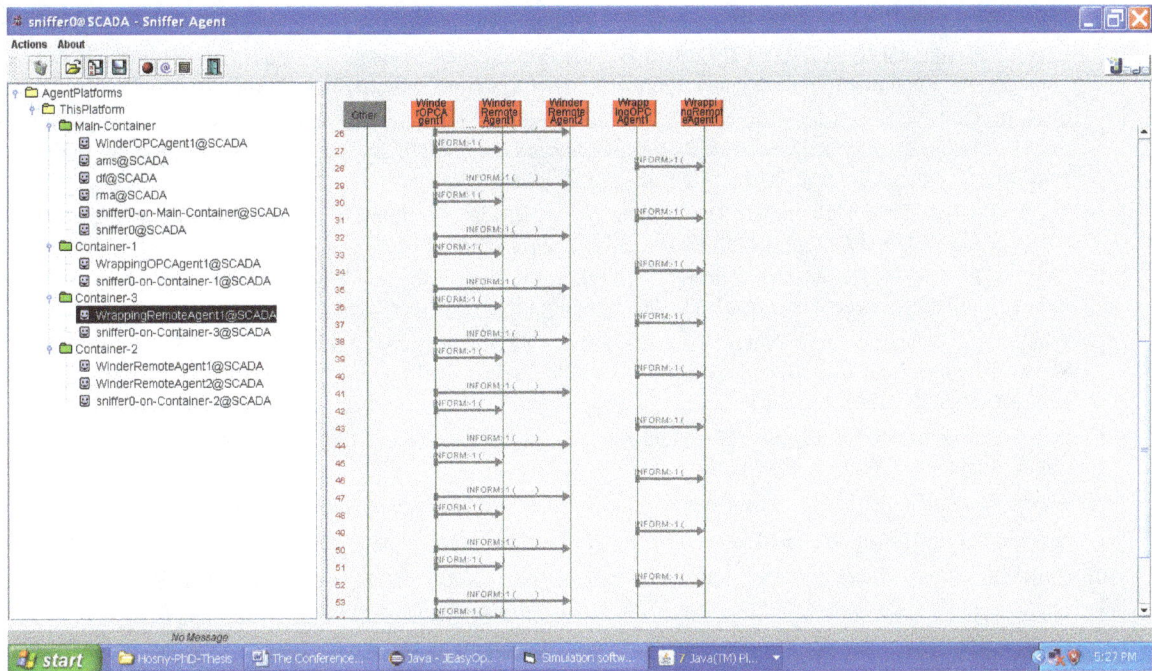

Figure 15. JADE's built-in sniffer tool, three normal jade agents, two of them subscribe to Winder OPCAgent and one subscribes to Wrapping OPCAgent.

as a result of subscription requests from the operator agents. The figure shows two OPC-Agents and three remote agents. Two remote operator agents subscribed to one of the two OPC-Agents and one subscribed to the other one. The number of OPC-Agents and remote agents can be increased easily by instantiating these agent types as required and anywhere in the LAN or the Internet. Note that it's not necessary for the OPC-Agents to have a GUI but the operator agents should have a user friendly GUI. In this experiment, a simple GUI is designed for both agent types.

6. Conclusion

Multi-agent systems propose solutions to highly distributed problems in dynamic and open computational domains. SCADA is one of these systems which are highly distributed, decentralized, open, and requires high degree of scalability. The agent-based approach should be adopted for developing flexible and scalable SCADA systems. In this research, the agent-based approach was adopted to develop a simple, flexible, and interoperable SCADA system. The achieved interoperability is realized by the adoption of the OPC technology for process communication, by this way it is possible to run the developed system with any type of control systems (*i.e.* PLC, DCS, CNC, etc.) independent of any vendor, and the same Java-OPC interface will be used with only some few modifications. By flexible, we mean that the system is able to adapt its dynamic working environment. The applicability of the agent-based approach is demonstrated by developing a practical project carried out in a paper mill. Furthermore, the proposed approach is cost-effective compared with custom SCADA packages. The future work will be the adoption of the agents' technology for building large-scale and highly distributed SCADA systems.

Acknowledgements

Hosny Abbas thanks the managers of Quena Paper Company (Egypt), Mr. Shazely Abd-El Azeem (General Manager) and Mr. Abd-El Hameid Omar (Manager of Automation Sector) for their encouragement and their acceptance to allow us to carry out this research in the company at their responsibilities and supervisory.

References

[1]   Chakrabarti, S., Kyriakides, E., Bi, T., Cai, D. and Terzija, V. (2009) Measurements Get Together. *IEEE Power and*

Energy Magazine, **7**, 41-49. http://dx.doi.org/10.1109/mpe.2008.930657

[2]   Buse, D.P. and Wu, Q.H. (2007) IP Network-Based Multi-Agent Systems for Industrial Automation: Information Management, Condition Monitoring and Control of Power Systems. Springer, New York.

[3]   Kota, R., Gibbins, N. and Jennings, N.R. (2009) Decentralised Structural Adaptation in Agent Organisations. In: Vouros, G., Artikis, A., Stathis, K. and Pitt, J., Eds., *Organized Adaption in Multi-Agent Systems*, Springer, Berlin Heidelberg, 54-71. http://dx.doi.org/10.1007/978-3-642-02377-4_4

[4]   Luck, M. (1999) From Definition to Deployment: What Next for Agent-Based Systems? *The Knowledge Engineering Review*, **14**, 119-124. http://dx.doi.org/10.1017/S0269888999142048

[5]   Oliveira, E., Fischer, K. and Stepankova, O. (1999). Multi-Agent Systems: Which Research for Which Applications. *Robotics and Autonomous Systems*, **27**, 91-106. http://dx.doi.org/10.1016/S0921-8890(98)00085-2

[6]   Woolbridge, M. and Jennings, N.R. (1995) Agent Theories, Architectures, and Languages: A Survey. In: Woolbridge, M. and Jennings, N.R., Eds., *Intelligent Agents*, Springer-Verlag, Berlin, 1-39. http://dx.doi.org/10.1007/3-540-58855-8_1

[7]   Di Marzo Serugendo, G., Gleizes, M.-P. and Karageorgos, A. (2011) Self-Organizing Software, from Natural to Artificial Adaptation. Springer, New York.

[8]   Weyns, D., *et al.* (2006) Multi-Agent Systems and Software Architecture. *AAMAS'06*, Hakodate, May 2006, 8-12.

[9]   Sargent, P. (1992) Back to School for a Brand New ABC. The Guardian Newspaper, 12 March 1992, 28.

[10]   Guilfoyle, C. and Warner, E. (1994) Intelligent Agents: The New Revolution in Software. Ovum Report.

[11]   Jennings, N.R. and Wooldridge, M. (1998) Applications of Intelligent Agents. Springer, Berlin.

[12]   Madejski, J. (2007) Survey of the Agent-Based Approach to Intelligent Manufacturing. *Journal of Achievements in Materials and Manufacturing Engineering*, **21**, 67-70.

[13]   McArthur, S.D.J., Catterson, V.M. and Hatziargyriou, N.D. (2007) Multi-Agent Systems for Power Engineering Applications—Part I: Concepts, Approaches, and Technical Challenges. *IEEE Transactions on Power Systems*, **22**, 1743-1752.

[14]   Shehory, O. (1998) Architectural Properties of Multi-Agent Systems, Technical Report CMU-RI-TR-98-28. The Robotics Institute, Carnegie Mellon University, Pittsburgh.

[15]   Foundation for Intelligent Physical Agents (FIPA) (2000) FIPA Agent Management Specification. http://www.fipa.org/specs/fipa00023/

[16]   Abbas, H.A. (2014) Future SCADA Challenges and the Promising Solution: The Agent—Based SCADA. *International Journal of Critical Infrastructures*, **10**, 307-333. http://dx.doi.org/10.1504/IJCIS.2014.066354

[17]   Abbas, H.A. and Mohamed, A.M. (2011) Review in the Design of Web Based SCADA Systems Based on OPC DA Protocol. *International Journal of Computer Networks*, **2**, 266-277.

[18]   Mohamed, A.M. and Abbas, H.A. (2011) Efficient Web Based Monitoring and Control System. *Proceedings of the 7th International Conference on Autonomic and Autonomous Systems*, Venice, 22-27 May 2011, 18-23.

[19]   Karnouskos, S. and Colombo, A.W. (2011) Architecting the Next Generation of Service-Based SCADA/DCS System of Systems. *Proceedings of the 37th Annual Conference on IEEE Industrial Electronics Society*, Melbourne, 7-10 November 2011, 359-364. http://dx.doi.org/10.1109/iecon.2011.6119279

[20]   Vale, Z.A., Morais, H., Silva, M. and Ramos, C. (2009) Towards a Future SCADA. *Proceedings of the Power & Energy Society General Meeting*, Calgary, 26-30 July 2009, 1-7.

[21]   Jennings, N.R. (2001) An Agent-Based Approach for Building Complex Software Systems. *Communications of the ACM*, **44**, 35-41. http://dx.doi.org/10.1145/367211.367250

[22]   Metzger, M. and Polakow, G. (2011) A Survey on Applications of Agent Technology in Industrial Process Control. *IEEE Transactions on Industrial Informatics*, **7**, 570-581. http://dx.doi.org/10.1109/TII.2011.2166781

[23]   OPC Foundation (2010) OPC DA 3.0 Specification [DB/OL].

[24]   Russell, J.P. (2001) Java Programming for Absolute Beginner. Prima Publishing, Roseville.

[25]   (2015) http://sourceforge.net/projects/jeasyopc/

[26]   Diaconescu, E. and Spirleanu, C. (2012) Communication Solution for Industrial Control Applications with Multi-Agents Using OPC Servers. *Proceedings of the 2012 International Conference on Applied and Theoretical Electricity (ICATE)*, Craiova, 25-27 October 2012, 1-6.

[27]   (2015) https://www.eclipse.org/

[28]   Bellifemine, F., Poggi, A. and Rimassi, G. (1999) JADE: A FIPA-Compliant Agent Framework. *Proceedings of the Practical Applications of Intelligent Agents and Multi-Agents*, April 1999, 97-108.

Appendix A: Java-OPC Interaction

The JEasyOPC project is established on LGPL (GNU Library or Lesser General Public License). The project can be downloaded from SourceForge.net as actual release (2.xx.xx) or night revision from SVN repository. The project is built on Eclipse 3.2.x Open Source IDE. The release is distributed as a zip-file for a quick download. In a zip-file, there are these important directories and files:

* *jeasyopc.jar*: the final library for usage in your application.
* *src.jar*: the source of library for a preview of library classes.
* *eclipse-project\JEasyOpc.zip*: zip-file with whole JEasyOPC project for Eclipse. There are all examples, JUnit tests, all sources!
* *Doc*: the directory includes documentation.
* *Resources*: the configuration files of JEasyOPC library. These resources have to be included in CLASSPATH of your project. There are all important information about usage of logging, internationalization and dll-library path (property library.path).

Interface with OPC servers:

```
Import javafish.clients.opc.JCustomOpc;
Import javafish.clients.opc.JEasyOpc;
Import javafish.clients.opc.JOpc;
Import javafish.clients.opc.asynch.AsynchEvent;
Import javafish.clients.opc.asynch.OpcAsynchGroupListener;
Import javafish.clients.opc.browser.JOpcBrowser;
Import javafish.clients.opc.component.OpcGroup;
Import javafish.clients.opc.component.OpcItem;
```

//Initialization

```
jopc_meas= new JEasyOpc(hostName, serverName, groupName+ "_meas");
```

//Reading

```
gotItem= jopc_meas.synchReadItem(statusGroup, (OpcItem)ItemToRead);
```

//Writing

```
jopc_command.synchWriteItem(commandGroup, item);
```

As an example, consider a JADE application contains two agents: one is connected through LAN to the OPC server and continuously read changed process variables from the server (call is OPC-Agent); the other is a remote agent interacts with the OPC-Agent by message passing to get the latest changed process variables. The following Java program presents a possible implementation of the OPC-Agent (to save paper size, exceptions handling is not included in the code):

```
package javafish.clients.opc;
// Jade Imports
import jade.core.*;
import jade.core.behaviours.TickerBehaviour;
import jade.lang.acl.ACLMessage;
// JEasyOPC imports
import javafish.clients.opc.component.*;
import javafish.clients.opc.exception.*;
public class Simple_OPC_Agent extends Agent
{
// OPC Declarations
private JEasyOpc jopc;
private OpcGroup group;
private OpcItem item1,item2;
private OpcItem responseItem1,responseItem2;
// Agent Setup function
protected void setup() {
// connecting to Siemens OPC.SimaticNet OPC server
jopc = new JEasyOpc("localhost", "OPC.SimaticNET", "JOPC1");
```

```
JOpc.coInitialize();
// OPC Group Creation
item1 = new OpcItem("s7:[@LOCALSERVER]db1,w0", true, "");
item2 = new OpcItem("s7:[@LOCALSERVER]db1,w2", true, "");
group = new OpcGroup("group1", true, 400, 0.0f);
group.addItem(item1);
group.addItem(item2);
jopc.addGroup(group);
// Starting the OPC Server
jopc.start();
addBehaviour(new TickerBehaviour(this, 1500) {
protected void onTick() {
responseItem1 = jopc.synchReadItem(group, item1);
responseItem2 = jopc.synchReadItem(group, item2);
System.out.println("item1="+ responseItem1.getValue());
System.out.println("item2="+ responseItem2.getValue());
ACLMessage msg1=new ACLMessage(ACLMessage.INFORM);
// Process the message
msg1.addReceiver(new AID("H2", AID.ISLOCALNAME));
msg1.setLanguage("English");
msg1.setContent(responseItem1.getValue()+responseItem2.getValue());
send(msg1);
}});}
```

In this example the OPC server is supposed to be *OPC. SimaticNET*, which is used to connect to Siemens control systems. Also the OPC server is supposed to be installed on the same machine as the sample client application which means that the communication between the client and the server takes place through COM, if the client and server are hosted by different machines then DCOM will be used. The remote agent is a simple Jade agent required to interact with the OPC-Agent by massage passing. First it might send only one starting message to the OPC-Agent and continuously, through a cyclic behavior, it receives messages from the OPC-Agent and retrieves the OPC data from these messages and prints them to the console. For more information about the application of JEasyOPC, interested readers are invited to read the documents and manuals of JEasyOPC.

7

Fuzzy Controller for Dual Sensors Cardiac Pacemaker System in Patients with Bradycardias at Rest

Basil Hamed[1], Abd Al Karim Abu Ras[2]

[1]Electrical Engineering Department, Islamic University of Gaza, Gaza, Palestine
[2]Electrical Engineer, Biomedical Equipment Co., Gaza, Palestine
Email: bhamed@iugaza.edu, aburasskarim@gmail.com

Abstract

Cardiovascular disease is defined as a heart rate that is less than 60 bpm. Implantable cardiac devices such as pacemakers are widely used nowadays. In this paper, design and implementation of the heart model can be controlled to be the heart of a patient suffering from a decrease in heart rate (Bradycardia). A system is designed to sense and calculate the heart rate per minute and it is considered as an input to the controller. The design and implementation of Mamdani fuzzy controller to generate electric pulses that mimic the natural pacing system of the heart maintains an adequate heart rate by delivering controlled, rhythmic electrical stimuli to the chambers of the patient heart. The proposed controller is tested by using Matlab/Simulink program.

Keywords

Pacemaker, Dual-Sensors, Heart Rate, Bradycardia, Fuzzy Controller

1. Introduction

Cardiovascular diseases are major causes of morbidity and mortality in the developed countries. One of the cardiac diseases, bradycardia, sometimes results in fainting, shortness of breath, and if severe enough, death. Bradycardia means a slow heart rate, which is usually defined as fewer than 60 bpm. This occurs because people with bradycardia may not be pumping enough oxygen to their own heart causing heart attack-like symptoms. Thus, early diagnosis and treatment of heart diseases can effectively prevent the sudden death of a patient [1]. It is well known that implantable cardiac devices such as pacemakers are widely used nowadays. They have become a therapeutic tool used worldwide with more than 250,000 pacemaker implants every year. A pacemaker is a medical device that uses electrical impulses, delivered by electrodes contacting the heart muscles, to regulate the beating of the heart. Its primary purpose is to treat bradycardia due to sinus node or atrioventricular conduction disorders and to maintain an adequate heart rate, either because the heart's native pacemaker is not fast enough, or there is a block in its electrical conduction system. It can help a person who has an abnormal heart rhythm resume a more active lifestyle [2]. Dual-sensors are used to avoid inappropriate rate increase and provide more accurate measurement of diagnostic data such that two sensors may compensate each other. Dual-sensors including accelerometer and QT interval adopted in this study provide activity signal, metabolic demand, and actual heart rate. An accelerometer placed in a pacemaker detects movement and patient's physical activity and generates an electronic signal that is proportional to physical activity. QT sensor type provides pacing rates more closely and specifically related to physical and mental stress requirements. The QT interval reflects the total duration of ventricular myocardial repolarization. It measures the interval between the pacing spike and the evoked T-wave as the sensor and this interval shortens with exercise. The target of this paper is to develop a pacemaker fuzzy controller for dual-sensors cardiac pacemaker system in patients with bradycardias at rest. It can automatically control the heart rate to accurately track a desired pre-set profile.

2. Related Works

o   In June 2012, W. V. Shi and M. C. Zhou [2] presented a survey of the body sensors applied in pacemakers, introduced new features and advances of modern pacemakers, and the advancement of varieties of body sensors incorporated in pacemakers with their rationales, features and applications. Using one sensor is not ideal to for heart rate adaptation. So combining different kinds of sensors (dual sensors) is better than one sensor for optimal rate adaptation

o   In 2011, Wei Vivien Shi and Meng Chu Zhou [1] designed Fuzzy PID Controllers for Dual-Sensor Pacing Systems in Patients with Bradycardias at Rest, the most important advantage of this method was to provide a quite satisfactory tracking of the desired heart rate profile. New model for heart, IECG signal, sensing system and Mamdani fuzzy controller will achieve a closer match between the actual heart rate and a desired profile.

o   In 2010, Xiaolin Zhou, Xin Zhu, Hui Wang, and Daming Wei [3] presented a Comparative Evaluation of Six Algorithms Using Simulated Electrocardiograms to measurement a QT Interval Prolongation. But relying only on one sensor is not ideal. So dual sensor would provide us better performance for sensing and heart rate adaptation.

o   In May/June 2006, S. A. P. Haddad, R. P. M. Houben, and W. A. Serdijin, [4] presented a brief overview of the history and development of circuit designs applied in pacemakers, the most important advantage of this work was to show the electrical operation of the heart, the history and development of cardiac pacing systems and some new features in modern pacemakers. Fuzzy Controllers for Dual-Sensor Pacing Systems would provide better tracking accuracy.

o   In September 2003, A. Ferro, C. Duilio, M. Santomauro, and A. Cuocolo [5] studied the role of heart rate on cardiac output (CO) at rest and during walk test in patients with dual-chamber pacemaker and depressed or normal left ventricular (LV) function. The importance is to the medical data sets for preset/desired heart rate profile as the reference input signal to our model.

o   In 2000, A. Wojtasik, Z. Jaworski, W. Kuzmicz, A. Wielgus, A. Walkanis, and D. Sarna [6] presented a study of several possible implementations of fuzzy logic controllers for rate-adaptive pacemakers. Most important advantage of this work shows fuzzy logic based control algorithm for adaptive pacemakers is technically feasible and can be implemented in several ways. However, there is a need for pacemakers with better control algorithms adapting the pacing rate to the physiological requirement of a particular patient. So design a fuzzy controller which can automatically control the heart rate (HR) to accurately track a desired pre-set HR profile will give us better performance.

3. Biomedical Background

The human heart occupies a small region between the third and sixth ribs in the central portion of the thoracic cavity of the body. The heart is divided by a tough muscular wall—the interatrial interventricular septum—into a somewhat crescent shaped right side and cylindrically shaped left side, each being one self-contained pumping station, but the two being connected in series [7]. The heart is actually two separate pumps as shown in **Figure 1**: a right heart that pumps blood through the lungs, and a left heart that pumps blood through the peripheral organs.

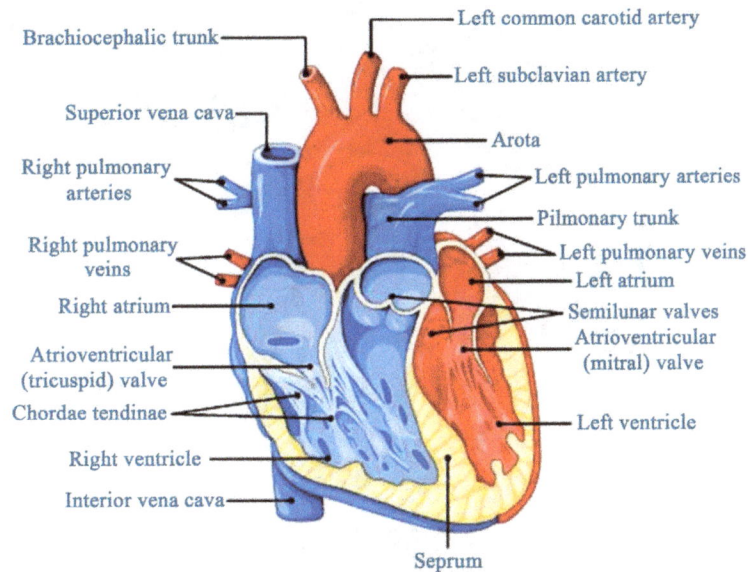

Figure 1. Heart structure.

In turn, each of these hearts is a pulsatile two-chamber pump composed of an atrium and a ventricle [8].

The heart is composed of three major types of cardiac muscle:

1) Atrial muscle,

2) Ventricular muscle,

3) Specialized excitatory and conductive muscle fibres.

The atrial and ventricular types of muscle contract in much the same way as skeletal muscle, except that the duration of contraction is much longer. Conversely, the specialized excitatory and conductive fibres contract only feebly because they contain few contractile fibrils; instead, they exhibit either automatic rhythmical electrical discharge in the form of action potentials or conduction of the action potentials through the heart, providing an excitatory system that controls the rhythmical beating of the heart [8].

There are two types of cardiac signals, one of them recorded from the chest, and the second one recorded from specific cardiac location inside the heart.

A. Pacemaker: is a device that generates electrical pulses and delivers them to the muscles of the heart (myocardium), in such a way as to cause those muscles to contract and the heart to beat. It is used to treat heart rhythms that are too slow, fast, or of any other irregularity. **Figure 2** illustrates the implantation of a pacemaker in a human body. A pacemaker helps a person who has an abnormal heart rhythm resume a more active lifestyle. Normally, small electrical pulses produced by a pacemaker can sustain a regular heartbeat. In case of deadly cardiac abnormalities, the pacemaker has to be adjusted to generate compulsive strong pulses assisting a patient to return to normal heartbeat [9].

B. Dual sensors: are used to avoid inappropriate rate increase and provide more accurate measurement of diagnostic data such that two sensors may compensate each other. During crosscheck both sensors can control each other and the pacing rate will only be changed if both or a predominant sensor agrees [1] [2]. Dual-sensors including accelerometer and QT interval adopted in this study provide activity signal, metabolic demand, and actual heart rate.

Rate Regulation with Dual Sensors: The dual-sensor pacemaker is expected to mimic the reaction of a healthy human heart to the levels of various body conditions, in other words to adapt the frequency of the heartbeat to the physiological needs. Dual-sensors including accelerometer and QT interval adopted in this study provide activity signal, metabolic demand, and actual heart rate. Acceleration is usually a good indicator of the physical activity requiring an increased heart rate [1]. This sensor is characterized as quick to respond but less physiologically accurate. On the other hand, QT interval rate response is characterized as physiological but it is slow to respond to the onset of moving. Consequently, to reduce these factors, two sensors are used to provide compensation, in that the fuzzy controller converges only when the system is stable. In the case of resting, for

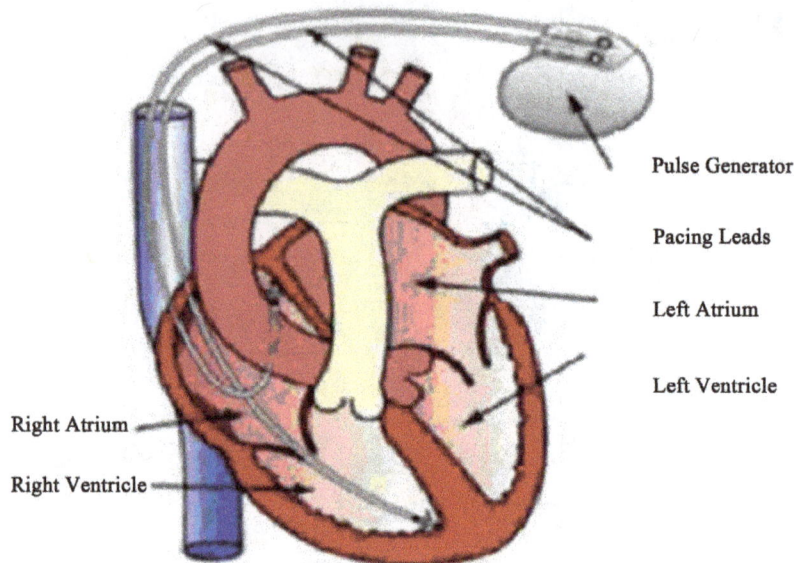

Figure 2. Implantation of a pacemaker.

instance, if the actual heart rate measured by dual-sensors is lower than preset normal rate for the particular patient, in order to assist the heartbeat regular, the pulses with adjustable pacing rate are generated by the pacemaker according to the fuzzy controller, such that the actual heart rate may track the preset desired heart rate in real-time.

In this paper the bradycardia at rest has been studied, so the Activity Sensor *State is Zero*. And its *response* will be the *same* as *desired profile value* = 80.

4. Designing FLC Dual Sensors Cardiac Pacemaker System

The system has heart model for IECG signal, adaptive sensing system for heart beats (dual sensors), and Mamdani fuzzy controller using for dual sensors cardiac pacemaker to achieve a closer match between the actual heart rate and a desired profile. As shown in the **Figure 3**.

A. **Heart Model:** consists of two basic parts, one is the natural pacemaker (SA) node, and the second is the cardiac muscle cell (myocyte).
o   The Sinoatrial (SA) node: is the normal pacemaker of the mammalian heart and generates the electrical impulse for the regular, rhythmic contraction of the heart [10].
o   The cardiac muscle cell: is the most physically energetic cell in the body, contracting constantly, without tiring, 3 billion times or more in an average human lifespan. By coordinating its beating activity with that of its 3 billion neighbours in the main pump of the human heart, over 7000 litres of blood are pumped per day, without conscious effort, along 100,000 miles of blood vessels [11]. **Figure 4** of the cardiac muscle cell underpins our understanding of how the electrical impulse, generated within the heart, stimulates coordinated contraction of the cardiac chambers.
B. **Sensing System/Simulation on Matlab**
    Stages for detecting heartbeat rate per minute
    In order to attenuate noise, the signal is passed through a band-pass filter. Subsequent processes are differentiation, squaring, Low-pass filter, saturation, quantizer, and a process for calculating mean value of the time between two qrs peaks, as shown in **Figure 5**.
C. **FLC for Dual-Sensor Cardiac Pacemaker System**
    FLC has one input which is: error and one output which is: pulses with adjustable pacing rate generated by the pacemaker (**Figure 6**).
    FLC Control Rules and Membership Functions
    Mamdani approach is used to implement FLC for dual-sensor cardiac pacemaker systems. FLC contains three basic parts: Fuzzification, Base rule, and Defuzzification.

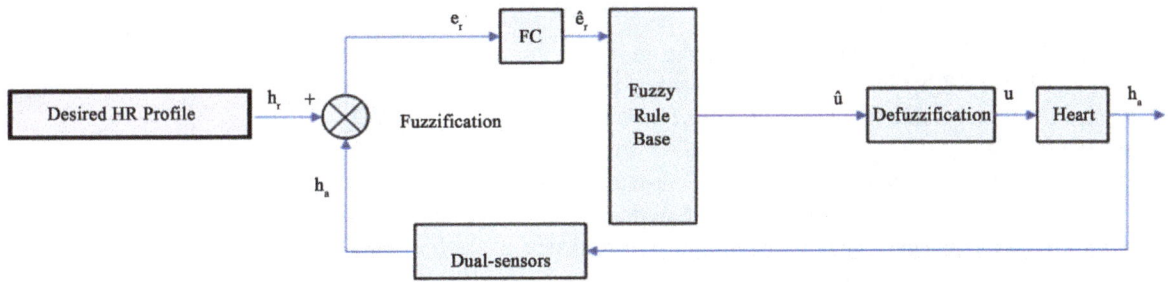

Figure 3. Dual-Sensor cardiac pacemaker system.

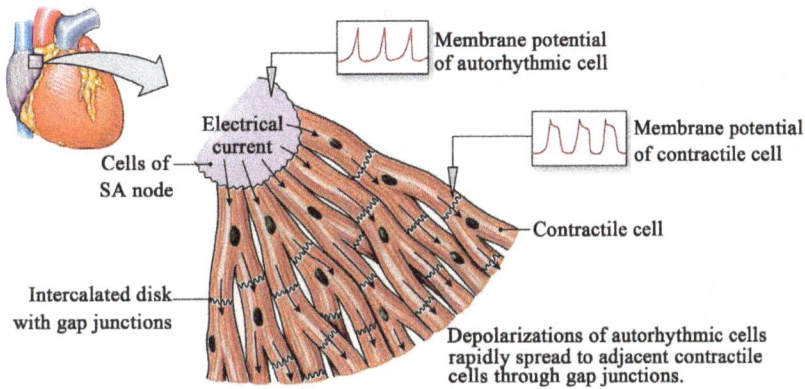

Figure 4. Cardiac muscle cell.

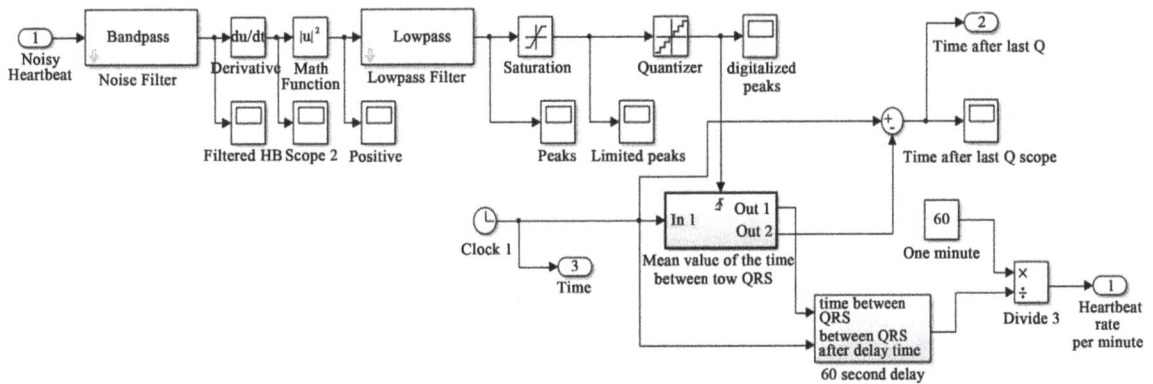

Figure 5. Testing the sensing system using Matlab/Simulink.

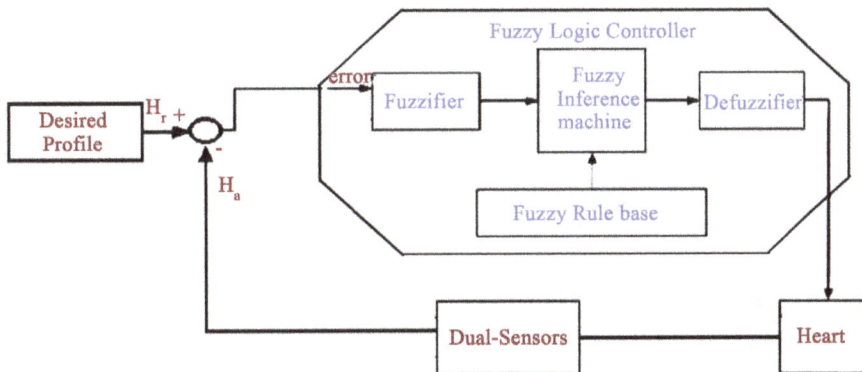

Figure 6. FLC controller for dual-sensor cardiac pacemaker system.

- **Fuzzification**

Mamdani FLC has one input which is error and one output which is pulses with adjustable pacing rate. **Figure 7(a)** and **Figure 7(b)** show the membership functions of fuzzy controller using Fuzzy Toolbox of Matlab software. Both membership functions have five Partitions VL, L, N, H, and VH.

- **Base Rule**

The fuzzy linguistic variables of the FLC system are illustrated in **Table 1**.

The negative values in the membership function are for the patient safety. If the patient heart rate rises above the normal rate suddenly then the controller will stop raising the heartbeat and try to decrease it to the normal rate. The rules for the proposed controller are:

1. If (error is VL) then (decision is VL).
2. If (error is L) then (decision is L).
3. If (error is N) then (decision is N).
4. If (error is H) then (decision is H).
5. If (error is VH) then (decision is VH).

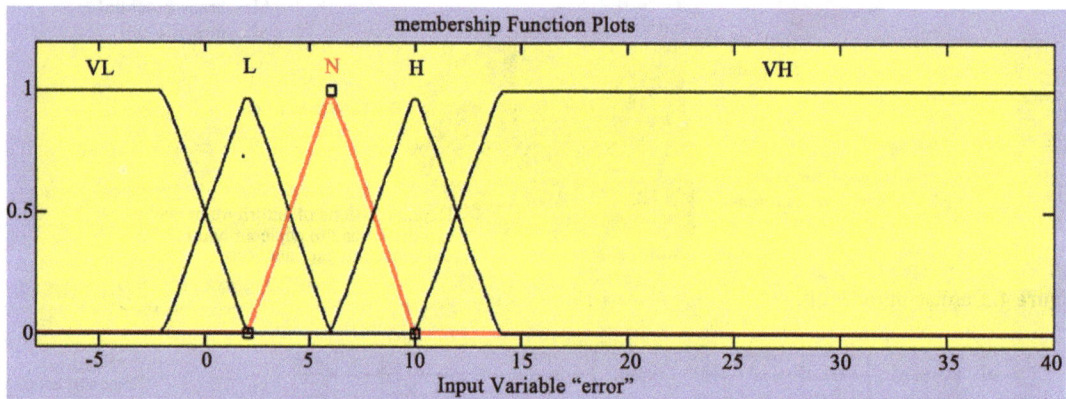

(a)

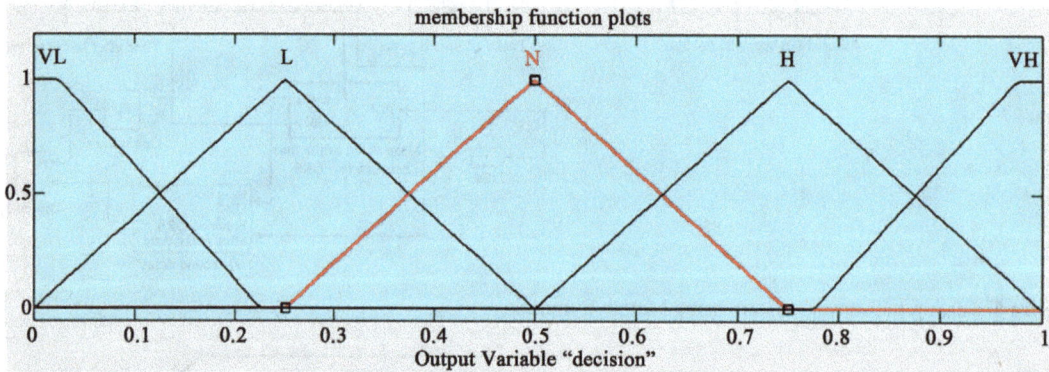

(b)

Figure 7. (a) Membership function of input error e; (b) Membership function of output.

Table 1. The linguistic variables.

VL	Very Low
L	Low
N	Normal
H	High
VH	Very High

- **Defuzzification**

Defuzzification method is the final stage of the fuzzy logic control. The FLC pacing rate is then defuzzified using the COA (Centroid) method for simplicity of implementation and it is faster than any defuzzification method.

$$COA = \frac{\sum_{Xmin}^{Xmax} X \cdot \mu(X)}{\sum_{Xmin}^{Xmax} \mu(X)}$$

D. Simulation on Matlab/Simulink for Dual-Sensor Cardiac Pacemaker System

Figure 8 illustrates the Simulink block diagram for the fuzzy controller dual-sensor cardiac pacemaker systems.

E. Case Study and Simulation Results

In this paper, the preset/desired heart rate profile as the reference input signal in the case study for particular patients presented are obtained using the medical data sets from [5].

Table 2 presents the characteristics of individual patient and the corresponding desired normal HR at rest.

The first 60 s is a delay time to allow the controller to understand the heart system, understand the patient heart and detect the patient heartbeats per minute while installs the pacemaker during the surgery. After discovering that the patient heartbeat per minute is under the normal rate, the controller must raise the heartbeats rate. During 30 s the controller begins to raise the patient heartbeats to the steady state heartbeats according to the desired heartbeats profile as shown in **Figure 9**.

- **Comparative Study**

The result of applying FLC to the pacemaker is compared with the work of Wei V Shi and Meng Chu Zhou; fuzzy logic controller, and fuzzy proportional-integral-derivative (FPID) [1].

Figure 10(a) and **Figure 10(b)**, show the effect of the proposed and compared controllers

As shown the proposed controller in this paper obtains the best results compared with other controllers (**Table 3**).

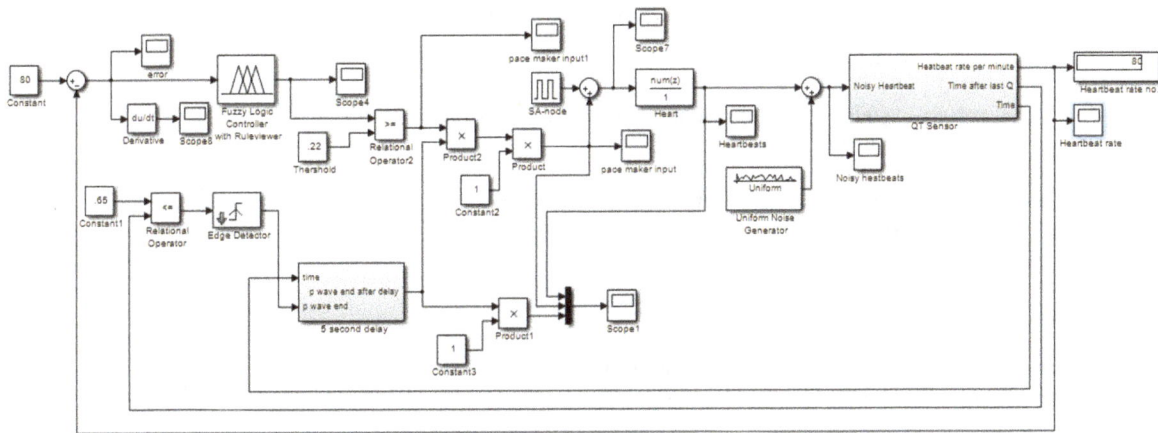

Figure 8. Testing FLC in the dual-sensor cardiac pacemaker system using Matlab/Simulink.

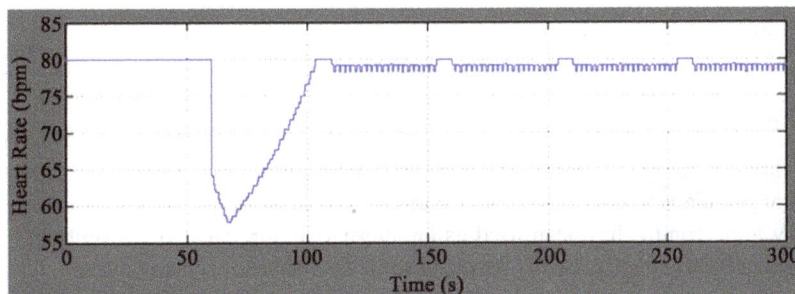

Figure 9. Pacemaker working mechanism & adjustable heartbeat rate.

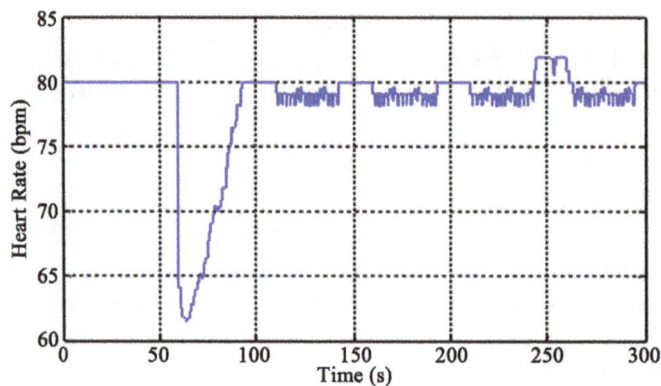

(a)

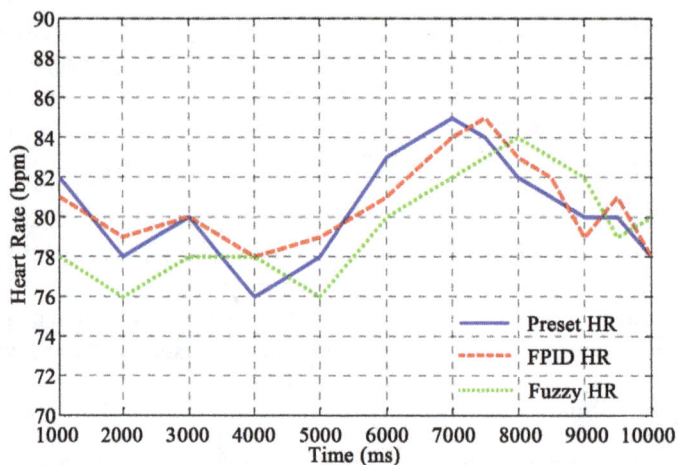

(b)

Figure 10. (a) Output of proposed FLC; (b) Output of FLC & FPID controller [1].

Table 2. Individual characteristics of the patient.

Age (year)	State	Preset HR (bpm)
50 - 65	at rest	81 ± 4

Table 3. Result comparison with FLC & FPID controllers in reference [1].

Controllers	rmse	Steady state error	Overshoot
Proposed FLC	1.0344	2.41%	$\cong 0$
FPID [1]	1.1902	2.63%	Not Clear
FLC [1]	2.3805	4.88%	Not Clear

5. Conclusion

Cardiovascular diseases are major causes of morbidity and mortality in the developed countries. One of the cardiac diseases, bradycardia, sometimes results in fainting, shortness of breath, and if severe enough, death. In recent years, fuzzy logic control has been used as an alternative approach to conventional process control techniques for biomedical equipment's. In this paper, new model for heart, sensing system, and Mamdani fuzzy controller is used to generate electric pulses that mimic the natural pacing system of the heart. The proposed model maintains an adequate heart rate by delivering controlled, rhythmic electrical stimuli to the chambers of the heart, and prevents human from being harmed by low heart rate. A comparative study shows that the proposed model with Mamdani FLC has a better response and demonstrates better performance than FPID controller.

References

[1] Shi, W.V. and Zhou, M.C. (2011) Design Fuzzy PID Controllers for Dual-Sensor Pacing Systems in Patients with Bradycardias at Rest. *IEEE International Conference on Systems*, *Man*, and *Cybernetics* (*SMC*), Anchorage, 9-12 October 2011, 1117-1122.

[2] Shi, W.V. and Zhou, M.C. (2012) Body Sensors Applied in Pacemakers: A Survey. *IEEE Sensors Journal*, **12**, 1817-1827. http://dx.doi.org/10.1109/JSEN.2011.2177256

[3] Zhou, X.L., Zhu, X., Wang, H. and Wei, D.M. (2010) Measurement of QT Interval Prolongation, a Comparative Evaluation of Six Algorithms Using Simulated Electrocardiograms. 2010 *IEEE Region* 10 *Conference TENCON* 2010, Fukuoka, 21-24 November 2010, 2077-2082.

[4] Haddad, S.A.P., Houben, R.P.M. and Serdijin, W.A. (2006) The Evolution of Pacemakers. *IEEE Engineering in Medicine and Biology Magazine*, **25**, 38-48. http://dx.doi.org/10.1109/MEMB.2006.1636350

[5] Ferro, A., Duilio, C., Santomauro, M. and Cuocolo, A. (2003) Walk Test at Increased Levels of Heart Rate in Patients with Dual-Chamber Pacemaker and with Normal or Depressed Left Ventricular Function. *European Heart Journal*, **24**, 2123-2132. http://dx.doi.org/10.1016/j.ehj.2003.09.007

[6] Jaworski, Z., Kuzmicz, W., Sadowski, M., Sama, D., Walkanis, A. Wielgus, A. and Wojtasik, A. (2000) VLSI Implementations of Fuzzy Logic Controllers for Rate-Adaptive Pacemakers. 1*st Annual International Conference on Microtechnologies in Medicine and Biology*, Lyon, 2000, 475-478.

[7] Bronzino, J.D. (2000) The Biomedical Engineering Handbook, Volume I. 2nd Edition, CRC Press, Boca Raton.

[8] Guyton, A.C. and Hall, J.E. (2006) Text Book of Medical Physiology. 11th Edition, Elsevier Saunders, Philadelphia, QP34.5.G9.

[9] Whittington, R.H., Giovangrandi, L. and Kovacs, G.T.A. (2005) A Closed Loop Electrical Stimulation System for Cardiac Cell Cultures. *IEEE Transactions on Biomedical Engineering*, **52**, 1261-1270. http://dx.doi.org/10.1109/TBME.2005.847539

[10] den Haan, A.D., Verkerk, A.O. and Tan, H.L. (2011) Creation of a Biopacemaker: Lessons from the Sinoatrial Node, Modern Pacemakers. In: Das, M.R., Ed., *Present and Future*, InTech, Morn Hill, Page. http://www.intechopen.com/books/modern-pacemakers-present-andfuture/creation-of-a-biopacemaker-lessons-from-the-sinoatrial-node

[11] Severs, N.J. (2000) The Cardiac Muscle Cell. *BioEssays*, **22**, 188-199. http://dx.doi.org/10.1002/(SICI)1521-1878(200002)22:2<188::AID-BIES10>3.0.CO;2-T

8

The Role of a Novel Discrete-Time MRAC Based Motion Cueing on Loss of Control at a Hexapod Driving Simulator

B. Aykent[1*], D. Paillot[1], F. Merienne[1], C. Guillet[1], A. Kemeny[1,2]

[1]CNRS Le2i Arts et Metiers ParisTech, Chalon sur Saone, France
[2]Technical Centre for Simulation, Renault, Guyancourt, France
Email: *b.aykent@gmail.com

Abstract

The objective of this paper is to present the advantages of Model reference adaptive control (MRAC) motion cueing algorithm against the classical motion cueing algorithm in terms of biomechanical reactions of the participants during the critical maneuvers like chicane in driving simulator real-time. This study proposes a method and an experimental validation to analyze the vestibular and neuromuscular dynamics responses of the drivers with respect to the type of the control used at the hexapod driving simulator. For each situation, the EMG (electromyography) data were registered from arm muscles of the drivers (flexor carpi radialis, brachioradialis). In addition, the roll velocity perception thresholds (RVT) and roll velocities (RV) were computed from the real-time vestibular level measurements from the drivers via a motion-tracking sensor. In order to process the data of the EMG and RVT, Pearson's correlation and a two-way ANOVA with a significance level of 0.05 were assigned. Moreover, the relationships of arm muscle power and roll velocity with vehicle CG (center of gravity) lateral displacement were analyzed in order to assess the agility/alertness level of the drivers as well as the vehicle loss of control characteristics with a confidence interval of 95%. The results showed that the MRAC algorithm avoided the loss of adhesion, loss of control (LOA, LOC) more reasonably compared to the classical motion cueing algorithm. According to our findings, the LOA avoidance decreased the neuromuscular-visual cues level conflict with MRAC algorithm. It also revealed that the neuromuscular-vehicle dynamics conflict has influence on visuo-vestibular conflict; however, the visuo-vestibular cue conflict does not influence the neuromuscular-vehicle dynamics interactions.

Keywords

Driving Simulator, EMG Analysis, Model Reference Adaptive Control, Discrete-Time Control, Loss

---

*Corresponding author.

of Control, Head Dynamics

1. Introduction

Multi-sensory datafusion: such as visual, auditory, haptic, inertial, vestibular, neuromuscular signals are of importance to represent a proper sensation (objectively) and so a perception (subjectively as cognition) in motion base driving simulators [1]-[10].

A use study of the physiological measurements (biofeedback methods) has been presented to estimate user interruptibility status by [3]. Heart Rate Variability (HRV) and Electromyogram (EMG) signals have been registered as users performed a diversity of assignments. Results have elicited high correlations for both HRV and EMG ($r = 0.96$ and $r = 0.85$ respectively) with user subjective reports for interruptibility [3].

Motion sickness has been discussed when a moving visual surround induces the illusion of self-rotation in [4] [11]. The vestibulo-ocular reflex and the occurrence of motion sickness are attached to the gravito-inertial force level according to [12].

Motion cueing for a 2 DOF (degrees of freedom) driving simulator has been examined by [6]. The theme of that research has been to test and compare performances of different washout algorithms applied to such sort of platform. The results have depicted that there has been no significant difference among those approaches [6]. The effects of different washout algorithms used for Stewart platforms (6 DOF) on subjective and objective ratings have been discussed in [2]. According to the simulator sickness test, closed-loop motion cueing algorithm; the subjects have reacted less stressfully (cold sweat) to the conditions, whereas they have behaved more stressfully in the conditions of the open-loop motion cueing (classical) algorithm. Regarding the visual sickness of the participants, closed-loop motion cueing algorithm has presented the most reasonable situation. Also concerning the mental pressure, the statistical distribution points out an agreeable experience for closed-loop motion cueing algorithm comparing to the open-loop motion cueing algorithm. Based upon the "modified simulator sickness questionnaire", the most realistic acceleration has been perceived by operating the motion platform with closed-loop motion cueing algorithm; where the most unpleasant steering has been coincided by exploiting the motion platform by open-loop motion cueing algorithm; and whereas the most agreeable condition has been experienced at closed-loop motion cueing algorithm with regards to perception of the pitch motion severity. Furthermore, the perception on curvature has been assessed as the most disagreeably during the attempts with classical washout algorithms [2].

Restituting the inertial cues on driving simulators play an important role to sustain a more proper functioning in proximity to the reality [13] [14]. Simulator sickness deals with whether this convergence is obtained or not, as being one of the main research issues for the driving simulators. Simulator sickness was assessed between moving base and fixed base simulators by [15]. However, there has been a very few publications of vehicle (visual)-vestibular cue conflict based approach and its correlation with the neuromuscular dynamics. This paper addresses a methodology in order to rate the loss of adhesion (LOA) as well as the agility and alertness level of the drivers as a correlated function of the vestibular cues with the EMG_{RMS} total power. Due to the restricted workspace, it is not possible to represent the vehicle dynamics permanently with scale one to one on the motion platform [6] [12] [16]-[18].

This research work was performed under the dynamic operations of the SAAM driving simulator as with a classical and a MRAC controlled tracking of the hexapod platform. The dynamic simulators' utilization scope diversifies from driver training to research purposes such as vehicle dynamics control, advanced driver assistance systems (ADAS) [19] [20]. The dynamic driving simulator SAAM (Simulateur Automobile Arts et Métiers) is made up of a hexapod platform system. It is exploited on a RENAULT Twin go 2 cabin with the original control instruments (gas, brake pedals, steering wheel). The visual system is handled by a 150° dome view. Multi-level real-time measuring techniques (XSens motion tracker, Biopac EMG (electromyography) device, Technoconcept postural stability platform) [1] are available, which are already used with numerous experiments/scenarios such as sinus steer test, NATO chicane, etc. The vehicle accelerations of translations (longitudinal X, lateral Y and vertical Z axes) as well as the vehicle accelerations of roll and pitch, which correspond to the vehicle dynamics, are taken into account for the control. Then the platform positions, velocities and accelerations were controlled and fed back to minimize the conflict between the vehicle and the platform levels [2].

The paper is organized as: Section 2 explains the proposed control approach for the dynamic platform of the driving simulator. Section 3 describes the used materials and methodology in order to analyze the multiple level data acquisition (neuromuscular via using EMG, vestibular through using motion tracking sensor and vehicle levels via SCANeR studio software). Section 4 discusses the results. And finally, Section 5 concludes up the paper.

2. Proposed Control Approach

This paper explores a comparative study between an open and a closed loop controlled platform to maintain the vehicle pursuing a chicane maneuver scenario (loss of control-LOC). For the evaluation and the validation procedure [3] [16] [18] [21]-[24], the scenario driven on the simulator SAAM with a classical motion cueing (open loop control) and a MRAC motion cueing (closed loop control) to describe the impact of the feedback control on LOC. The results from a case study were illustrated in the scope of this research with real time controls of the platform at a longitudinal velocity of 60 km/h. This research surveys the following hypotheses:

-Neuromuscular (EMG_{RMS} total power) and vehicle dynamics (lateral displacement of the vehicle CG (center of gravity)) interaction indicates the limit of LOC in the driving simulation experiments.

- If they are positively significant correlated, it means that an avoidance loss of control (LOC) is possible to occur.
- If they are negatively significantly correlated, it shows that the drivers are prone to experience a loss of control (LOC) phenomenon.

-Vehicle dynamics approach (lateral displacement of the vehicle CG): Lateral displacement area decreases when the MRAC algorithm is used (avoidance of LOC).

-Vestibular (roll velocity perception threshold—RVT) and neuromuscular dynamics (EMG_{RMS} mean total power and EMG_{RMS} maximum total power) interaction gives the characteristics of the driver: If they are positively correlated; perception of agility, in other words alertness level of the drivers and avoidance of LOC increase when the MRAC motion cueing is used.

2.1. Motion Cueing Algorithm

The proposed classical motion cueing algorithm's sketch is indicated with continuous lines and arrows where the model reference controlled motion cueing algorithm's sketch was drawn with discrete lines and arrows (**Figure 1**).

The proposed MRAC motion cueing algorithm here uses the same filters, gains (**Figure 1** and **Table 1**) like used in the classical algorithm to compare the effects of the model reference adaptive control of the dynamic driving simulator. The main idea for Model Reference Adaptive Control (MRAC, which is also known as an

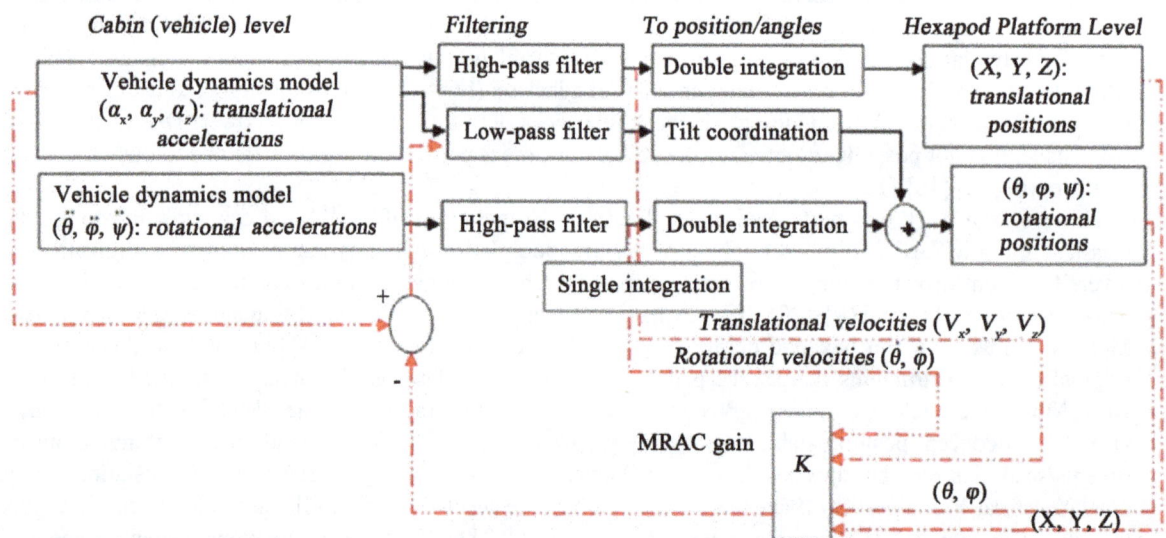

Figure 1. Classical and MRAC motion cueing algorithms sketches.

Table 1. Classical motion cueing algorithm parameters [25].

Symbol	Longitudinal	Lateral	Roll	Pitch	Yaw
2nd order LP cut-off frequency (Hz)			0.3	0.7	
2nd order LP damping factor			0.3	0.7	
1st order LP time constant (s)	0.1	0.1			0.1
2nd order HP cut-off frequency (Hz)	0.5	0.5			2
2nd order HP damping factor	1	1			1
1st order HP time constant (s)	2	2			2

MRAS or Model Reference Adaptive System) is to build a closed-loop controller with parameters that can be updated to change the response of the system. The output of the system is compared to a desired response from a reference model. The control parameters are updated based on this error. The goal is for the parameters to converge to ideal values that cause the plant response to match the response of the reference model. It can focus on the continuous-time case and also on discrete-time design. A discrete-time MRAC was referred in this paper. The objective is to regulate the output minimized (platform-vehicle levels' sensed acceleration difference minimization). The system (dynamic driving simulator) is subject to disturbances and is driven by controls [26]-[30].

Table 2 illustrates the constraints of the dynamic driving simulator SAAM, which was used in the real-time dll plugin in the SCANeR studio software for the both motion cueing algorithms. For the longitudinal, lateral and vertical displacements, the used gains were 0.2, 0.2 and 0.22 respectively.

The design step searches a state-feedback law that minimizes the cost function via applying this logic. **Figure 2** illustrates the research method for the whole group of the subjects used in this article.

Both of the motion control algorithms (classical and MRAC) were integrated at the dynamic driving simulator SAAM with a "dll plugin" which were created with Microsoft Visual 2008 C++ used in SCANeR studio version 1.1.

2.2. Control Problem

We considered adiscrete-time MIMO (multiple-input multiple-output) system described by [27]-[30]

$$x(k+1) = Ax(k) + Bu(k), \quad y(k) = Cx(k) \qquad (1)$$

with $A \in \mathbb{R}^{n \times n}$, $B \in \mathbb{R}^{n \times M}$ and $C \in \mathbb{R}^{M \times n}$ being unknown and constant parameter matrices and $x(k) \in \mathbb{R}^n$, $u(k) \in \mathbb{R}^M$ and $y(k) \in \mathbb{R}^M$ being the system state, input and output vector signals.

2.2.1. Control Objective

The control objective is to design a state feedback control signal $u(k)$ in Equation (1) such that all the closed-loop signals remain bounded and the system output signal $y(k)$ tracks a given reference output $y_m(k) \in \mathbb{R}^M$ that is generated from the reference model system

$$y_m(k) = W_m(z)r(k) \qquad (2)$$

$Wm(z)$ is an $M \times M$ transfer matrix, and $r(k)$, an M-dimensional real array, is a bounded reference input signal [27] [30] [31].

2.2.2. Assumptions

To begin the real-time controller design which is implemented in driving simulator as "dll plugin", we assume [27]:

(A1) All zeros of $G_i(z) = C_i(zI - A_i)^{-1} B_i, i = 1, 2, \cdots, N$ lie within the unit circle in the z-plane;

(A2) $G_i(z), i = 1, 2, \cdots, N$, have full rank, there is a left interactor matrix $\xi m(z)$ for all $G_i(z), i = 1, 2, \cdots, N$

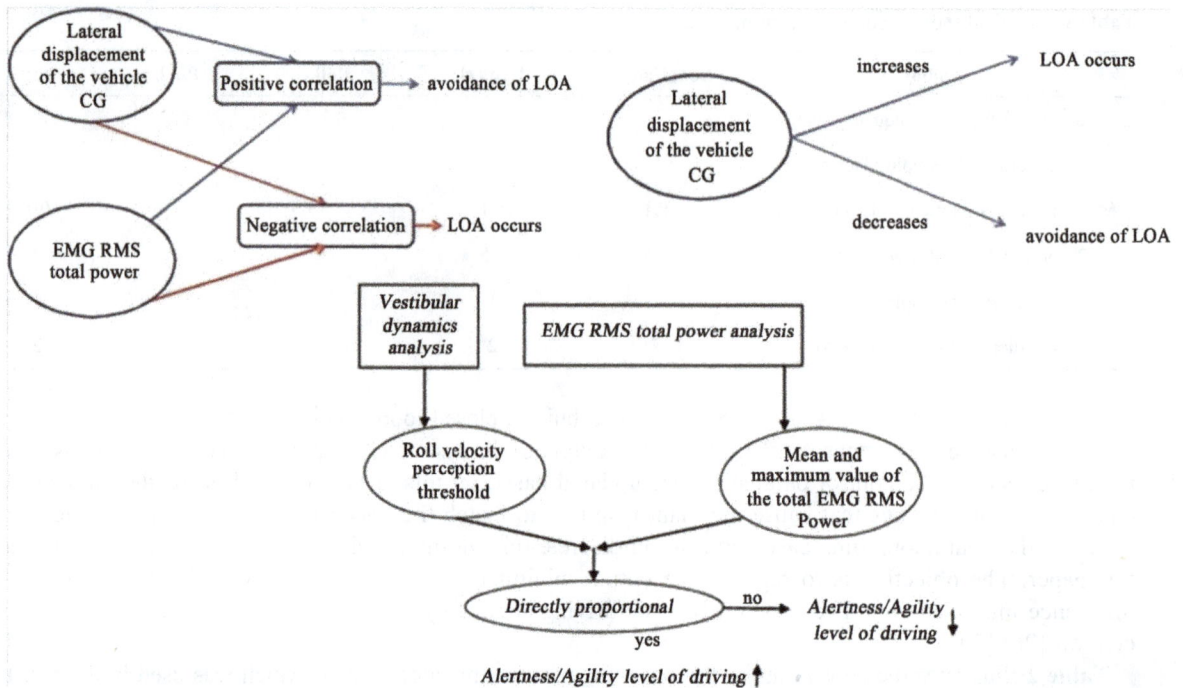

Figure 2. Hypotheses: Our hypotheses are made up of three parts. 1) We stated that if the lateral displacement of the vehicle CG used inside the driving simulator and the power spent by arm muscles (from flexor carpi radialis) are positively correlated, it is an indicator of avoidance of LOA (LOC). Inversely, if they are negatively correlated, it is an objective metrics of occurrence of LOA. 2) We also declared that if the lateral displacement of the vehicle CG increases LOA occurs and in adverse case LOA decreases. 3) According to our third hypothesis, if the head (vestibular) level roll velocity perception threshold and the arm muscles power dissipation are positively correlated (cues conflict decrease) the alertness/agility level of the drivers increase too. We checked this criterion with respect to the real-time mean and maximum values of the total EMG_{RMS} power measured from the arm muscles. Mean EMG_{RMS} power is corresponding to the whole scenario whereas the maximum value of the EMG_{RMS} power is referring to the sudden change mostly depending on the sudden change of the road curvatures, *i.e.* LOA.

Table 2. Limits of each degree of freedom (DOF) for the SAAM driving simulator [25].

DOF	Displacement	Velocity	Acceleration
Pitch	±22 deg	±30 deg/s	±500 deg/s²
Roll	±21 deg	±30 deg/s	±500 deg/s²
Yaw	±22 deg	±40 deg/s	±400 deg/s²
Heave	±0.18 m	±0.30 m/s	±0.5 g
Surge	±0.25 m	±0.5 m/s	±0.6 g
Sway	±0.25 m	±0.5 m/s	±0.6 g

and the reference system transfer matrix $W_m(z) = \xi_m^{-1}(z)$;

(A3) All leading minors $\Delta_j, j = 1, 2, \cdots, M,$ of the high frequency gain matrix Kp are nonzero and their signs are known.

2.2.3. State Feedback for State Tracking

For a state feedback for state tracking design, the controller structure is

$$u(k) = K_1^T(k)x(k) + K_2(k)r(k) \tag{3}$$

where $K_1(k) \in \mathbb{R}^{n \times M}$ and $K_2(k) \in \mathbb{R}^{M \times M}$ are parameter matrices updated from some adaptive laws, so that the plant state vector signal $x(k)$ can asymptotically track a reference state vector signal $x_m(k)$ generated from a chosen reference system

$$x_m(k+1) = A_m x_m(k) + B_m r(k) \tag{4}$$

where $A_m \in \mathbb{R}^{n \times n}$ is stable and $B_m \in \mathbb{R}^{n \times M}$. For such an adaptive control design, the matching conditions $K_1(k)$ and $K_2(k)$ are the estimates of the nominal K_1^* and K_2^* which satisfy the conditions of matching

$$C_i\left(zI - A_i - B_i K_1^{*T}\right)^{-1} B_i K_2^* = W_m(z), \quad K_2^{*-1} = K_P \tag{5}$$

where $K_P = \lim_{z \to \infty} \xi_m(z) G_i(z)$ is the high frequency gain matrix of $G_i(z)$. The existence of K_1^* and K_2^* is guaranteed under the nominal system condition: (A4) (A, B) is stabilizable and (A, C) is observable.

2.2.4. Tracking Error Equation
Substituting the control law Equation (3) in Equation (1), we obtain

$$x(k+1) = \left(A + BK_1^{*T}\right)x(k) + BK_2^* r(k) + B\left(\left(K_1^T(k) - K_1^{*T}\right)x(k) + \left(K_2(k) - K_2^*\right)r(k)\right), \quad y(k) = Cx(k) \tag{6}$$

In view of the reference model Equation (2), matching equations Equation (5) and Equation (6), the output tracking error $e(k) = y(k) - y_m(k)$ is

$$e(k) = W_m(k) + K_P\left[\tilde{\theta}^T \omega\right](k) + Ce^{\left(A + BK_1^{*T}\right)(k)} x(0) \tag{7}$$

where $Ce^{\left(A + BK_1^{*T}\right)(k)} x(0)$ converges to zero exponentially and

$$\tilde{\theta}(k) = \theta(k) - \theta^* \tag{8}$$

$$\theta(k) = \left[K_1^T(k), K_2(k)\right]^T \in \mathbb{R}^{(n+M) \times M} \tag{9}$$

$$\theta^* = \left[K_1^{*T}, K_2^*\right]^T \tag{10}$$

$$\omega(k) = \left[x^T(k), r^T(k)\right]^T \tag{11}$$

2.2.5. LDS Decomposition
In this section, we present the design and analysis of an adaptive scheme based on the LDS decomposition of the high frequency gain matrix K_P.

To design an adaptive parameter update law, it is crucial to develop an error model in terms of some related parameter errors and the tracking error $e(k) = y(k) - y_m(k)$.

2.2.6. Error Model
Neglecting the term $Ce^{\left(A + BK_1^{*T}\right)(k)} x(0)$, we obtain from Equation (7) and Equation (2)

$$\xi_m(z)[e](k) = K_P \tilde{\theta}^T(k) \omega(k) \tag{12}$$

To deal with the uncertainty of the high frequency gainmatrix K_P, we use its LDS decomposition

$$K_P = L_s D_s S \tag{13}$$

where $S \in \mathbb{R}^{M \times M}$ with $S = S^T > 0$, L_s is an $M \times M$ unit triangular matrix, and

$$D_s = \text{diag}\left\{s_1^*, s_2^*, \cdots, s_M^*\right\} = \text{diag}\left\{\text{sign}[\Delta_1]\gamma_1, \cdots, \text{sign}\left[\frac{\Delta_M}{\Delta_{M-1}}\right]\gamma_M\right\} \tag{14}$$

such that $\gamma_i > 0, i = 1, \cdots, M$, where γ is called adaptive gain, and s_2^* is the second term in the diagonal of the matrix D_s of the LDS decomposition [27] [28], substituting Equation (13) in Equation (12) yields

$$L_s^{-1} \xi_m (z)[e](k) = D_s S \tilde{\theta}^{\mathrm{T}}(k) \omega(k) \tag{15}$$

To parameterize the unknown L_s, θ_0^* is introduced in Equation (16)

$$\theta_0^* = L_s^{-1} - I = \begin{bmatrix} 0 & 0 & \cdots & 0 \\ \theta_{21}^* & 0 & \cdots & 0 \\ \theta_{31}^* & \theta_{32}^* & \cdots & 0 \\ & \cdots & \cdots & \\ \theta_{M-11}^* & & 0 & 0 \\ \theta_{M1}^* & & \theta_{MM-1}^* & 0 \end{bmatrix}, \text{ the dimension of } \theta^* \text{ is } \mathbb{R}^{M \times M} \tag{16}$$

Then, it yields (17):

$$\xi_m(z)[e](k) + \theta_0^* \xi_m(z)e(k) = D_s S \tilde{\theta}^{\mathrm{T}}(k)\omega(k) \tag{17}$$

A filter was designed $h(z) = 1/f(z)$, where $f(z)$ is a stable monic polynomial of degree equals to the degree of $\xi m(z)$, operating both sides of Equation (17) by $h(z)$IM leads to

$$\overline{e}(k) + \left[0, \theta_2^{*\mathrm{T}} \eta_2(k), \theta_3^{*\mathrm{T}} \eta_3(k), \cdots, \theta_M^{*\mathrm{T}} \eta_M(k) \right] = D_s S h(z) \left[\tilde{\theta}^{\mathrm{T}} \omega \right](k) \tag{18}$$

where

$$\overline{e}(k) = \xi_m(z)h(z)[e](k) = \left[\overline{e}_1(k), \cdots, \overline{e}_M(k) \right] \tag{19}$$

$$\eta_i(k) = \left[\overline{e}_1(k), \cdots, \overline{e}_{i-1}(k) \right]^T \in R^{i-1}, i = 2, \cdots, M \tag{20}$$

$$\theta_i^* = \left[\theta_{i1}^*(k), \cdots, \theta_{ii-1}^*(k) \right]^{\mathrm{T}} \in R^{i-1}, i = 2, \cdots, M \tag{21}$$

Based on this parameterized error equation, we reach the estimation error signal

$$\varepsilon(k) = \left[0, \theta_2^{\mathrm{T}} \eta_2(k), \theta_3^{\mathrm{T}} \eta_3(k), \cdots, \theta_M^{\mathrm{T}} \eta_M(k) \right] + \psi(k)\xi(k) + \psi(k)\xi(k) + \overline{e}(k) \tag{22}$$

where $\theta_i(k), i = 2, 3, \cdots, M$ are the estimates of θ_i^*, and $\psi(k) - \psi^*$ are related parameter errors.

2.2.7. Adaptive Laws
Within the estimation error model Equation (22), the chosen adaptive laws [27]:

$$\dot{\theta}_i(k) = \frac{\Gamma_{\theta_i \epsilon_{i(k)} \eta_i(k)}}{m^2(k)}, \ i = 2, 3, \cdots, M \tag{23}$$

$$\dot{\theta}^{\mathrm{T}}(k) = -\frac{D_s \epsilon(k) \zeta^{\mathrm{T}}(k)}{m^2(k)} \tag{24}$$

$$\dot{\psi}(k) = -\frac{\Gamma_{\epsilon(k)\xi^{\mathrm{T}}(k)}}{m^2(k)} \tag{25}$$

$$\omega(k) = \left[x^{\mathrm{T}}(k), r^{\mathrm{T}}(k) \right]^{\mathrm{T}} \tag{26}$$

where the signal $\epsilon(k) = \left[\epsilon_1(k), \epsilon_2(k), \cdots, \epsilon_M(k) \right]$ is computed from Equation (22), $\Gamma_{\theta_i} = \Gamma_{\theta_i^{\mathrm{T}}} > 0, \ i = 2, 3, \cdots, M$ and $\Gamma = \Gamma^{\mathrm{T}} > 0$ are adaptation gain matrices and

$$m(k) = 1 + \zeta^{\mathrm{T}}(k)\zeta(k) + \xi^{\mathrm{T}}(k)\xi(k) + \sum_{i=2}^{M} \left(\eta_i^{\mathrm{T}}(k)\eta_i(k) \right)^{1/2} \tag{27}$$

is a standard normalization signal, where $\zeta(k) = h(z)[\omega](k)$ [27].

3. Methods

We consider the dynamic model of the hexapod simulator described by Equation (28) which is a three state variable (x). The control inputs are (u) the pitch angle (θ_v), the roll angle (ϕ_v) and the yaw angle (ψ_v) of the vehicle model. Our MRAC model was given in Equation (29) (HP: high pass filtered motion, LP: low pass filtered motion) which was with a sampling interval of $T = 1/60$ seconds to obtain the discrete-time motion cueing algorithms implemented in our dll plugin.

$$x = \left[x(k+1), y(k+1), z(k+1) \right]^{\mathrm{T}}, \tag{28}$$

$$u = \left[\theta_v(k), \phi_v(k), \psi_v(k) \right]^{\mathrm{T}}$$

$$\begin{bmatrix} x(k+2) \\ y(k+2) \\ z(k+2) \end{bmatrix} = \begin{bmatrix} \dfrac{-5.0625 + 2 \cdot x(k+1)_{HP}}{x(k+2)_{HP}+1} & 0 & 0 \\ 0 & \dfrac{-5.0625 + 2 \cdot y(k+1)_{HP}}{y(k+2)_{HP}+1} & 0 \\ 0 & 0 & \dfrac{-5.0625 + 2 \cdot z(k+1)_{HP}}{z(k+2)_{HP}+1} \end{bmatrix} \cdot \begin{bmatrix} x(k+1) \\ y(k+1) \\ z(k+1) \end{bmatrix}$$

$$+ \begin{bmatrix} -\sin^{-1}\left(\dfrac{x(k+2)_{LP}}{9.81} \right) & & \\ & -\sin^{-1}\left(\dfrac{y(k+2)_{LP}}{9.81} \right) & \\ & & -\sin^{-1}\left(\dfrac{z(k+2)_{LP}}{9.81} \right) \end{bmatrix} \cdot \begin{bmatrix} \theta_v(k) \\ \phi_v(k) \\ \psi_v(k) \end{bmatrix}$$

$$- \begin{bmatrix} \dfrac{10 \cdot x(k)_{HP} + 6.3167}{x(k+2)_{HP}+1} & & \\ & \dfrac{10 \cdot y(k)_{HP} + 6.3167}{y(k+2)_{HP}+1} & \\ & & \dfrac{10 \cdot z(k)_{HP} + 6.3167}{z(k+2)_{HP}+1} \end{bmatrix} \cdot \begin{bmatrix} x(k) \\ y(k) \\ z(k) \end{bmatrix} \tag{29}$$

3.1. Subjects

Twenty-six healthy participants took place in the experiments (4 females, 22 males) with a mean age of 28.9 ± 5.8 years old and a driving license holding with a mean experience of 9.7 ± 6.6 years.

3.2. Protocol

Figure 3 shows the trajectory of the chicane maneuver that we used in this experiment protocol (Here $W = 1$ m and $L = 1.5$ m). The vehicle velocity during the simulator experiments was chosen as constant at 60 km/h. The same scenario was driven at the classical and the MRAC motion cueing algorithms in order to compare the biomechanical interactions (head level dynamics and vehicle level dynamics interaction, neuromuscular dynamics and vehicle level dynamics and lastly head level dynamics and neuromuscular dynamics) of the participants.

Vestibular level dynamics of the participants refer to the head movements of them (see **Figure 4**). It was measured via a XSens motion tracking sensor. Vehicle level dynamics indicate the visual cues which come from the surroundings of the vehicle when it is driven at the simulator as real-time.

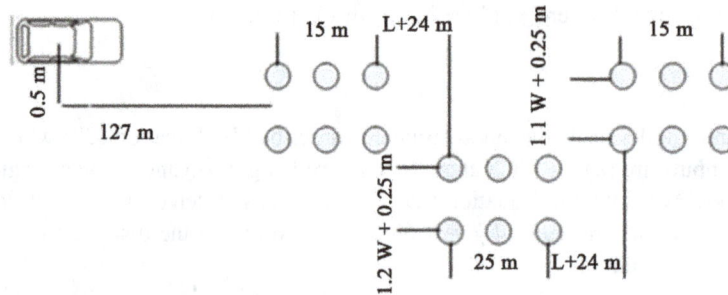

Figure 3. Tested trajectory of a chicane maneuver.

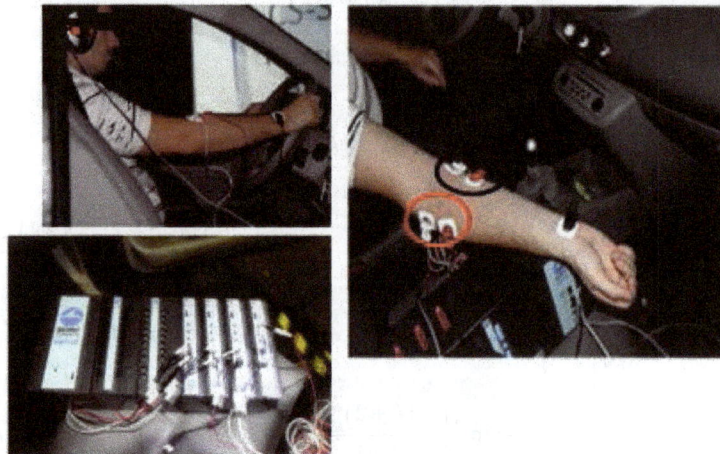

Figure 4. Head movement and EMG analysis.

3.3. Data Analysis

Multi-level data acquisition was performed at two levels as follows:

3.3.1. Vestibular Level Data Acquisition (through Sensor)

Such as the roll, pitch, yaw angles and rates as well as the accelerations in X,Y and Z. Quaternions have been used, since they are simpler to compose and to avoid singularity for angular calculations, so-called the problem of gimbal lock compared to Euler angles. The application domains of quaternions can be counted as computer graphics, computer vision, robotics, navigation, flight dynamics [32] and orbital mechanics of satellites [33]. Because we have dealt with the hexapod driving simulator in real-time, we have used quaternions. The data are calibrated due to three dimensional quaternion orientation. The sampling rate for the data registration during the sensor measurements is 20 Hz. For the calibrated data acquisition, the alignment reset has been chosen which simply combines the object and the heading resets at a single instant in time. This has the advantage that all coordinate systems can be aligned with a single action.

3.3.2. Electromyography (via Biopac System)

Electromyography (EMG) is an evaluation method of the electrical activity produced by musculoskeletal system. EMG is performed using an instrument called an electromyograph, to realize a record called an electromyogram. An electromyography detects the electrical potential generated by muscle cells [34] when these cells are electrically or neurologically activated. The signals can be analyzed to detect and identify medical abnormalities, muscle activation level, and recruitment order or to analyze the biomechanics of human or animal movements [34].

By using the Biopac systems, several frequency and time domain techniques could be used for data reduction of EMG signals [21].

For this study, it was chosen to deal with the EMG_{RMS} (root mean square: which is a product of longitudinal, lateral and vertical dynamics related dissipated power) power analysis (V^2/Hz in unit) in time domain, in order to investigate their associations with RVT (°/s in unit, which is an indicator of the conflict in dynamics).

- EMG_{RMS} mean total power yields the average power of the power spectrum within the epoch [21].
- EMG_{RMS} total power is equal to the sum of power at all frequencies of the power spectrum within the epoch [21].
- Epoch corresponds to how many time steps (Δt) a whole time series signal is divided into [21].

For the calibration of the electromyography, a gain of 1000 was used. And the **Figure 4** depicts the data acquisition during the experiments; the electrical activities of the muscles were registered by a non-invasive surface EMG method through two analog channels. The signals were collected with 10 Hz for low cut-off, 500 Hz for high cut-off frequencies (a band-pass filter with a frequency range: 10 - 500 Hz).

Electrodes in black circle were connected to flexor carpi radialis muscle where the electrodes in red circle were connected to brachioradialis muscles. We measured and saved the electrical activity changes on the brachioradialis and flexor carpi radialis muscles. In this paper, we explained the results which were taken from the muscle 'flexor carpi radialis' (at right hand side, **Figure 4**).

4. Results and Discussion

4.1. Vestibular and Neuromuscular Dynamics Interaction with Vehicle Dynamics

Figure 5 corresponds to the vehicle velocity during the experiments with driver for the EMG-RV analysis done in **Figure 6**.

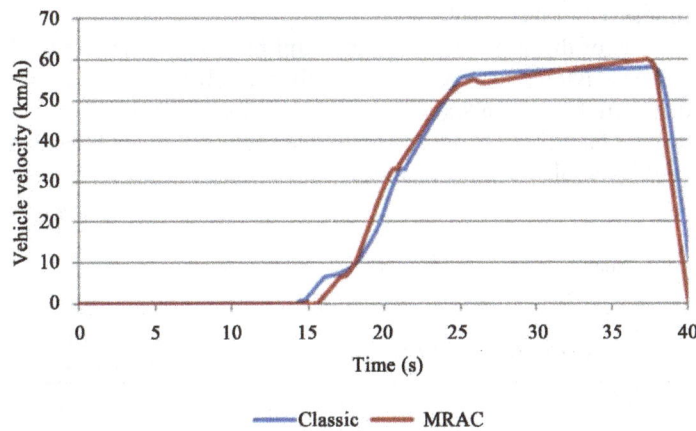

Figure 5. Vehicle velocity (km/h).

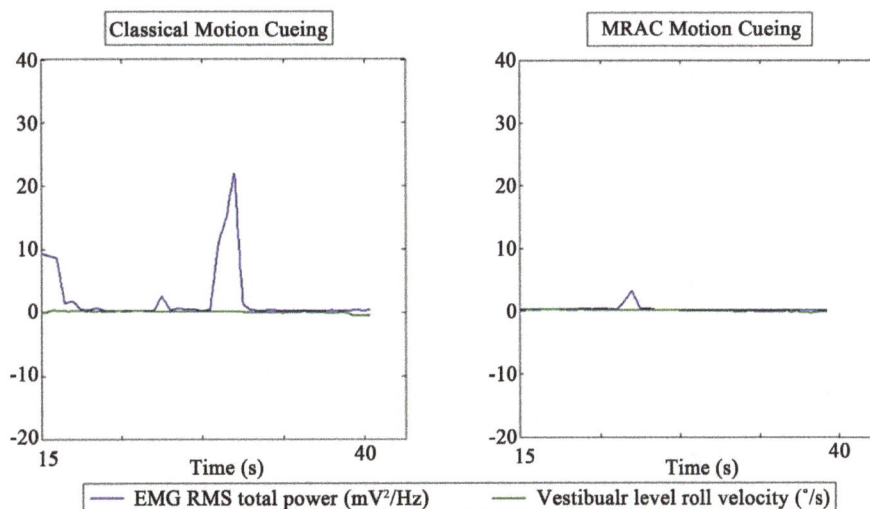

Figure 6. EMG_{RMS} total power (mV^2/Hz) and vestibular roll velocity (°/s).

Figure 6 indicates a real-time measurement of EMG_{RMS} total power-head roll velocity for one out of the twenty six subjects who participated in the experiments.

If **Figure 3**, **Figure 5** and **Figure 6** are evaluated together it is possible to characterize the neuromuscular dynamics (via EMG analysis), vestibular dynamics more clearly.

It can be seen that the discrepancy has been decreased between the EMG_{RMS} total power and the roll velocity at vestibular level by using MRAC motion cueing (see **Figure 6**). This figure also proves less contradicting cues from the arm muscular and the vestibular dynamics system, in other words less LOA (loss of adhesion) of the vehicle or more agility and alertness levels of the driver in the lateral dynamics.

Table 3 summarizes the correlation of the arm muscle (flexor carpi radialis) power and the head level roll velocity from a sensor attached to the right ear of the drivers via a headphone (**Figure 4**) with the vehicle CG (center of gravity) lateral displacement for the classical motion cueing algorithm situation. From **Table 3**, it is seen that there have been significant correlation between vestibular roll velocity and vehicle CG lateral displacement except for Subject 1 and Subject 13. Furthermore, merely the subject 6 has yielded a positive correlation between the vestibular level roll velocity and the arm muscles power. Apart from subject 6, they have demonstrated a negative correlation, which actually represents an increased level of the visuo-vestibular cue conflict.

According to **Table 3**, it can be concluded that there have been significant correlations between arm muscle powers and vehicle CG lateral displacements for all the thirteen subjects at the classical motion cueing case. Moreover, a negative correlation has been resulted between the arm muscles power and the vehicle CG lateral displacements for the subjects 1, 3, 7, 8 and 13 for the classical algorithm, whereas a positive correlation has been obtained for the rest of the subjects. These correlations show us that for the 8 subjects the vehicle has been driven with the less controlloss out of the 13 subjects.

Table 4 gives the correlation of the arm muscle (flexor carpi radialis) power and the head level roll velocity with the vehicle CG (center of gravity) lateral displacement for the MRAC motion cueing algorithm case. From **Table 4**, it can be seen that there have been significant correlation between vestibular roll velocity and vehicle CG lateral displacement except for Subject 8. Furthermore, the subject 3, 6 and 8 have yielded a positive correlation between the vestibular level roll velocity and the arm muscles power. Apart from these subjects, they have demonstrated a negative correlation.

Table 3. Vestibular-arm muscle dynamics interaction for classical motion cueing.

	Classical motion cueing Correlation of vestibular roll velocity-vehicle CG lateral displacement		Classical motion cueing Correlation of EMG RMS total power for arm muscles-vehicle CG lateral displacement	
	r	*p*	*r*	*p*
Subject 1	−0.0252	0.6025	−0.7730	0.0000***
Subject 2	−0.5375	0.0000***	0.4004	0.0000***
Subject 3	−0.5815	0.0000***	−0.1068	0.0244*
Subject 4	−0.2872	0.0000***	0.1714	0.0003***
Subject 5	−0.6761	0.0000***	0.5870	0.0000***
Subject 6	0.3348	0.0000***	0.4380	0.0000***
Subject 7	−0.4486	0.0000***	−0.2023	0.0000***
Subject 8	−0.3638	0.0000***	−0.1652	0.0007***
Subject 9	−0.1770	0.0003***	0.1994	0.0000***
Subject 10	−0.6624	0.0000***	0.1413	0.0051**
Subject 11	−0.6265	0.0000***	0.1931	0.0001***
Subject 12	−0.7544	0.0000***	0.2197	0.0000***
Subject 13	−0.0691	0.1448	−0.1083	0.0220*

*Means one zero after the point ".", **means two zeros after the point ".", ***means three and more than three zeros after the point ".".

Table 4. Vestibular-arm muscle dynamics interaction for MRAC motion cueing.

	MRAC motion cueing _Correlation of vestibular roll velocity-vehicle CG lateral displacement_		**MRAC motion cueing** _Correlation of EMG RMS total power for arm muscles-vehicle CG lateral displacement_	
	r	*p*	*r*	*p*
Subject 1	−0.2241	0.0000***	0.3925	0.0000***
Subject 2	−0.6283	0.0000***	0.3586	0.0000***
Subject 3	0.4826	0.0000***	0.0896	0.0618
Subject 4	−0.5073	0.0000***	−0.3667	0.0000***
Subject 5	−0.4519	0.0000***	0.0969	0.0182*
Subject 6	0.1222	0.0155*	0.4443	0.0000***
Subject 7	−0.1934	0.0001***	0.1651	0.0008***
Subject 8	0.0368	0.4506	0.1650	0.0007***
Subject 9	−0.1310	0.0063**	0.0667	0.1658
Subject 10	−0.5408	0.0000***	0.1386	0.0064**
Subject 11	−0.1222	0.0198*	0.0036	0.9456
Subject 12	−0.6599	0.0000***	0.3367	0.0000***
Subject 13	−0.4904	0.0000***	0.3724	0.0000***

*Means one zero after the point ".", **means two zeros after the point ".", ***means three and more than three zeros after the point ".".

According to **Table 4**, it can be concluded that there have been significant correlations between arm muscle powers and vehicle CG lateral displacements apart from the subjects 3, 9 and 11 at the MRAC motion cueing condition. Moreover, a negative correlation has been coincided between the arm muscles power and the vehicle CG lateral displacements only for the subject4 for the MRAC algorithm, whereas a positive correlation has been occurred for the rest of the subjects. These correlations show us that only for the 1subject (subject 4) the vehicle has been driven with a propensity of control loss out of the 13 subjects, when uniquely the correlation between the arm muscles power and the vestibular level roll velocity are taken into account.

From **Table 3**, **Table 4** and **Figure 7**, it is seen that loss of adhesion (LOA) of the vehicle causes a visuo-vestibular cues conflict (if the correlation of vestibular roll velocity with vehicle CG lateral displacement is negative) however visuo-vestibular conflict is not always followed by a LOA (if the correlation of EMG_{RMS} total power with vehicle CG lateral displacement is negative).

Figure 8 summarizes our findings about the relationships of multi sensory (vestibular, neuromuscular, vehicle (visual)) cues with motion sickness incidence as a metrics for lateral dynamics in terms of sensory cue conflict theory [35] [36] depending on **Table 3** and **Table 4** respectively for this study. According to this figure; when there is no conflict at all in cues, no motion sickness is occurred. As there is visuo-vestibular cue conflict, it results as a moderate level of motion sickness. Eventually when LOA is observed, the motionsickness gets higher levels.

Table 5 illustrates the lateral displacement area (Equation (30) in m·s) [19] [20] [37] (see **Figure 9**) under the vehicle CG lateral displacement (Y_{CG}) from the time series graphs where t is time

$$\text{Lateral displacement area} = \int_o^t Y_{CG} \mathrm{d}t \qquad (30)$$

According to **Table 5**, it can be summed up that apart from the subjects 6, 7, 8 and 9 the loss of adhesion has decreased; in other words agility or alertness level of the drivers have increased for the classical algorithm comparing to MRAC algorithm which make the vehicle maintain on the desired route. In contrast for the rest of the subjects, the agility level of the drivers has increased, in other words LOC of the vehicle has decreased with MRAC motion cueing. For the subjects 6, 7, 8 and 9 (4 subjects out of 13 subjects) the classical motion cueing algorithm had the higher level of loss of control (LOC) of the vehicle in the dynamic driving simulator. For the

Visuo-vestibular sensory conflict does not always cause
loss of adhesion

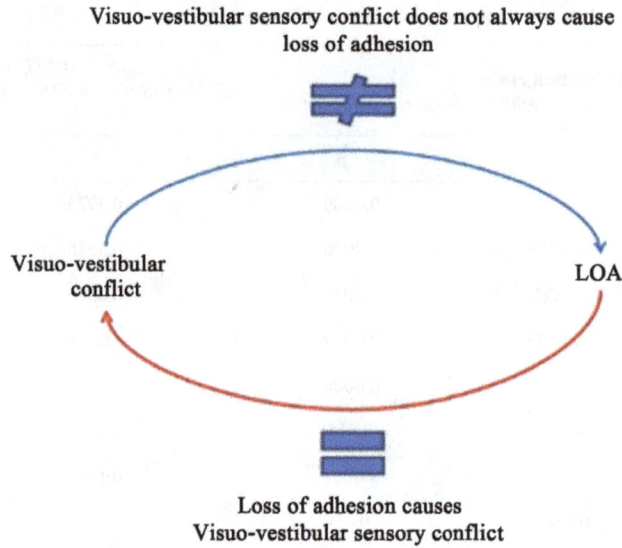

Visuo-vestibular   LOA
conflict

Loss of adhesion causes
Visuo-vestibular sensory conflict

Figure 7. Vestibular-neuromuscular-vehicle dynamics interaction.

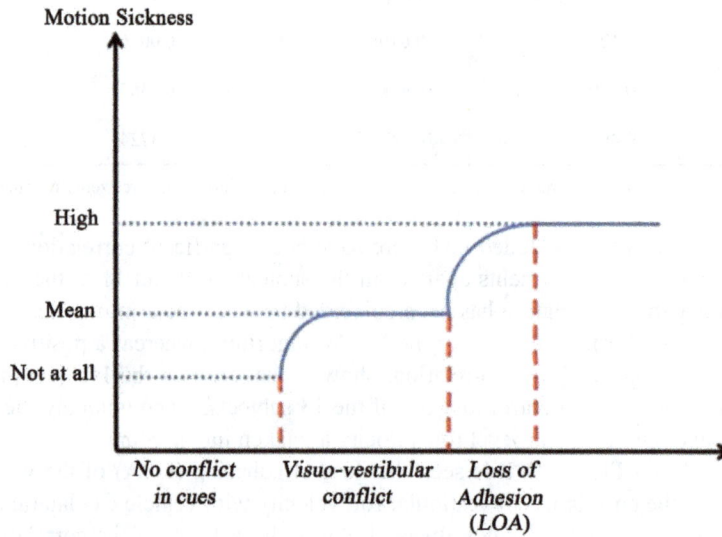

Figure 8. Relationships of multi sensory cues conflict with motion sickness.

Figure 9. Lateral displacement area.

rest of the subjects (9 subjects out of 13 subjects), the MRAC motion cueing algorithm brought a more controll-able vehicle in the dynamic simulator as operated in real-time. This shows that the MRAC motion cueing algorithm (closed loop control) provides a less LOA and LOC comparing to the classical motion cueing algorithm (open loop control).

Table 5. Lateral displacement area for both algorithms.

	Classical motion cueing *Lateral displacement area* (m·s)	MRAC motion cueing *Lateral displacement area* (m·s)	LOA change (%) from Classical to MRAC algorithm
Subject 1	3.2166×10^4	1.6569×10^4	48.4% of decrease
Subject 2	1.5717×10^4	1.4588×10^4	7.1% of decrease
Subject 3	2.1987×10^4	1.7697×10^4	19.5% of decrease
Subject 4	1.6291×10^4	1.5022×10^4	7.8% of decrease
Subject 5	1.3508×10^4	1.2317×10^4	8.8% of decrease
Subject 6	1.0828×10^4	1.3160×10^4	21.5% of increase
Subject 7	1.3432×10^4	1.3630×10^4	1.4% of increase
Subject 8	1.4279×10^4	1.4479×10^4	1.4% of increase
Subject 9	1.3804×10^4	1.5422×10^4	11.7% of increase
Subject 10	1.2864×10^4	1.2560×10^4	2.36% of decrease
Subject 11	1.4529×10^4	1.1474×10^4	21% of decrease
Subject 12	1.2642×10^4	1.1966×10^4	5.35% of decrease
Subject 13	1.7150×10^4	1.5708×10^4	8.4% of decrease

4.2. Vestibular and Neuromuscular Dynamics Interaction

After having completed the data evaluation by individual as above, the overall data analysis has been studied for the thirteen subject group for each motion cueing (both for the classical and the MRAC motion cueing algorithms).

The overall data analysis was done by using a two-way ANOVA (Analysis of Variance) and Pearson's correlation with an $\alpha = 0.05$. In order to accomplish the statistical analysis, we took
-   the vestibular roll velocity into account, which were measured from the right ear level of the participants during the real-time simulator experiments through the motion tracking sensor.
-   the EMGRMS total power from the muscle 'flexor carpi radialis'.

The two-way ANOVA was applied to identify the level of significance as between subjects' principle test. The Pearson's correlation was computed to clarify the correlation of the vestibular level roll velocity threshold with the EMG_{RMS} power dissipation of the arm muscles.

4.2.1. Two-Way ANOVA

The two-way ANOVA tests were used to check the influence of one type of response dynamics (spent power from the arms) of the drivers on the other type of response dynamics (vestibular roll velocity) of them as an interaction metrics.

By studying the output of the two-way ANOVA for the proposed classical motion cueing algorithm, we see that there is no evidence of a significant interaction effect ($F = 3.74$, $p = 0.085 > 0.05$) between those two types of response dynamics of the drivers. We therefore can conclude that no interaction was obtained between the EMG_{RMS} mean power and the vestibular roll velocity threshold. The test for the main effect of the vestibular roll velocity perception threshold ($F = 45.17$, $p < 0.0001$) shows a significant roll velocity perception threshold effect on the EMG_{RMS} maximum power level. Finally, the test for the main effect of the EMG_{RMS} mean power ($F = 1360.89$, $p < 10^{-9}$) tells us there is an evidence to conclude that the EMG_{RMS} mean power has a significant effect on the EMG_{RMS} maximum power level.

By investigating the output of the two-way ANOVA for the proposed model reference adaptive control motion cueing algorithm, we see that there is no evidence of a significant interaction effect ($F = 0.43$, $p = 0.528 > 0.05$) between those two types of response dynamics of the drivers. We therefore cannot conclude that an interactionis found between the EMG_{RMS} mean power and the roll velocity threshold. The test for the main effect of the vestibular roll velocity perception threshold ($F = 9.22$, $p = 0.01412 < 0.05$) shows a significant roll velocity perception threshold effect on the EMG_{RMS} maximum power level. Finally, the test for the main effect of the EMG_{RMS} mean power (F = 232.44, $p < 10^{-6}$) tells us there is an evidence to conclude that the EMG_{RMS} mean power has a significant effect on the EMG_{RMS} maximum power level.

4.2.2. Pearson's Correlation

Having searched the relationships of the roll velocity thresholds with the EMG_{RMS} mean and maximum total power, we recognized that if we drive the same scenario with the MRAC, it shows a more alerted (agile) mode compared to the classical motion cueing algorithm. Because, the correlation coefficients (r) between the roll velocity thresholds and the EMG_{RMS} mean/maximum total power are positive for the MRAC and negative for the classical motion cueing algorithm.

We described the RV (roll velocity) at "vestibular" level [5] [6] [19] [20]. MATLAB/Simulink applications were used to process the RVT data (*maximum head level roll velocity at low frequent motion independent from the direction*): Vestibular level roll velocity signals were conditioned with a 1st order Butterworth low-pass filter at 5 Hz.

Root-mean square (RMS) of EMG (EMG_{RMS}) (mV) values were computed based on a total range of motion during the driving phase of the simulator experiments [18] [29], using the following Equation (31):

$$EMG_{RMS} = \sqrt{\frac{1}{T} \int_{t}^{t+T} EMG^2(t)\,dt} \qquad (31)$$

where t is the onset of signal time and T is the duration of RMS averaging [38].

For the frequency domain analysis used in this study to determine the spent power/energy by the arm muscles of the drivers, a Fast Fourier Transformation algorithm was used to calculate the power spectrum of EMG signals (mV) [34] [38] with the following Equation (32):

$$PSD(f) = \left| FFT(x_n) \right|^2 \qquad (32)$$

where x_n is a set of consecutive EMG signals with the specific number of epochs, PSD is power spectral density in mV^2/Hz [34] [38].

In this article:

- EMG_{RMS} total power indicates the dissipated power (times series data) during the driving simulation experiments (Equation (31) and Equation (32)). It yields the sum of the power at all frequencies of the power spectrum within the epoch.
- EMG_{RMS} maximum total power refers to the peak values (one value point) obtained from the dissipated energy during the driving simulation experiments (Equation (31) and Equation (32)).
- EMG_{RMS} mean total power indicates the mean of the sum of the dissipated power (one value point) at all frequencies during the driving simulation experiments (Equation (31) and Equation (32)).
- Roll velocity perception threshold gives *maximum vestibular level roll velocityat low frequent motion independent from the direction*: Those signals were conditioned with a 1st order Butterworth low-pass filter at 5 Hz.

Figure 10 and **Figure 11** illustrate the difference of the both motion cueing algorithms in terms of RVT (°/s) to EMG_{RMS} maximum total power (mV^2/Hz) as well as to EMG_{RMS} mean total power (mV^2/Hz) changes. In order to assess the relationships, we used MATLAB. The vertical axes were illustrated in the logarithmic scale in order to show the data points better.

Due to **Figure 12**, it is clearly seen that some of the participants reached extreme outliers (red stars) in terms of roll velocity thresholds by driving the classical motion cueing algorithms while the RVT for the MRAC motion cueing algorithm showed no extreme outliers. In addition, the discrepancy of the RVT for classical motion cueing algorithm (from −3.1°/s to 2.4°/s, it makes a discrepancy of 5.5°/s) is greater than the one (from −0.8°/s to 1.1°/s, it makes a discrepancy of 1.9°/s) in MRAC motion cueing algorithm, which also specifies a higher level of vestibular sensory conflict.

According to Pearson's correlation, it is shown that the RVT and the EMG_{RMS} mean power are positively correlated ($r = 0.228$, $p = 0.453$) for the MRAC algorithm and they are negatively correlated ($r = -0.167$, $p = 0.586$) for the classical algorithm.

According to Pearson's correlation, it is shown that the RVT and the EMG_{RMS} maximum power are positively

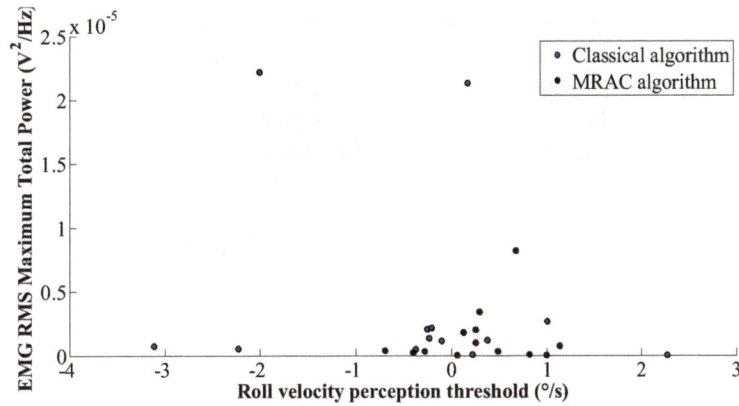

Figure 10. RVT and EMG RMS maximum total power analysis relationships for all subjects.

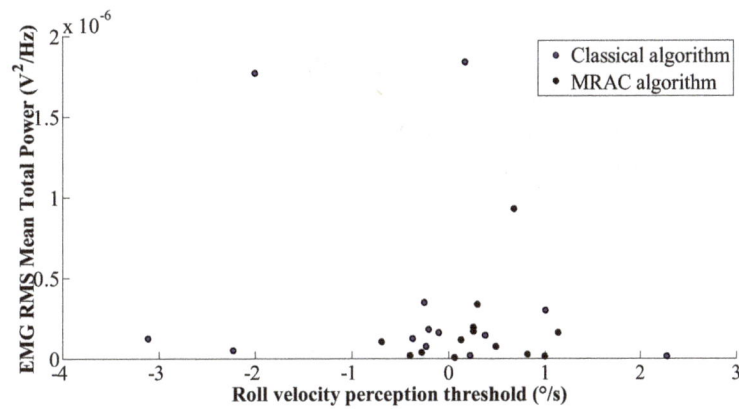

Figure 11. RVT and EMG RMS mean total power analysis relationships for all subjects.

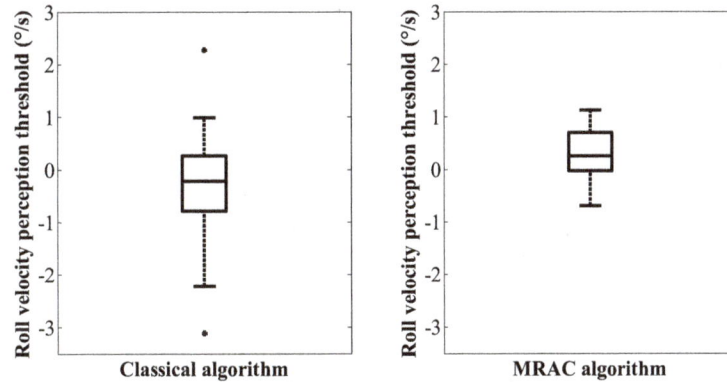

Figure 12. RVT comparison for the classical and the MRAC motion cueing algorithms for all subjects.

correlated ($r = 0.192$, $p = 0.531$) for the MRAC motion cueing algorithm and they are negatively correlated ($r = -0.178$, $p = 0.560$) for the classical motion cueing algorithm. (**Figure 10** and **Figure 11**).

5. Conclusion

Head roll velocity, arm muscle and vehicle dynamics interaction were investigated. For the classical motion cueing the discrepancy of the curves (vestibular-arm muscle dynamics) has increased (see **Figure 6**). Concluding up from **Table 5**; 9 out of the 13 subjects have obtained a lower amount of surface according to the lateral displacement of the vehicle CG with the MRAC algorithm. This indicates an improvement in LOA (loss of adhesion) at the MRAC motion cueing comparing to the classical motion cueing (69.2% of decrease in LOA, 30.8 % of increase in LOA). Due to **Table 3**, 5 out of 13 subjects have shown a propensity on LOA and decrease in agility/alertness level of the drivers at the classical algorithm (proximity to a LOA incidence is 38.5%). **Table 4** has illuminated an inclination to an incidence of LOA for merely 1 subject (propensity to a LOA is 7.7%).

We also investigated the vestibular sensed roll velocity perception threshold (impulse effect dynamics: high frequent motion) with the total power spent by the arm muscles. It gave an idea about the driver's behaviours as "alertness (agility)".

Having a closed loop control of the hexapod platform (a MRAC motion cueing strategy) supplied more alertness (39.5% increase in agility: $r = 0.228$, $p = 0.453$ for the MRAC algorithm and $r = -0.167$, $p = 0.586$ for the classical algorithm in terms of RVT-EMG$_{RMS}$ mean total power. And also 37% increase in agility: $r = 0.192$, $p = 0.531$ for the MRAC algorithm and $r = -0.178$, $p = 0.560$ for the classical algorithm in terms of RVT-EMG$_{RMS}$ maximum total power), that helps the dynamic simulator be driven more controllably, compared to an open loop control of the hexapod platform (a classic motion cueing strategy) with the same filters and gains (**Figure 1** and **Table 1**).

We thus cannot conclude that there is an interaction between the EMG$_{RMS}$ means power and the roll velocity threshold. The test for the main influence of the vestibular roll velocity perception threshold indicates a significant roll velocity perception threshold influence on the EMG$_{RMS}$ maximum power level. Eventually, the test for the main effect of the EMG$_{RMS}$ mean power gives us an evidence to conclude that there is a significant EMG$_{RMS}$ mean power effect on the EMG$_{RMS}$ maximum power level for both classical and MRAC motion cueing algorithms.

As a conclusion, the MRAC motion cueing strategy optimized the dynamic simulator condition with respect to the classical motion cueing strategy, so that it resulted as an improved situation for the drivers in terms of the "avoidance of LOA" and improving the "motion sickness" depending on sensory conflict theory between neuromuscular-vehicle (visual) cues, between vestibular-vehicle (visual) cues.

It also proved that the neuromuscular-vehicle dynamics conflict has influence on visuo-vestibular conflict; however, the visuo-vestibular cue conflict does not influence the neuromuscular-vehicle dynamics interactions.

As prospective work, we would like to evaluate the sickness regarding the pitch velocity/acceleration perception threshold as well as the neuromuscular-vestibular reaction time relationships of the drivers on various road scenarios, under different controls of the hexapod platform with correlations in inertial, vestibular, neuromuscular cues.

Acknowledgements

Arts et Métiers Paris Tech built up the SAAM driving simulator with the partnership of Renault.

References

[1]   Angelaki, D.E., Gu, Y. and DeAngelis, G.C. (2009) Multisensory Integration: Psychophysics, Neurophysiology, and Computation. *Current Opinion in Neurobiology*, **19**, 452-458. http://dx.doi.org/10.1016/j.conb.2009.06.008

[2]   Aykent, B., Paillot, D., Merienne, F., Fang, Z. and Kemeny, A. (2011) Study of the Influence of Different Washout Algorithms on Simulator Sickness for Driving Simulation Task. *Proceedings of the ASME* 2011 *World Conference on Innovative Virtual Reality WINVR*2011, Milan, 331-341. http://dx.doi.org/10.1115/WINVR2011-5545

[3]   Chen, D., Hart, J. and Vertegaal, R. (2007) Towards a Physiological Model of User Interruptability. *IFIP International Federation for Information Processing, INTERACT* 2007, 439-451.

[4]   Dichgans, J. and Brandt, T. (1973) Optokinetic Motion Sickness and Pseudo-Coriolis Effects Induced by Moving Visual Stimuli. *Acta Otolaryngologica*, **76**, 339-348. http://dx.doi.org/10.3109/00016487309121519

[5]   DiZio, P. and Lackner, J.R. (1989) Perceived Self-Motion Elicited by Postrotary Head Tilts in a Varying Gravitoinertial Force Background. *Perception & Psychophysics*, **46**, 114-118. http://dx.doi.org/10.3758/BF03204970

[6]   Nehaoua, L., Arioui, H., Espié, S. and Mohellebi, H. (2006) Motion Cueing Algorithms for Small Driving Simulator. *IEEE International Conference in Robotics and Automation* (*ICRA*06), Orlando.

[7]   Kemeny, A. (2014) From Driving Simulation to Virtual Reality. *Proceedings of the* 2014 *Virtual Reality International Conference*, 32.

[8] Benson, A.J., Hutt, E.C. and Brown, S.F. (1989) Thresholds for the Perception of Whole Body Angular Movement about a Vertical Axis. *Aviation, Space, and Environmental Medicine*, **60**, 205-213. http://psycnet.apa.org/psycinfo/1989-21186-001

[9] Guedry Jr., F.E. (1964) Visual Control of Habituation to Complex Vestibular Stimulation in Man. Bureau of Medicine and Surgery, Project MR005.13-6001, Subtask 1 Report No. 95, NASA Order No. R-93, US Naval School of Aviation Medicine, US Naval Aviation Medical Center, Pensacola, 1-16. http://informahealthcare.com/doi/abs/10.3109/00016486409121398

[10] Guedry, F.E. and Montague, E.K. (1961) Quantitative Evaluation of the Vestibular Coriolis Reaction. *Aviation, Space, and Environmental Medicine*, **32**, 487-500. http://eurekamag.com/research/025/329/025329418.php

[11] Wiederhold, B.K. and Bouchard, S. (2014) Sickness in Virtual Reality. *Advances in Virtual Reality and Anxiety Disorders*, 35-62. http://dx.doi.org/10.1007/978-1-4899-8023-6_3

[12] DiZio, P. and Lackner, J.R. (1988) The Effects of Gravitoinertial Force Level and Head Movements on Post-Rotational Nystagmus and Illusory After-Rotation. *Experimental Brain Research*, **70**, 485-495. http://dx.doi.org/10.1007/BF00247597

[13] Kolasinski, E.M. (1995) Simulator Sickness in Virtual Environments. Army Project Number 2O262785A791, Education and Training Technology.

[14] Asadi, H., Mohammadi, A., Mohamed, S. and Nahavandi, S. (2014) Adaptive Translational Cueing Motion Algorithm Using Fuzzy Based Tilt Coordination. *Neural Information Processing*, Springer International Publishing, Berlin, 474-482.

[15] Curry, R., Artz, B., Cathey, L., Grant, P. and Greenberg, J. (2002) Kennedy SSQ Results: Fixed- vs Motion-Based FORD Simulators. *Proceedings of Driving Simulation Conference*, Paris, 11-13 September 2002, 289-300.

[16] Kim, M.S., Moon, Y.G., Kim, G.D. and Lee, M.C. (2010) Partial Range Scaling Method Based Washout Algorithm for a Vehicle Driving Simulator and Its Evaluation. *International Journal of Automotive Technology*, **11**, 269-275. http://dx.doi.org/10.1007/s12239-010-0034-0

[17] MOOG FCS, 6 DOF Motion System (2006) Motion Drive Algorithm (MDA) Software Tuning Manual Version 1.0. Document No: LSF-0468, Revision: A.

[18] Siegler, I., Reymond, G., Kemeny, A. and Berthoz, A. (2001) Sensorimotor Integration in a Driving Simulator: Contributions of Motion Cueing in Elementary Driving Tasks. *Proceedings of Driving Simulation Conference*, Sophia-Antipolis, 5-7 September 2001, 21-32.

[19] Meywerk, M., Aykent, B. and Tomaske, W. (2009) Einfluss der Fahrdynamik-regelung auf die Sicherheit von N1-Fahrzeugen bei unterschiedlichen Bela-dungszuständen. Teil 1: Grundlagen, Unfallstatistik, Abstütz- und Beladungs-einrichtung, Fahrzeugdatenermittlung. FE 82.329/2007. http://bast.opus.hbz-nrw.de/frontdoor.php?source_opus=354&la=de

[20] Meywerk, M., Aykent, B. and Tomaske, W. (2009) Einfluss der Fahrdynamik-regelung auf die Sicherheit von N1-Fahrzeugen bei unterschiedlichen Bela-dungszuständen. Teil 2: Fahrversuche und Fahrsimulatorversuche. FE 82.329/ 2007. http://bast.opus.hbz-nrw.de/frontdoor.php?source_opus=355&la=de

[21] AcqKnowledge® 4 (2011) Software Guide For Life Science Research Applications, Data Acquisition and Analysis with BIOPAC MP Systems Reference Manual for AcqKnowledge® 4.2 Software &MP150 or MP36R Hardware/ Firmware on Windows® 7 or Vista or Mac OS® X 10.4-10.6. 357-358.

[22] Benson, A.J. (1990) Sensory Functions and Limitations of the Vestibular Systems. In: Warren, R. and Wertheim, A.H., Eds., *Perception and Control of Self-Motion*, Laurence Erlbaum Associates, Hinsdale, 145-170.

[23] Kemeny, A. and Panerai, F. (2003) Evaluating Perception in Driving Simulation Experiments. *Trends in Cognitive Sciences*, **7**, 31-37. http://dx.doi.org/10.1016/S1364-6613(02)00011-6

[24] Pick, A.J. (2004) Neuromuscular Dynamics and the Vehicle Steering Task. Ph.D. Dissertation, St Catharine's College, Cambridge University Engineering Department, Cambridge.

[25] Aykent, B., Merienne, F., Paillot, D. and Kemeny, A. (2013) Influence of Inertial Stimulus on Visuo-Vestibular Cues Conflict for Lateral Dynamics at Driving Simulators. *Journal of Ergonomics*, **3**, 1-7. http://omicsgroup.org/journals/influence-of-inertial-stimulus-on-visuo-vestibular-cues-conflict-for-lateral-dynamics-at-driving-simulators-2165-7556.1000113.pdf

[26] Duarte, M.A. and Ponce, R.F. (1997) Discrete-Time Combined Model Reference Adaptive Control. *International Journal of Adaptive Control and Signal Processing*, **11**, 501-517. http://dx.doi.org/10.1002/(SICI)1099-1115(199709)11:6<501::AID-ACS448>3.0.CO;2-G

[27] Guo, J., Liu, Y. and Tao, G. (2009) Multivariable MRAC with State Feedback for Output Tracking. *American Control Conference*, Hyatt Regency Riverfront, PaperWeA18.5, St. Louis, 10-12 June 2009, 592-597.

[28] Ioannou, P.A. and Sun, J. (1995) Robust Adaptive Control. Prentice-Hall Inc., Upper Saddle River, 313-408.

[29] Maiti, D., Guo, J. and Tao, G. (2011) A Discrete-Time Multivariable State Feedback MRAC Design with Application to Linearized Aircraft Models with Damage. *American Control Conference*, San Francisco, 29 June-1 July 2011, 606-611.

[30] Tao, G. and Ioannou, P.A. (1993) Model Reference Adaptive Control for Plants with Unknown Relative Degree. *IEEE Transactions on Automatic Control*, **38**, 976-982. http://dx.doi.org/10.1109/9.222314

[31] Tao, G. (2003) Adaptive Control Design and Analysis. John Wiley and Sons, New York. http://dx.doi.org/10.1002/0471459100

[32] Katz, A. (1996) Computational Rigid Vehicle Dynamics. Krieger Publishing Co., Malabar.

[33] Kuipers, J.B. (1999) Quaternions and Rotation Sequences: A Primer with Applications to Orbits, Aerospace, and Virtual Reality. Princeton University Press, Princeton.

[34] Mesh Electromyography (2015) National Library of Medicine—Medical Subject Headings, Mesh. http://www.ncbi.nlm.nih.gov/mesh/68004576

[35] Oman, C. (1989) Sensory Conflict in Motion Sickness: An Observer Theory Approach. NASA, Ames Research Center, Spatial Displays and Spatial Instruments, 15 p.

[36] Oman, C. (1990) Motion Sickness: A Synthesis and Evaluation of the sensory Conflict Theory. *Canadian Journal of Physiology and Pharmacology*, **68**, 294-303. http://dx.doi.org/10.1139/y90-044

[37] Electronic Code of Federal Regulations (2015) Title 49: Transportation Part 571—Federal Motor Vehicle Safety Standards Subpart B—Federal Motor Vehicle Safety Standards § 571.126 Standard No. 126; Electronic Stability Control Systems. US Government Printing Office, Electronic Code of Federal Regulations. http://www.ecfr.gov/cgi-bin/text-idx?SID=6099cd0521e2251f35642c554b1c3001&node=pt49.6.571&rgn=div5#se49. 6.571_1126

[38] Fu, W., Liu, Y., Zhang, S., Xiong, X.J. and Wei, S.T. (2012) Effects of Local Elastic Compression on Muscle Strength, Electromyographic, and Mechanomyographic Responses in the Lower Extremity. *Journal of Electromyography and Kinesiology*, **22**, 44-50. http://dx.doi.org/10.1016/j.jelekin.2011.10.005

9

Adaptive Control for a Class of Systems with Output Deadzone Nonlinearity

Nizar J. Ahmad[1], Ebraheem K. Sultan[1], Mohammed Q. Qasem[1], Hameed K. Ebraheem[1], Jasem M. Alostad[2]

[1]Faculty of Electronic Engineering Technology, College of Technological Studies, The Public Authority for Applied Education and Training (PAAET), Kuwait City, Kuwait
[2]Faculty of Computer Science, College of Basic Education, The Public Authority for Applied Education and Training (PAAET), Kuwait City, Kuwait
Email: nj.ahmad@paaet.edu.kw, ek.sultan@paaet.edu.kw, mq.qasem@paaet.edu.kw, hk.ebraheem@paaet.edu.kw, jm.alostad@paaet.edu.kw

Abstract

This paper presents a continuous-time adaptive control scheme for systems with uncertain non-symmetrical deadzone nonlinearity located at the output of a plant. An adaptive inverse function is developed and used in conjunction with a robust adaptive controller to reduce the effect of deadzone nonlinearity. The deadzone inverse function is also implemented in continuous time, and an adaptive update law is designed to estimate the deadzone parameters. The adaptive output deadzone inverse controller is smoothly differentiable and is combined with a robust adaptive nonlinear controller to ensure robustness and boundedness of all the states of the system as well as the output signal. The mismatch between the ideal deadzone inverse function and our proposed implantation is treated as a disturbance that can be upper bounded by a polynomial in the system states. The overall stability of the closed-loop system is proven by using Lyapunov method, and simulations confirm the efficacy of the control methodology.

Keywords

Adaptive Inverse Control, Output Deadzone, Hard Nonlinearity

1. Introduction

The problem of deadzone nonlinearity has been addressed by many researches with great success by utilizing adaptive control methods to eliminate the undesirable effects on the output of a plant [1]-[5]. Demonstrated in

Figure 1 is the effect of deadzone on the output of a plant for a pure sinusoidal input trajectory. The majority of earlier investigations to this problem focus on the problem where the nonlinearity is located at the input of the plant as an actuator problem [1] [2]. In an actuator deadzone, the control effort is within the span of the nonlinearity which makes it somewhat easier to reduce or eliminate its deleterious effects before it enters the dynamics of the system to be controlled. As a matter of fact, several papers present a two structure control schemes that can be designed to handle deadzone as well as other requirements for plant performance criteria [3]. On the other hand, output deadzone, which is physically inherent in some sensors that measure output signals of a plant, is a more complicated problem. The control effort has to eliminate the deleterious effect of the deadzone nonlinearity whilst going through the complicated dynamics of the plant. Therefore, whatever added control requirements enforced on the designer due to disturbances or noise affecting the plant, will further complicated the task. One of the earliest investigations of output nonlinearities such as deadzone was presented by [4]. Their proposed methodology was based on output matching control which involved the design of an adaptive deadzone inverse used to reshape the input reference trajectory to negate the effect of the deadzone. The parameters of the deadzone were adaptively estimated by designing an error function utilizing the output to observe plants states. The implementation was quiet complex in design and implemented in discrete time. In [5], an output feedback design was analysed for robustness and was developed using input to state stability (ISS) small gain tools. The combination of observer and controller design was proved to be essential when handling output nonlinearities. An adaptive compensation scheme without constructing a dead-zone inverse was presented in [6]. The proposed adaptive method requires only the information of bounds of the deadzone slopes and treats the time-varying input coefficient as a system uncertainty. The new control scheme ensures bounded-error trajectory tracking and assures the boundedness of all the signals in the adaptive closed loop. Tian Ping *et al.* utilized the integral-type Lyapunov function to design an adaptive compensation term for the upper bound of the residual and optimal approximation error as well as the dead-zone disturbance [7]. It was demonstrated that the closed-loop control system was semi-globally uniformly bounded. In [8], an inverse deadzone function was incorporated in control system driven from a mathematical model of a deadzone in pneumatic servo valves. Tests were performed out using controllers with and without dead zone compensation to comparison validated the efficacy of the method. In [9], a somewhat earlier work was presented in discrete time which successfully achieved reduction

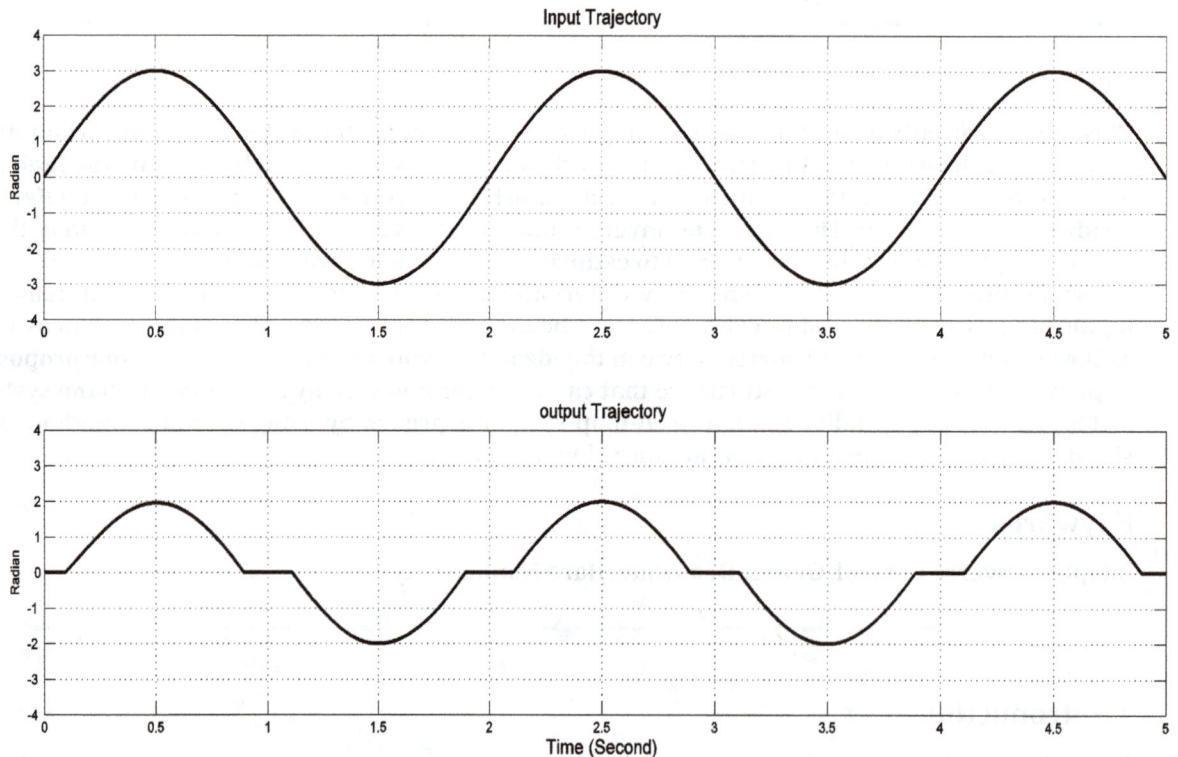

Figure 1. The distortion effect of output deadzone nonlinearity on a sinusoidal of signal.

of the tracking error in plants with output deadzone nonlinearity while ensuring the global boundedness stability. The paper presented by Jing Zhoua *et al.* introduced a smooth approximation to the deadzone model which allowed them to employ back stepping technique [10]. In their approach, no knowledge was assumed of the uncertainty's and the deadzone's parameters. It is shown that the proposed controller not only can guarantee global stability, but also can achieve excellent transient performance. It is worthwhile to note that other non-classical control methods, such as fuzzy logic or neural network, have been presented by several researchers to reduce the effect of a deadzone nonlinearity [11]-[14]. For example, Wallace and Max used an adaptive fuzzy controller for nonlinear systems subject to dead-zone input. The boundedness of all closed-loop signals and the convergence properties of the tracking error are proven using Lyapunov stability theory and Barbalat's lemma [15].

Motivated by the success in producing successful results in handling input deadzone, we present an extended method to reduce the errors caused by output deadzone nonlinearity. The proposed method relies on the premise that by pre-shaping the input trajectory to mimic an inverse form of the deadzone nonlinearity, the combined effect will reduce if not completely eliminating the effect of output deadzone.

In this paper, a new continuous time robust adaptive output deadzone inverse controller (RAODI) is used in conjunction with a conventional model reference adaptive control to counter the distortions cause by output deadzone. The ideal deadzone inverse controller is approximated by an infinitely differentiable implementation to insure asymptotic tracking and minimized error generation. The overall stability of the system under the proposed scheme will be proven analytically and demonstrated by simulation to a practical application. The structure of the paper starts with a brief presentation of the dynamics of an output deadzone nonlinearity that defines various parameters and its effect on the output of a system are presented in Section 2. Meanwhile, the proposed control methodology is presented and its analytical proof using the Lyapunov argument is shown in Section 3. Consequently, an illustrative example of a model reference adaptive control scheme combined with the inverse control method is presented and followed by simulation results in Section 4.

2. The dynamics of Output Deadzone Nonlinearity

A common representation of a non-symmetrical deadzone nonlinearity, shown in **Figure 1**, can be described as follows

$$DZ(y) = \begin{cases} m(x - d_r), & \text{if } x > d_r \\ 0, & \text{if } -d_l < x < d_r \\ m(x + d_l), & \text{if } x < -d_l \end{cases} \qquad (1)$$

where $DZ(y)$ denotes the output of deadzone function, $x(t)$ the output of a plant, m is the slope of the lines, $(d_r - d_l)$ is the width of the deadzone distance, and $u(t)$ is the input of the plant block as shown in **Figure 2**. Although the width of the deadzone spacing is assumed not to be exactly known, an upper bounds on it is given by

$$|d_r - d_l| \le d_M \qquad (2)$$

where d_M is a positive scalar. Output deadzone may also be written as

$$DZ(y) = x - sat_d(x) \qquad (3)$$

where $sat_d(u)$ represents a non-symmetrical saturation function given by

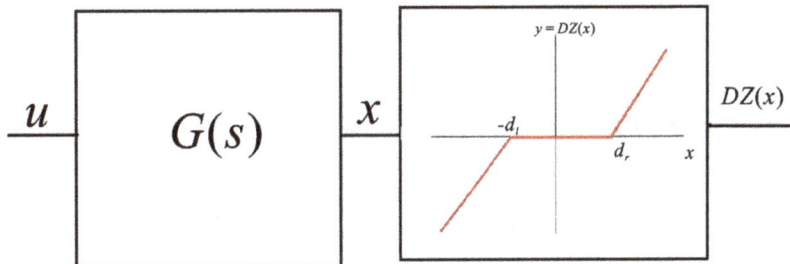

Figure 2. Non-symmetric deadzone nonlinearity as a function of a plant output signal.

$$sat_d(x) = \begin{cases} d_r, & \text{if } x > d_r \\ x, & \text{if } d_l < x < d_r \\ -d_l, & \text{if } x < -d_l \end{cases} \tag{4}$$

By defining a logical switching operator

$$\chi_r = \begin{cases} 1 & \text{if } x > 0 \\ 0 & \text{otherwise} \end{cases} \tag{5}$$

$$\chi_l = \begin{cases} 1 & \text{if } x < 0 \\ 0 & \text{otherwise} \end{cases} \tag{6}$$

Then, the dynamics of the non-symmetrical deadzone presented in (3) can be rewritten as follows

$$y = DZ(x) = x(t) - \chi_l d_l - \chi_r d_r = x(t) - d^{\mathrm{T}} \overline{\chi} \tag{7}$$

where $x(t)$ is the. Meanwhile, thelogical indicators, $\overline{\chi} = [\chi_r \chi_l]$ can be implemented by utilizing the definition of a sign function given as

$$\text{sgn}(x_d) = \begin{cases} 1 & x_d > 0 \\ -1 & x_d \leq 0 \end{cases}. \tag{8}$$

To obtain a smoothly differentiable implementation of (8), we replace it with a

$$\text{sgn}(x_d) \approx \tanh(k_s \cdot x_d). \tag{9}$$

with $k_s > 0$ appropriately selected with high value for fast switching applications.
Hence, rewriting Equation (5) and Equation (6) as

$$\chi_r = \frac{1 + \tanh(k_s \cdot x_d)}{2} \tag{10}$$

$$\chi_l = \frac{1 - \tanh(k_s \cdot x_d)}{2} = (1 - \chi_r). \tag{11}$$

To proceed with the design of the compensator the following assumptions are required:
(A1) The deadzone parameters $d_r > 0$ and $-d_l < 0$.
(A2) The deadzone parameters d_r and d_l are bounded as follows:

$$d_l \in [d_l \min, d_l \max] \text{ and } d_r \in [d_r \min, d_r \max].$$

(A3) Without any loss of generality the slope of the deadzone m is positive and is set to 1.
Assumption (A1) and (A2) are the actual physical attributes of a real industrial deadzone and is adopted in [16]. Therefore, the saturation function given by (4) is physically bounded

$$\|sat(x)\| = \|d^{\mathrm{T}} \overline{\chi}\| \leq d_M. \tag{12}$$

3. Robust Adaptive Controller Design

Considering the following nonlinear systems with input deadzone nonlinearity described as

$$\dot{x} = Ax + f(x) + B\{(u) + \psi(x)\}$$
$$y = DZ(x) \tag{13}$$

where the matrices A and B are given by

$$A = \begin{pmatrix} 0 & 1 & \cdots & 0 \\ 0 & 0 & 1 & 0 \\ \vdots & & \ddots & \vdots \\ 0 & 0 & \cdots & 0 \end{pmatrix} \quad B = \begin{pmatrix} 0 \\ \vdots \\ 0 \\ 1 \end{pmatrix}.$$

Meanwhile, the unmeasurable disturbances represented as $\psi(x)$ and $f(x)$ are assumed to be bounded by a known p^{th} order polynomial in the states [17]:

$$\|\psi(x)\| \le \sum_{k=0}^{p} \varsigma_k \|x\|^k \tag{14a}$$

$$\|f(x)\| \le \sum_{k=0}^{p} \varsigma_k \|x\|^k . \tag{14b}$$

The desired reference model is given by

$$\dot{x}_d = Ax_d + B\{Kx_d + r\}, \tag{15}$$

where $K \in R^{1 \times n}$ and r is a reference signal. By reshaping the desired reference model in a way to produce a deadzone inversed version of it will reduce the effect of the deadzone. Tracking the reshaped copy of the reference model will force the output of the deadzone nonlinearity to track the original desired reference signal. The adaptive output deadzone inverse compensator can be deduced from (7) as

$$x_d^* = \widehat{DI}(x_d) = x_d + \chi_l \hat{d}_l + \chi_r \hat{d}_r = x_d + \hat{d}^T \overline{\chi}, \tag{16}$$

where $\hat{d}^T = \left[\hat{d}_r \hat{d}_l\right]$ is the adaptively estimated values of the exact deadzone spacing $d = \left[d_r^* d_l^*\right]$. The adaptive inverse dynamics may be determined by differentiating (16) as follows

$$x_{1d}^* = x_d + \hat{d}^T \overline{\chi}$$

$$x_{2d}^* = \dot{x}_{1d}^* = \dot{x}_d + \dot{\hat{d}}^T \overline{\chi} + \hat{d}^T \dot{\overline{\chi}}$$

$$x_{3d}^* = \ddot{x}_d^* = \ddot{x}_d + \ddot{\hat{d}}^T \overline{\chi} + 2\dot{\hat{d}}^T \dot{\overline{\chi}} + \hat{d}^T \ddot{\overline{\chi}}$$

$$\vdots$$

$$x_{nd}^* = x_d^{*(n)} = x_d^{(n)} + \sum_{k=0}^{n} \binom{n}{k} \cdot \hat{d}^{T(k)} \cdot \overline{\chi}^{(n-k)}$$

Consequently, we can utilize (15) to construct the inverse deadzone model reference as

$$\dot{x}_d^* = Ax_d^* + B\left\{K \cdot \left(x_d + \sum_{k=0}^{n} \binom{n}{k} \cdot \hat{d}^{T(k)} \cdot \overline{\chi}^{(n-k)}\right) + r\right\}. \tag{17}$$

Hence, the states tracking error dynamics $\tilde{x} = x - x_d^*$ may be written as follows

$$\dot{\tilde{x}} = A\tilde{x} + B\left\{u + \psi(\tilde{x}) - K \cdot \left(x_d - \sum_{k=0}^{n} \binom{n}{k} \cdot \hat{d}^{T(k)} \cdot \overline{\chi}^{(n-k)}\right) - r\right\}, \tag{18}$$

where r is the desired reference signal. Equation (18) is written compactly as

$$\dot{\tilde{x}} = A\tilde{x} + B\left\{u + \psi(\tilde{x}) - Kx_d^* - r\right\}. \tag{19}$$

where dynamics of x_d^* are given by (17).

By defining the output tracking error $\epsilon(t) = y - x_d$ an adaptive update law for $\hat{d}^T$ can be written as

$$\dot{\hat{d}} = -\sigma \epsilon(t) \overline{\chi} \tag{20}$$

Once again, by ensuring that the plant states $x(t)$ tracking $x_d^*(t)$ will cause

$$y(t) = DZ\left(x_d^*(t)\right) = DZ\left(\widehat{DI}(x_d(t))\right) = x_d + \epsilon(t) \tag{21}$$

where $\epsilon(t)$ is the output mismatch error caused by the difference between the exact deadzone parameter and the estimated one is expressed as $\tilde{d} = d^* - \hat{d}$. To parameterize $\epsilon(t)$, we utilize Equation (7) to get

$$\epsilon(t) = y(t) - y^*(t) = x_d(t) - d^T \overline{\chi} - x_d(t) + \hat{d}^T \overline{\chi} \tag{22}$$

or simply written as

$$\epsilon(t) = \tilde{d}^T \overline{\chi}. \tag{23}$$

where $\tilde{d}^T$ the deadzone parameters estimation error is

$$\tilde{d}^T = \begin{bmatrix} d_r^* - \hat{d}_r \\ d_l^* - \hat{d}_l \end{bmatrix}, \tag{24}$$

Therefore, the deadzone effect noted by the term $d^T \overline{\chi}$ in (7) can be cancelled by simply ensuring that the system's states vector $x(t)$ track the inverse dynamics of the desired trajectory $x_d(t)$. To achieve proper tracking and global bounded stability of the overall system, we propose the following RAODI controller:

$$u_d(t) = -\alpha B^T P \tilde{x} - \hat{\beta} B^T P \tilde{x} + K x_d^* + r \tag{25}$$

where $\alpha > 0$, $\tilde{x} = x - x_d^*$, and P is the positive definite symmetric solution of the Algebraic Riccati equation (ARE). Moreover, the adaptation law for $\hat{\beta}$ is given by

$$\dot{\hat{\beta}} = \Gamma \left\| B^T P \tilde{x} \right\|, \quad \Gamma > 0. \tag{26}$$

The properties of the controller (25) are stated in the following theorem:

Theorem. For the plant described by (13) with input deadzone (1), and the RAODI control law (25) along with the adaptive update laws (22) and (26) will ensure the closed-loop stability and boundedness of tracking error, hence reducing the effects of deadzone on the control law driving the system dynamics and ensures bounded output tracking.

Proof. Using the following positive definite control Lyapunov function

$$V = \tilde{x}^T P \tilde{x} + \frac{\Gamma^{-1}}{2} \tilde{\beta}^2 + \frac{\sigma^{-1}}{2} \tilde{d}^2 \tag{27}$$

Differentiating along the trajectories of the system and substituting for the closed loop dynamics given by (19) yields

$$\begin{aligned} \dot{V} &= \dot{\tilde{x}}^T P \tilde{x} + \tilde{x}^T P \dot{\tilde{x}} + \Gamma^{-1} \tilde{\beta} \dot{\hat{\beta}} + \sigma^{-1} \tilde{d} \dot{\hat{d}} \\ &= \left(A\tilde{x} + B\{u + \psi(x) - Kx_d^* - r\} \right)^T P \tilde{x} \\ &\quad + \tilde{x}^T P \left(A\tilde{x} + B\{u + \psi(x) - Kx_d^* - r\} \right) + \Gamma^{-1} \tilde{\beta} \dot{\hat{\beta}} + \sigma^{-1} \tilde{d} \dot{\hat{d}} \end{aligned} \tag{28}$$

Applying the robust controller given in (25) into (28) gives

$$\begin{aligned} \dot{V} &= \left(A\tilde{x} + B\{-\alpha B^T P \tilde{x} - \hat{\beta} B^T P \tilde{x} + \psi(x)\} \right)^T P \tilde{x} \\ &\quad + \tilde{x}^T P \left(A\tilde{x} + B\{-\alpha B^T P \tilde{x} - \hat{\beta} B^T P \tilde{x} + \psi(x)\} \right) + \Gamma^{-1} \tilde{\beta} \dot{\hat{\beta}} + \sigma^{-1} \tilde{d} \dot{\hat{d}} \end{aligned} \tag{29}$$

Collecting terms and simplifying

$$\dot{V} = \tilde{x}^T \left(A^T P + PA \right) \tilde{x} + 2\tilde{x}^T PB \left(\{-\alpha B^T P \tilde{x} - \hat{\beta} B^T P \tilde{x} + \psi(x)\} \right) + \Gamma^{-1} \tilde{\beta} \dot{\hat{\beta}} + \sigma^{-1} \tilde{d} \dot{\hat{d}}$$

$$\dot{V} = \tilde{x}^T \left(A^T P + PA - 2\alpha PBB^T P \right) \tilde{x} + 2\tilde{x}^T PB \left(\{-\hat{\beta} B^T P \tilde{x} + \psi(x)\} \right) + \Gamma^{-1} \tilde{\beta} \dot{\hat{\beta}} + \sigma^{-1} \tilde{d} \dot{\hat{d}} \tag{30}$$

$$\dot{V} = \tilde{x}^T \left(A^T P + PA - 2\alpha PBB^T P \right) \tilde{x} - \hat{\beta} B^T P \tilde{x} \tilde{x}^T PB + 2\tilde{x}^T PB (\psi(x)) + \Gamma^{-1} \tilde{\beta} \dot{\hat{\beta}} + \sigma^{-1} \tilde{d} \dot{\hat{d}} \tag{31}$$

The first term can be simplified by solving the Algebraic Reccati Equation given by

$$A^T P + PA - 2\alpha PBB^T P = -Q \tag{32}$$

which gives

$$\dot{V} = -\tilde{x}^T Q \tilde{x} - \hat{\beta} \left\| \tilde{x}^T PB \right\|^2 + 2\tilde{x}^T PB \left(\psi(x) \right) + \Gamma^{-1} \tilde{\beta} \dot{\hat{\beta}} + \sigma^{-1} \tilde{d}^T \dot{\hat{d}} \tag{33}$$

Replacing the adaptation law (23) and replacing $\hat{\beta} = \tilde{\beta} + \beta^*$ in (31) yields

$$\dot{V} = -\tilde{x}^T Q \tilde{x} - \left(\tilde{\beta} + \beta^* \right) \left\| \tilde{x}^T PB \right\|^2 + 2\tilde{x}^T PB \left(\psi(x) \right) + \tilde{\beta} \left\| \tilde{x}^T PB \right\|^2 + \sigma^{-1} \tilde{d}^T \dot{\hat{d}} \tag{34}$$

$$\dot{V} = -\tilde{x}^T Q \tilde{x} - \beta^* \left\| \tilde{x}^T PB \right\|^2 + 2\tilde{x}^T PB \psi(x) + \sigma^{-1} \tilde{d}^T \dot{\hat{d}} \tag{35}$$

Substituting the adaptive update law (7) $\dot{\hat{d}} = -\sigma \epsilon(t) \overline{\chi}$ makes the fourth term in

$$\dot{V} = -\tilde{x}^T Q \tilde{x} - \beta^* \left\| \tilde{x}^T PB \right\|^2 + 2\tilde{x}^T PB \psi(x) - \tilde{d}^T \epsilon(t) \overline{\chi} \tag{36}$$

Utilizing Equation (23) for output tracking error $\epsilon(t) = \tilde{d}^T \overline{\chi}$.

$$\dot{V} = -\tilde{x}^T Q \tilde{x} - \beta^* \left\| \tilde{x}^T PB \right\|^2 + 2\tilde{x}^T PB \psi(x) - \left(\tilde{d}^T \overline{\chi} \right)^2 \tag{37}$$

Renders the last term negative. For the third term, we utilize the general inequality $2ab \le a^2 + b^2$ the third term in (37) can be bounded as

$$\left\| 2\tilde{x}^T PB \psi(x) \right\| \le \varsigma \left\| \tilde{x}^T PB \right\| + \varsigma^{-1} \left\| \tilde{x} \right\| \tag{38}$$

Applying this bound to (37)

$$\dot{V} \le -\left(\lambda_{\min}(Q) - \varsigma^{-1} \gamma \right) \left\| \tilde{x} \right\| - \beta^* \left\| \tilde{x}^T PB \right\|^2 - \left(\tilde{d}^T \overline{\chi} \right)^2 \tag{39}$$

By choosing the degree of freedom ς satisfying the condition $\varsigma < \dfrac{\lambda_{\min}(Q)}{\gamma}$ and choosing β^* to be greater than ς ensures all terms of $\dot{V}$ negative.

4. Illustrative Example & Simulations

To illustrate the efficacy of the proposed compensator a second order sinusoidal desired reference model is selected for tracking. Simulations of the system in (22) under the adaptive control law (23) and (24) have been performed for a sinusoidal reference trajectory given by $x_d(t) = 3\sin(\pi t)$ represented by a second order model. The actual plant is also chosen to be a second order system simulating a rotational gear with deadzone resulting form the spacing between its meshing teeth.

$$\begin{aligned} \ddot{\theta}_m + k_1 \dot{\theta}_m + k_2 \theta_m &= u(t) \\ \theta_l = DZ(\theta_m) &= \theta_m - sat(\theta_m), \end{aligned} \tag{40}$$

where $\begin{bmatrix} \theta_m & \dot{\theta}_m \end{bmatrix}^T$ represent the driving motor angle and velocity respectively; $\begin{bmatrix} k_1 k_2 \end{bmatrix}^T$ represent the viscous friction and the electromotive force constant; and θ_l represents the output load angle. By defining the state vector $\begin{bmatrix} x_1 x_2 \end{bmatrix}^T$ to represent $\begin{bmatrix} \theta_m \dot{\theta}_m \end{bmatrix}$, then the system under investigation can be represented in space state form as

$$\begin{aligned} \dot{x} &= Ax + B \left\{ k^T x + u(t) \right\} \\ y &= DZ(x) = x - sat(x). \end{aligned} \tag{41}$$

where the matrices A and B along with the gain k are given by

$$A = \begin{bmatrix} 0 & 1 \\ 0 & 0 \end{bmatrix}, B = \begin{bmatrix} 0 \\ 1 \end{bmatrix}, k = \begin{pmatrix} k_1 \\ k_2 \end{pmatrix} \tag{42}$$

Meanwhile, the desired reference model to be tracked at the output for the overall system may be rewritten as

$$\dot{x}_d = Ax_d + B\left\{\gamma k^{\mathrm{T}} x_d - 3\pi^2 \sin(\pi t)\right\}$$

$$y = x_d,$$

(43)

where $\gamma > 0$ used to insure the stability of the desired tracked model. In the case of meshing gears, the dead-zone spacing parameter can be easily predetermined and measured. The reference point is chosen to be at the center of the deadzone spacing. Hence, define $d^* = d_r^* = -d_l^* = 1$ with $\hat{d}$ being the adaptation that estimates-its value as given by Equation (20). Therefore, the adaptive deadzone inverse trajectory written as follows

$$x_d^* = \widehat{DI}(x_d) = x_d + \hat{d}\bar{\chi}.$$

(44)

The proposed controller is given by

$$u_d(t) = -\alpha B^{\mathrm{T}} P\tilde{x} - \hat{\beta} B^{\mathrm{T}} P\tilde{x} + Kx_d^* + r$$

(45)

where the first term is the conventional PD-controller, the second term is the robust adaptive controller, and the third term is the adaptive deadzone inverse one.

$$\dot{\hat{\beta}} = \Gamma\left\|B^{\mathrm{T}} P\tilde{x}\right\|, \ \Gamma > 0.$$

(46)

Meanwhile, the initial value of $\hat{d}$ is set to be zero and no prior knowledge of its values is needed. The exact value of the simulated deadzone parameter is set to $d^* = 1$. For all other simulated parameters refer to **Table 1**.

Figure 3 shows the output trajectory $y_o = \theta_l$ for the system under RAODI control is presented and is compared to the trajectory tracking of the system under adaptive without the inverse (in dotted blue), and a PD-controller (dashed red). The system performance is shown with the black solid line while the performance of a regular PD controller is shown in dotted red line. Clearly, the output of the system under RAODI outperforms the system with a conventional PD controller. The deadzone spacing effect is practically eliminated and the tracking error is held to a small negligible amount.

The improvement in reducing the effect of output deadzone on the output signal is demonstrated in **Figure 4** where the error $(y_o - x_d) = \theta_l - \theta_d$ is plotted in solid line as apposed to the same error for the system under a PD controller plotted in dotted red line. In addition, in **Figure 4**, the **dashed blue line** reflects the output tracking error for the system without the use of inverse deadzone modifier. The error without the deadzone inverter is much larger than the improved performance due to RAODI controller. The system state $x_1 = \theta_m(t)$ tracking performance (solid) verses the deadzone inverted trajectory $x_{1d} = \theta_d$ for the system under RAODI control is presented in **Figure 5**, with **Figure 6** demonstrating the state tracking error $\epsilon(t) = \theta_l - \theta_d$ for the system under the proposed control scheme. The second state $x_2 = \omega(t)$ tracking performance and its error $\epsilon_2 = \omega - \omega_d$ are presented in **Figure 7** and **Figure 8**, respectively. In addition, **Figure 9** and **Figure 10** show the evolution of the adaptations $\hat{\beta}$ and $\hat{d}$ confirming their bounded stability. Meanwhile, the adaptive controller effort $u_d(t)$ is shown in **Figure 11**.

Table 1. Parameters utilized in the example.

	Systems Physical Attributes		
	Parameter	**Value**	**Unit**
1	k_p	40	Gain Constant
2	k_v	13	Gain Constant
3	γ	100	Gain Constant
4	d_r^*	1.0	radian
5	d_l^*	−1.0	radian
6	α	1.0	N.m/rad
7	Γ	100	Gains
8	J	1	$\frac{\mathrm{V}}{\mathrm{rad}} \cdot s^{-2}$
9	ω_d	π	rad/s

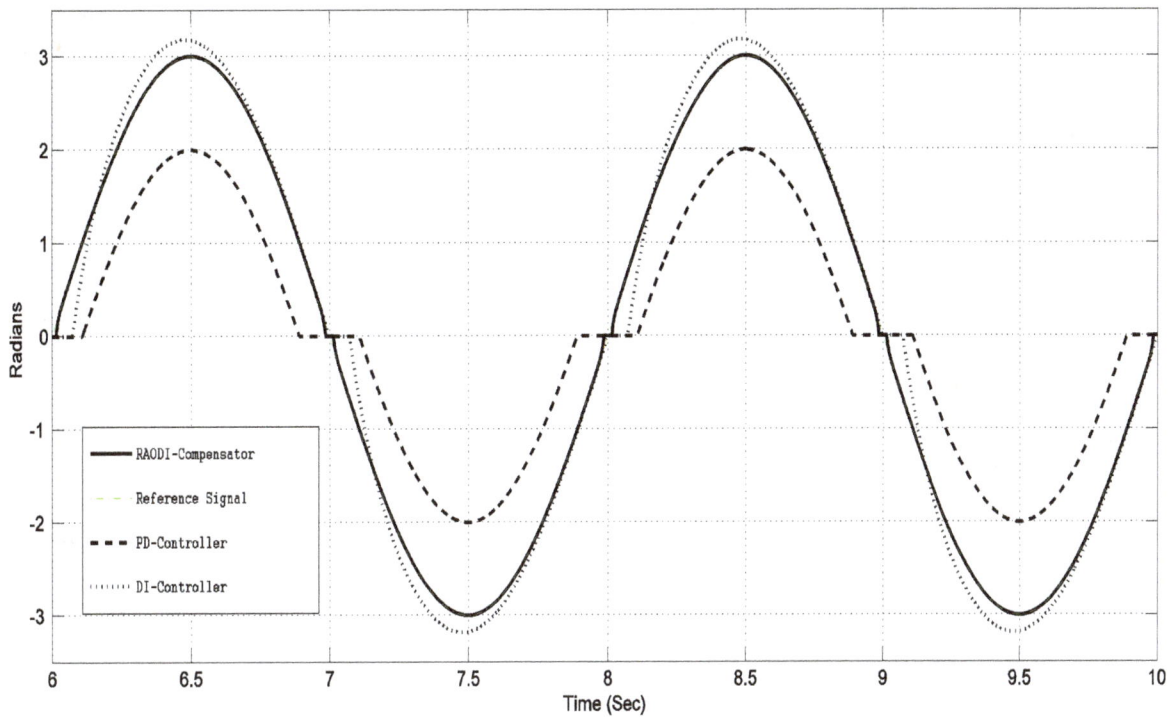

Figure 3. The output trajectory y_o (**black-solid**) for the system under RAODI control vs. the performance of an adaptive controller (**blue-dotted**), and a PD-controller (**red-dashed**).

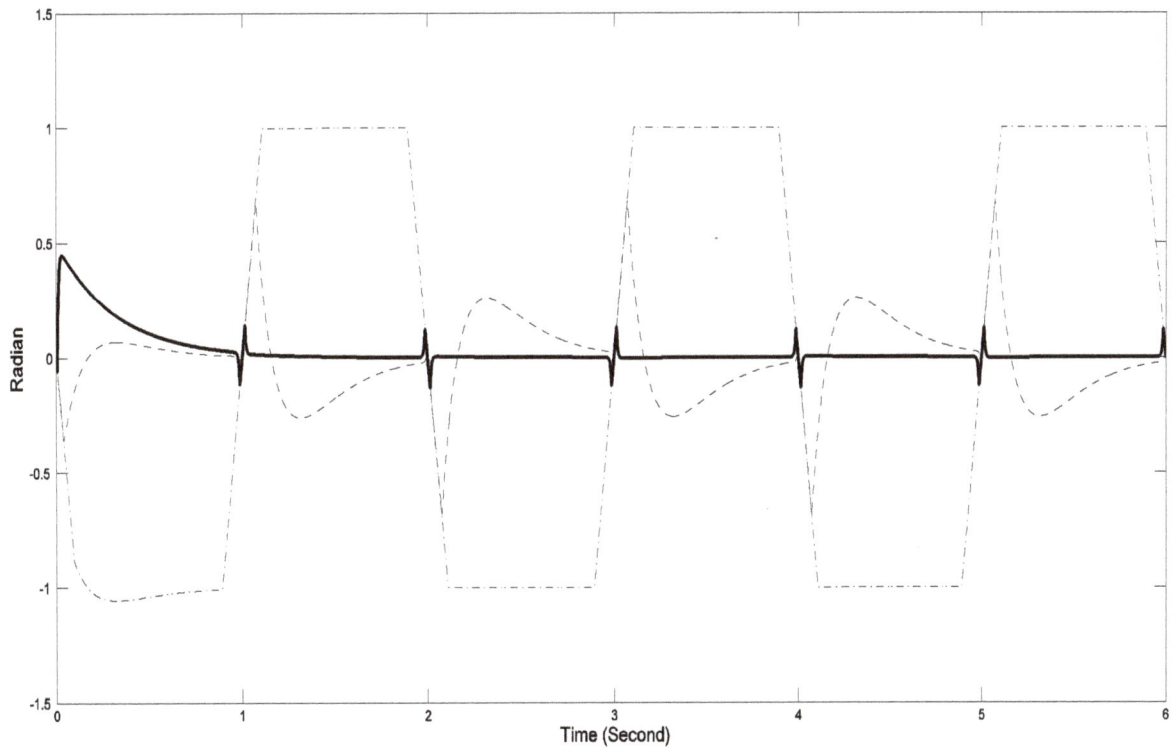

Figure 4. The output tracking error y_o (solid) for the system under RAODI vs. the tracking error of the system under a PD-controller (red-dashed). The **dashed blue line** reflects the output tracking error for the system without the use of inverse deadzone modifier.

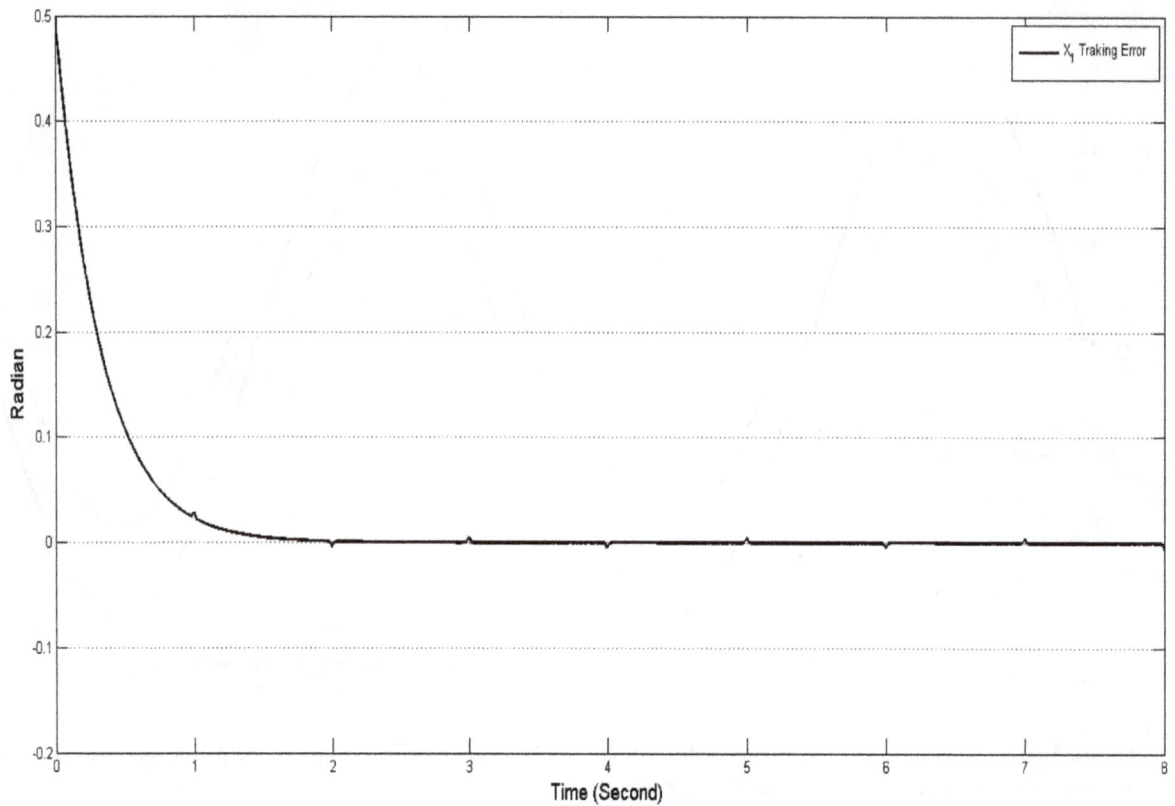

Figure 5. The system state $x_1 = \theta(t)$ tracking performance (solid) verses the deadzone inverted trajectory $x_{1d} = \theta_d$ for the system under RAODI control (red-dashed).

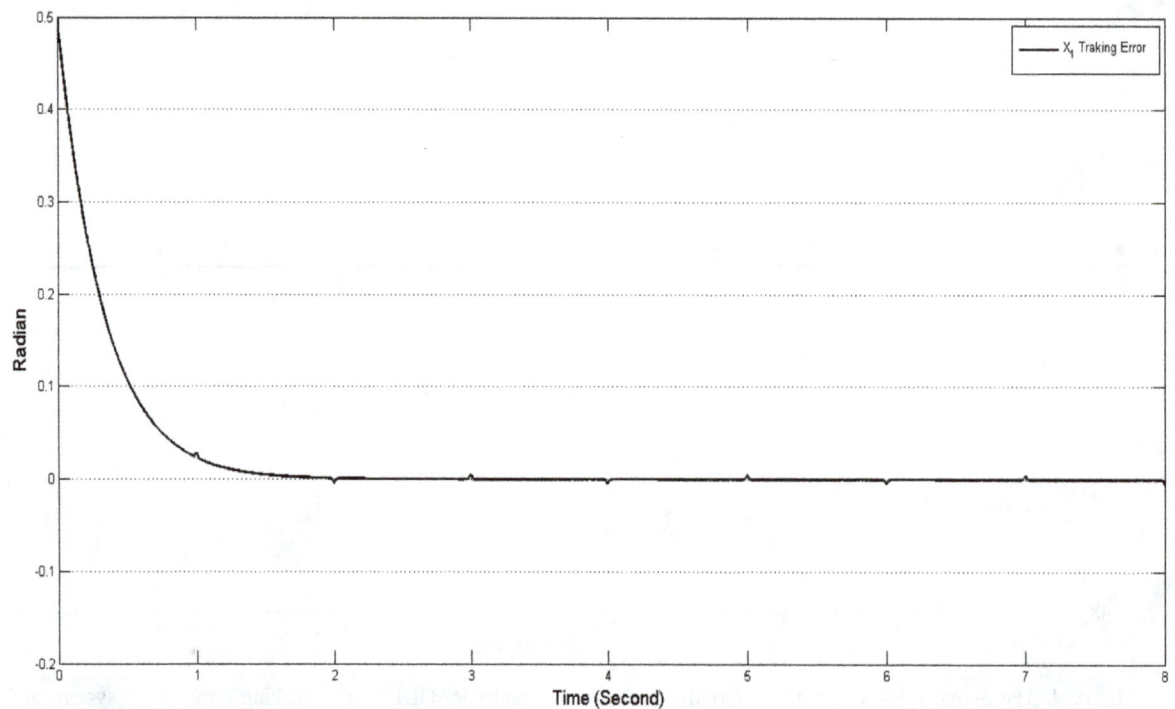

Figure 6. The state tracking error $\epsilon(t) = x_1 - x_{1d}$ for the system under RAODI control.

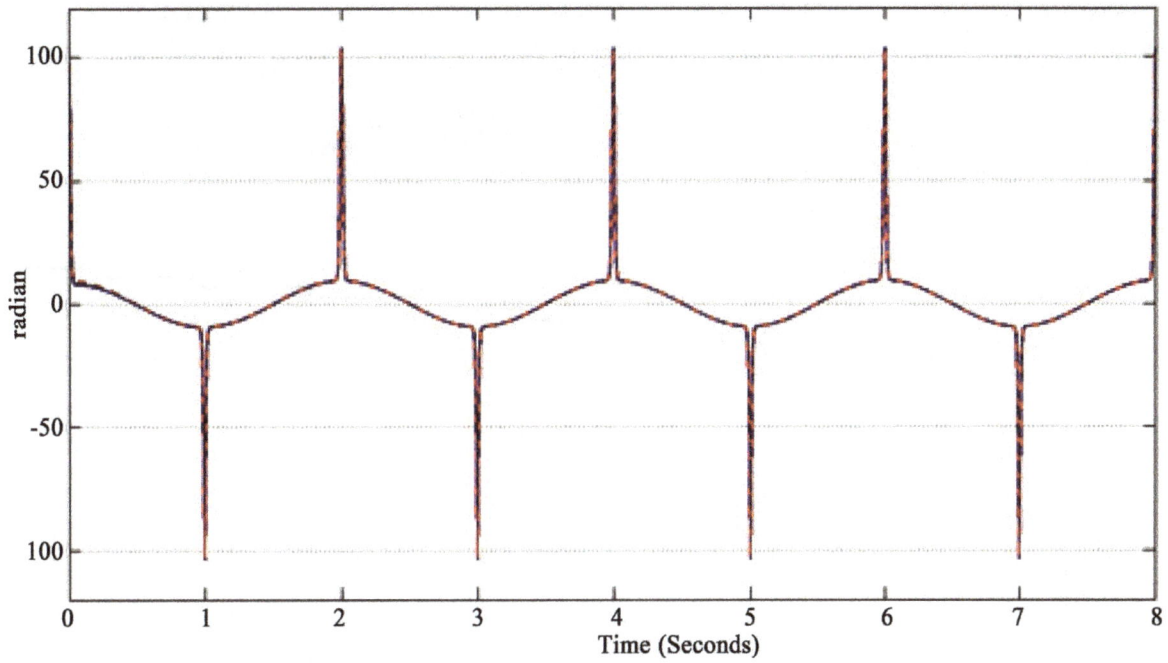

Figure 7. The system state $x_2 = \omega(t)$ tracking performance (solid) verses the inverted deadzone trajectory $x_{2d} = \omega_d$ for the system under RAODI control (red-dashed).

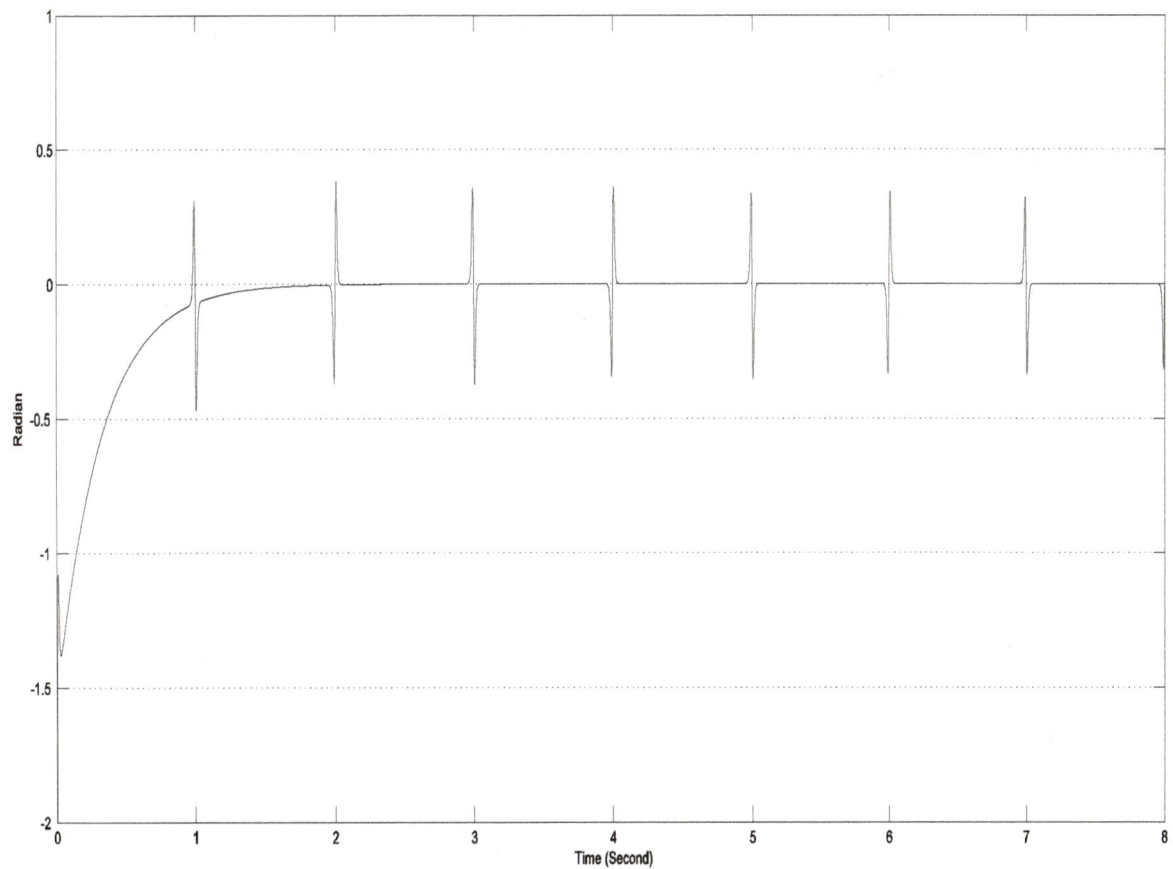

Figure 8. The second state error $\epsilon_2 = x_2 - x_{2d}$ for the system under RAODI control.

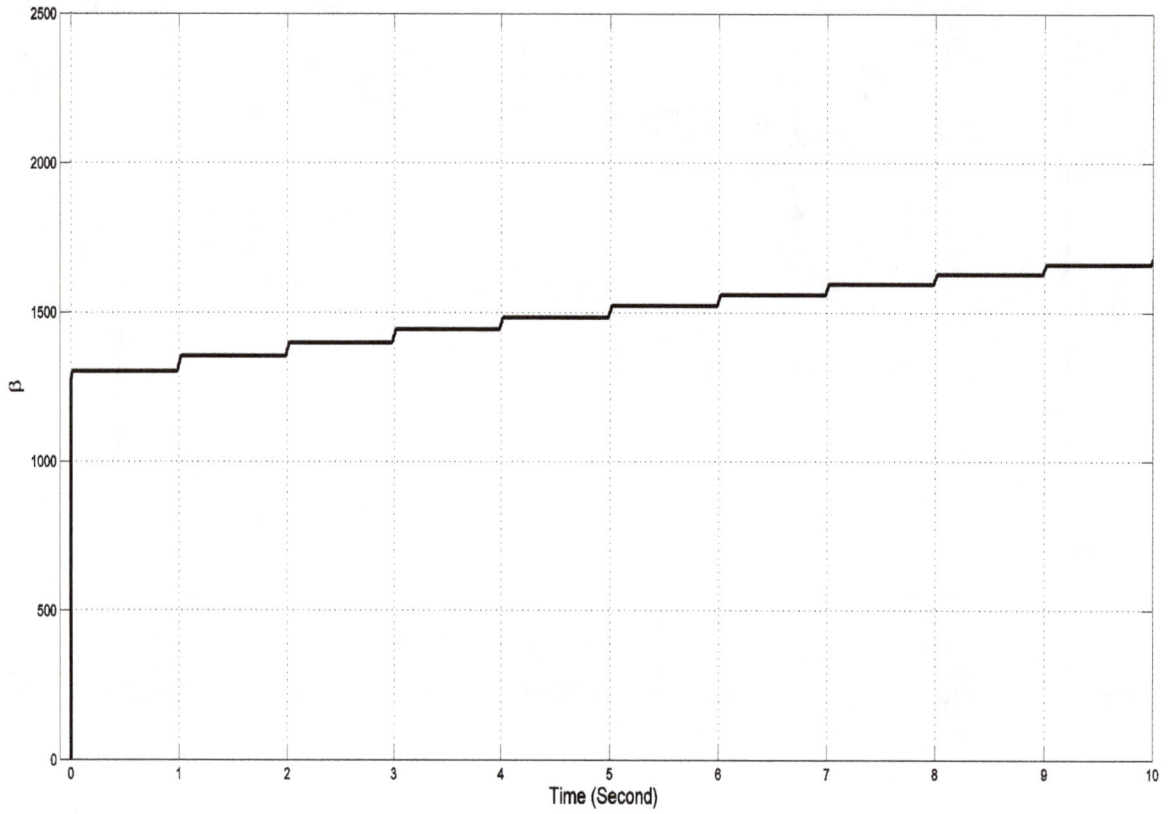

Figure 9. The evolution of the robust adaptation $\hat{\beta}$.

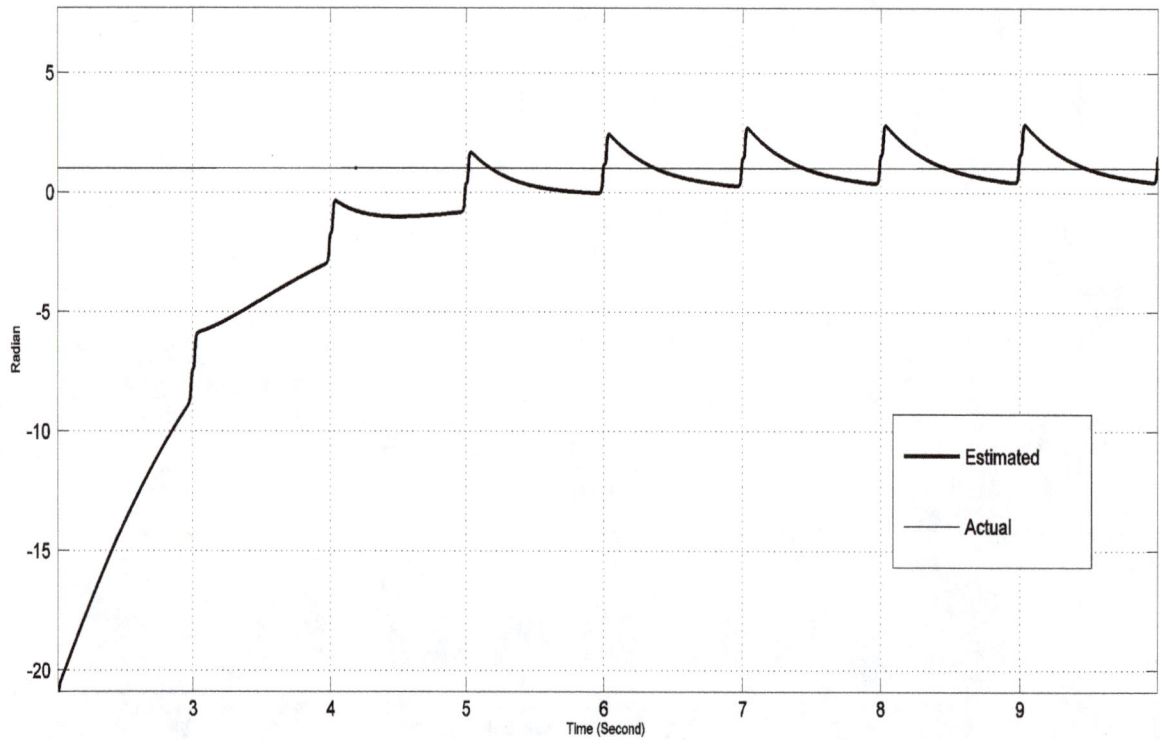

Figure 10. The evolution of the adaptation $\hat{d}$ estimating the actual $d^* = 1.0$ radian.

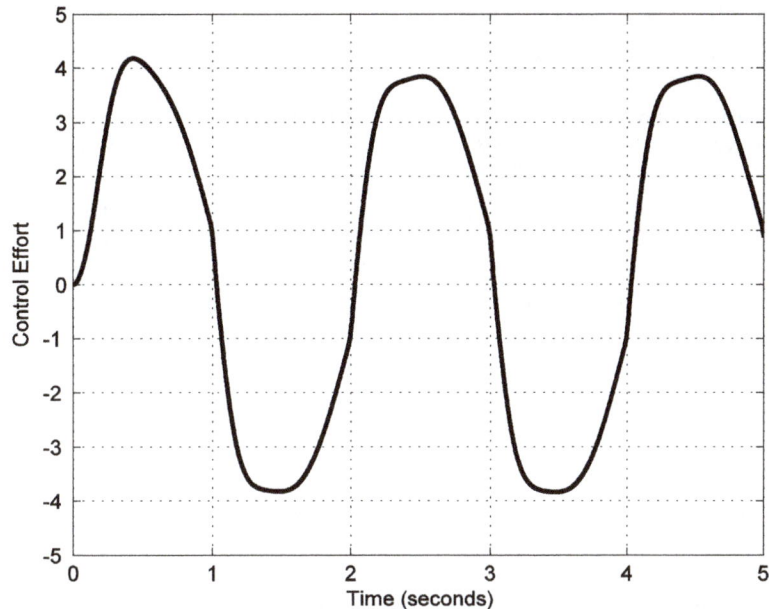

Figure 11. Evolution of the control.

5. Conclusion

In this paper, an adaptive inverse deadzone controller is compared with a robust adaptive controller for systems with output deadzone nonlinearity. Both controllers have been shown to effectively stabilize a second order system, and achieve bounded input bounded output (BIBO) tracking. The proposed deadzone inverse controller has greatly improved the performance of the system over the robust controller. The deadzone inverse controller was implemented in continuous time and was used to modify a desired model reference to mimic an inverse deadzone trajectory. The RAODI is smoothly differentiable and can easily be combined with any of the advanced control methodologies. The stability of the closed-loop system has been proven by using Lyapunov arguments and simulations results confirm the efficacy of the control methodology.

Acknowledgements

This work is supported by the Public Authority for Applied Education and Training (PAAET) Kuwait grant number TS-14-03.

References

[1]   Ahmad, N.J., Alnaser, M.J. and Alsharhan, W.E. (2013) Asymptotic Tracking of Systems with Non-Symmetrical Input Deadzone Nonlinearity. *International Journal of Automation and Power Engineering*, **2**, 287-292.

[2]   Tao, G. and Kokotovic, P. (1995) Discrete-Time Adaptive Control of Systems with Unknown Dead-Zones. *International Journal of Control*, **61**, 1-17. http://dx.doi.org/10.1080/00207179508921889

[3]   Ahmad, N.J., Ebraheem, H.K., Alnaser, M.J. and Alostath, J.M. (2011) Adaptive Control of a DC Motor with Uncertain Deadzone Nonlinearity at the Input. 2011 *Chinese Control and Decision Conference* (*CCDC*), Mianyang, 23-25 May 2011, 4295-4299. http://dx.doi.org/10.1109/CCDC.2011.5968982

[4]   Toa, G. and Kokotovic, P. (1996) Adaptive Control of Systems with Actuator and Sensor Nonlinearities. John Wiley & Sons, Inc., New York.

[5]   Arcak, M. and Kokotovic, P.V. (2000) Robust Output Feedback Design Using a New Class of Nonlinear Observer. *Proceedings of the 39th Conference on Decision and Control*, Sydney, December 2000, 778-783.

[6]   Ibrir, S., Xie, W.F. and Su, C.-Y. (2007) Adaptive Tracking of Nonlinear Systems with Non-Symmetric Dead-Zone Input. *Automatica*, **43**, 522-530. http://dx.doi.org/10.1016/j.automatica.2006.09.022

[7]   Zhang, T.-P., Zhou, C.-Y. and Zhu, Q. (2009) Adaptive Variable Structure Control of MIMO Nonlinear Systems with Time-Varying Delays and Unknown. *International Journal of Automation and Computing*, **6**, 124-136.

http://dx.doi.org/10.1007/s11633-009-0124-5

[8]   Andrighetto, P.L. and Bavaresco, D. (2008) Dead Zone Compensation in Pneumatic Servo Systems. *ABCM Symposium Series in Mechatronics*, **3**, 501-509.

[9]   Recker, D.A. and Kokotovic, P.V. (1993) Indirect Adaptive Nonlinear Control of Discrete-Time Systems Containing a Deadzone. *Proceedings of the 32nd Conference on Decision and Control, San Antonio*, 15-17 December 1993, 2647-2653. http://dx.doi.org/10.1109/CDC.1993.325676

[10] Zhou, J., Er, M.J. and Wen, C.Y. (2005) Adaptive Control of Nonlinear Systems with Uncertain Dead-Zone Nonlinearity. *44th IEEE Conference on Decision and Control, and the European Control Conference*, Seville, 12-15 December 2005, 797-801.

[11] Betancor-Martin, C.S., Montiel-Nelson, J.A. and Vega-Martinez, A. (2014) Deadzone Compensation in Motion Control Systems Using Model Reference Direct Inverse Control. 2014 *IEEE 57th International Midwest Symposium on Circuits and Systems* (*MWSCAS*), College Station, 3-6 August 2014, 165-168. http://dx.doi.org/10.1109/MWSCAS.2014.6908378

[12] Jang, J.O., Chung, H.T. and Jeon, G.J. (2005) Saturation and Deadzone Compensation of Systems Using Neural Network and Fuzzy Logic. *Proceedings of the* 2005 *American Control Conference*, **3**, 1715-1720. http://dx.doi.org/10.1109/ACC.2005.1470215

[13] Chang, C.-Y., Hsu, K.-C., Chiang, K.-H. and Huang, G.-E. (2006) An Enhanced Adaptive Sliding Mode Fuzzy Control for Positioning and Anti-Swing Control of the Overhead Crane System. *IEEE International Conference on Systems, Man and Cybernetics*, **2**, 992-997.

[14] Kong, X.Z. and Zang, F.Y. (2009) Study on the Intelligent Hybrid Control for Secondary Regulation Transmission System. *IEEE International Conference on Automation and Logistics*, Shenyang, 5-7 August 2009, 726-729. http://dx.doi.org/10.1109/ICAL.2009.5262830

[15] Bessa, W.M. and Barrêto, R.S.S. (2010) Adaptive Fuzzy Sliding Mode Control of Uncertain Nonlinear Systems. *Revista Controle & Automação*, **21**, 117-126.

[16] Lewis, F.L., Tim, W.K., Wang, L.-Z. and Li, Z.X. (1999) Deadzone Compensation in Motion Control System Using Adaptive Fuzzy Logic Control. *IEEE Transactions on Control Systems Technology*, **7**, 731-742. http://dx.doi.org/10.1109/87.799674

[17] Jain, S. and Khorrami, F. (1995) Robust Adaptive Control of a Class of Nonlinear Systems: State and Output Feedback. *Proceeding of the* 1995 *American Control Conference*, Seattle, June 1995, 1580-1584. http://dx.doi.org/10.1109/acc.1995.529773

10

Design of the Control System of Adaptive Balancing Device Based on PLC

Sha Zhu, Yuexiang Li, Yu Wang, Yonghui Wang

School of Petroleum Engineering, Harbin Institute of Petroleum, Harbin, China
Email: 563334464@qq.com, lyxalhy1017425@qq.com, wangyude1387@163.com, 37473103@qq.com

Abstract

Due to the well condition and the un-expected imbalance movement of the pumping unit in use, the energy consumes a lot. The existing balancing equipment cannot adjust and monitor the pumping units in real time. Therefore this paper introduces the new adaptive balancing equipment—fan-shaped adaptive balancing intelligent device, projects a design of such control system based on PLC, and determines the principle of the control system, the execution software and the design flow. Site commissioning effect on Daqing Oilfield shows this fan-shaped adaptive balancing intelligent device can effectively adjust and monitor the pumping unit in real time, the balance even adjusts from 0.787 to 0.901, and integrated energy saving rate is 14.2%. It is approved that this control device is professionally designed, with strong compatibility, and high reliability.

Keywords

Pumping Unit, Fan-Shaped Adaptive Balancing Intelligent Device, Balance Adjustment, Energy Saving, PLC, Remote Control

1. Introduction

The existing pumping units are commonly biased compound balanced pumping units, the mechanism of which is adding stationary counterbalance to the tail of pumping unit, thus effectively combining crank-balance and biased compound balanced facility to improve the balance of pumping units. However, problems are that the balance of pumping units cannot be adjusted in real time, and such balance structure also consumes lots of steel, etc. [1]-[3]. To solve these problems, the researchers at home and abroad have been keening on the development of auto-adaptive balancing intelligent units in recent years. And over the years of improvement, researchers have eventually developed new integrated energy saving pumping units and increased integrated balancing technology [4], such as a biased downward barbell type pumping unit, double horse head pumping unit, hydraulic pumping unit, and the balanced pumping device [5], etc. Researchers have also added auto-balancing device to the existing pumping unit to realize intelligent control of auto-balancing facility [6] [7]. Ji Xiao-Ke and his partners [8] have developed a mobile balancing intelligent pumping unit controlled by computer cabinet; Wang Xue-Ling *et al*. [9] put forward a pusher auto-balancing device, both of which above are controlled by RTU system. Deng Si-Ming and his colleagues [10] have designed a balance transfer mobile structure, driven by servo control system. The existing adaptive balancing units are mostly based on overall digitalized control; hence their integrated and complex bodies need more funds.

Aiming at a new auto-adaptive balancing device—the fan-shaped adaptive balancing intelligent device, this paper presents a program of a design of the control system based on PLC. The control system of the device has strong compatibility and high reliability. Site adjustment effect of the device on Da-Qing Oilfield shows that the device has solved the real time control problems of adaptive balancing pumping unit, achieved the purpose of little investment, strong suitability, energy saving and consumption reduction of the pumping unit, and provided a good design scheme for the research of digitalized pumping units.

2. The Structure Model of Adaptive Balancing Intelligent Equipment

Fan-shaped adaptive balancing intelligent device is a balancing component at the rear of the walking beam. It mainly includes: balance boom, sector gear, step-balancing motor, swing arm, swing arm pinion, counterbalance.

The working principle of the device is (as shown in **Figure 1**): At the rear of the beam, a gear sector is fixed through the balance arm, the both sides of the gear sector connect two swing arms through a fixed axle, one end of the swing arm is fixedly connected to the fixed axle, the other end is connected to the pinion by regulation axes, sector gear and swing-arm pinion sit between two swing-arms, swing-arms engage sector gear and swing-arm pinion through the fixed axle and regulation axes separately, and counterbalance is fixed in connection with the pinion and swing arms. When the balance exceeds the set value, the step-balancing motor then drives swing-arm pinion and makes the cunterbalance to move, so that the beam maintains the best balance and finally achieves balance adjustment during the oil pumping. The working principle is shown in **Figure 1**.

3. Design of the Control System of Adaptive Balancing Intelligent Device

3.1. Control Principle

Fan-shaped adaptive balancing intelligent device is operated by a strong stability gearing. The main actuator of the control system for adaptive and real-time balance monitoring include: Programmable Logic Controller (PLC), Power-collection Module, Main Motor, Inverter, Angular-displacement Sensor, Load Sensor, Step-balancing Motor, Counterbalance, GPRS Module, command input and display devices. Of the above components, PLC is the main controller, which determines the trackway of the pumping unit over the angular displacement sensor and load sensor, confirms balance condition by calculating balance through current collected in the current collecting module, and simultaneously drives step-balancing motor to adjust the counterbalance in realizing adaptive balance. The main controller connects to a wireless module, which carries out information exchange with remote control system of oil fields. The inverter connects to the main motor and step-balancing motor controls counterbalance. The control logic flow chart is shown in **Figure 2**.

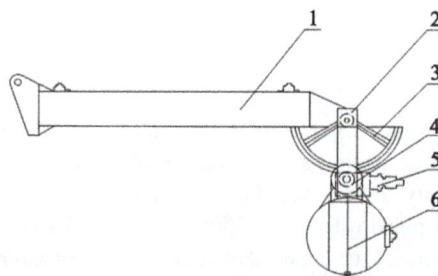

Figure 1. The mechanical system of auto-adaptive balancing intelligent equipment. 1: Balance Boom; 2: Swing Arm; 3: Sector Gear; 4: Swing-arm Pinion; 5: Step-balanceing Motor; 6: Cunterbalance.

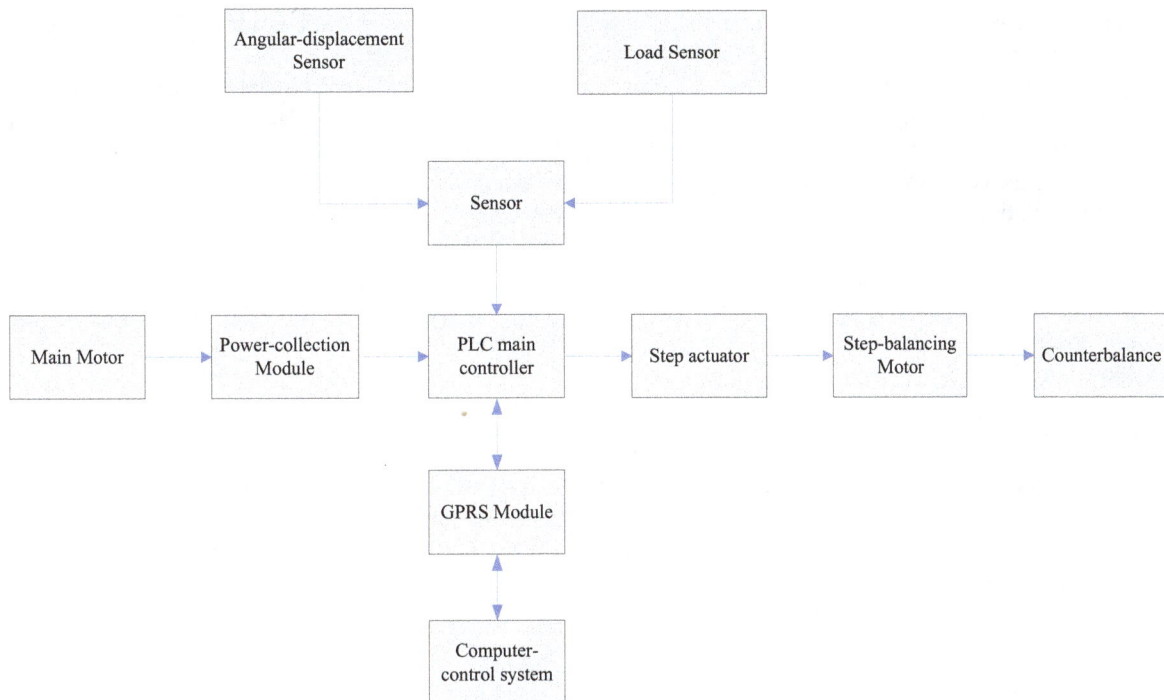

Figure 2. The control logic flow chart of adaptive balancing intelligent device.

3.2. Hardware Design

Main controller selects Programmable Logic Controller (PLC), which is in use in industry for 60 years. Even though the operation ability of Omron CP1E Collection CP1E-NA20DR-A is not as good as embedded development board, this brand is sufficient enough to meet the functional requirements of auto-adaptive balancing intelligent equipment. And after half a century of verification, PLC's reliability has been fully recognized by various industries [11].

Angular-displacement sensor chooses Moon Instruments collection SMA36, which measures the rotational position of an object by a linear relationship between current output and axis circumflex angle. Load Sensor uses Moon Instruments collection SMA35GDDC-B, which is designed for polished rod load test of pumping unit on the oil field. All the selected sensors have good performances with a good calmer output, two-wire type regardless positive and negative pole in case of wrongly wiring, and mounting and dismounting conveniently and safely.

Step-balancing Motor picks Billions Stars collection 86BYG250H (including step actuator), which is suitable for mechanical engineering equipment and can operate steadily in low-speed. It can also monitor the pulse signal in real-time with advanced science and technology.

Main-core action devices use Schneider of France and Siemens of Germany, including circuit breaker, contactor, relay and so on. They can enhance the safety and reliability of the control loop and reduce the chance of motor burning accidents.

Except the main power line, in order to avoid field personnel wrongly wiring and burning of devices, all other necessary wiring port and peripheral signal cable interfaces must be in air plug butt docking form. This makes installation much more convenient and efficient.

3.3. Software Design

"Balance Beam Pumping Unit Evaluation Criteria" of the fan-shaped adaptive balancing device is determined by the current method, which is deemed equal balance when up-stroke peak torque of the crank shaft of gearbox is as same as that of down-stroke [12]-[14]. Since both the motor input current and power and crankshaft torque of gearbox are approximately in direct proportion, usually the equilibrium ratio (ER) of the main motor peak current in up-down stroke is compared instead of using the peak torque of the crank shaft [15]. When ER of the main motor peak current in up-down stroke is equal to 1, the pumping unit is considered balance. Usually oil-fields specify that current ER ranging from 0.8 to 1.1 is considered balance.

VB6.0 is used for compiling the main control system. When the fan-shaped adaptive balancing device is at work, PLC analyzes up-down strokes based on the changing signal of the angular displacement sensor. At the same time, the current acquisition module gathers the maximum up-down stroke current from main motor and calculates the equilibrium ratio (ER) of the main motor peak current in up-down stroke, which equals to the current calculating of the balance of the pumping unit. After comparing with the set value, a command is sent to drive the step-balancing motor to adjust the counterbalance position of the sector gear by using the swing arm, until the actual operating current ratio (the degree of balance) equals to the set value. Fan-shaped adaptive balancing intelligent process is implemented in this way. The wireless module GPRS, which carries out information exchange with remote control system of oil fields, is connected to PLC. The control system flow chart is shown in **Figure 3**.

4. Field Experiments

In February 2015, we carried out an experiment at the well site of the 2 oil brigade Daqing Oilfield, China. The fan-shaped adaptive balancing intelligent device is installed on several 10-type pumping units to conduct balance adjustment. The experimental results are shown in **Table 1**. It is apparent that the experimental effect with the fan-shaped adaptive balancing intelligent device is better than with original pumping units.

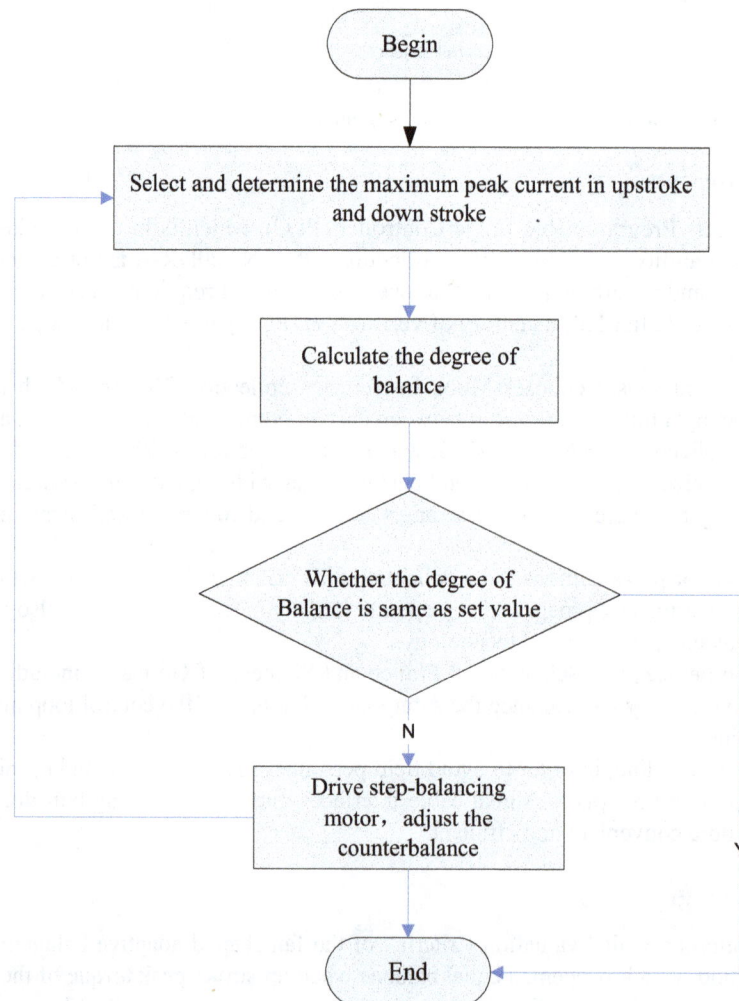

Figure 3. The control system flow chart of the fan-shaped adaptive balancing intelligent device.

Table 1. The experimental results.

Pumping stroke (m)		3	Pumping speed (r/min)	6
Balance mode	Peak current (A)		Degree of balance	Integrated energy-saving rate
	Upstroke	Downstroke		
Crank balance	45	35	0.787	14.2%
Self-intelligent balance	35.79	32.24	0.901	

After the fan-shaped adaptive balancing intelligent device is installed, degree of balance of pumping unit that is in unbalanced state is adjusted to 0.901, which drops in the preset range of 0.8 - 1.1. Thus the balance of pumping units is improved effectively. In addition, integrated energy-saving rate of the pumping units can reach around 14.2%, better than original pumping units.

Computer control system displays the pumping unit balance, peak current and other operating parameters to achieve manual and automatic balance function adjustment. This device reduces labor intensity, and improves work efficiency.

5. Conclusion

This paper proposes solutions based on PLC for the actual control requirement of the fan-shaped adaptive balancing intelligent device for pumping unit, gives the overall design scheme, introduces the implementation details of hardware and software, issues a command to balance motor automatically, achieves automatic adjustment function, maintains the balance of the device in an optimum range, reduces the peak current, and achieves the purpose of energy saving. Site commissioning results show that the system design is reasonable, reliable, and provides references to similar construction machinery control system.

References

[1]   Zhu, X.-M. (2014) A Summary of Energy-Saving Technology of Beam Pumping Unit. Inner Mongolia Petrochemical Industry, No. 17, 117-118.

[2]   Peng, H.-T. (2014) Energy-Saving Analysis and Design Calculation of the Down-deviation Compound Balance Beam Pumping Unit. Northeast Petroleum University, Daqing.

[3]   Li, X.-Y. (2012) Research on Energy Saving Control of Beam Pumping Units. Northeastern University, Shenyang.

[4]   Liang, H.-B., Yi, L.-N. and Meng, Q.-W. (2011) Summary on the Energy-Saving Technical Transformation of Beam Pumping Unit. *Applied Energy Technology*, No. 2, 4-7

[5]   Ma, B.-S. (2015) Application Effect of the New Balance Beam Pumping Unit of Analysis. *Chemical Engineering & Equipment*, No. 4, 69-73.

[6]   Ji, X.-K. (2013) The Research of Pumping Balance Technology on Domestic and Foreign. *Petroleum Geology and Engineering*, **27**, 121-124.

[7]   Liang, Z.-B. (2014) Design of Automatic Balance Beam Pumping Unit. *Mechanical Engineers*, No. 10, 206-207.

[8]   Ji, X.-K. and Xu, L. (2013) Development and Application of a New Type of Digital Pumping Unit. *China Petroleum Machinery*, **41**, 96-99.

[9]   Wang, X.-L. and Xue, Z.-J. (2014) Development and Application of Self-Balancing Intelligent Pumping Unit. *China Petroleum Machinery*, **42**, 82-85.

[10]   Deng, S.-Z. and Zhang, D.-X. (2014) Design and Application on Auto-Balancing Device of Walking-Beam Pumping Unit. *Journal of Oil and Gas Technology*, **36**, 231-233.

[11]   Shen, F.-Q. and Jiang, J. (2015) Control System Design of Crushing Vehicle mounted Machines Based on PLC. Electric Drive, **45**, 74-76.

[12]   Wan, B.-L. (1988) Oil Mechanical Design and Calculations. Petroleum Industry Press, Beijing.

[13]   Wu, Y.-J., Liu, Z.-J. and Zhao, G.x. (1994) Pumping Unit. Petroleum Industry Press, Beijing.

[14]   Qin, W. and Xi, X. (2012) Design of Beam Pumping Unit Balancing Adjustment. *Device Advanced Materials Research*, **524-527**, 1368-1371. http://dx.doi.org/10.4028/www.scientific.net/AMR.524-527.1368

[15]   Liang, H.-B. and Wang, J.-X. (2013) Study on Balance Principle of Beam-pumping Unit. *Oil Field Equipment*, **42**, 16-19.

11

Controlling Speed of DC Motor with Fuzzy Controller in Comparison with ANFIS Controller

Aisha Jilani[1], Sadia Murawwat[1], Syed Omar Jilani[2]

[1]Electrical Engineering Department, Lahore College for Women University, Lahore, Pakistan
[2]Electrical Engineering Department, University of Lahore, Lahore, Pakistan
Email: ai_jilani@yahoo.com, sadia.murawwat@hotmail.com, so_jils@hotmail.com

Abstract

Machines have served the humanity starting from a simple ceiling fan to higher industrial applications such as lathe drives and conveyor belts. This research work aims at providing an appropriate software based control system because it provides computer featured applications, prevents rapid signal loss, reduces noise while also significantly improves the steady state and dynamic response of the motor. In this research paper, we have worked on DC motors due to its significant advantages over other types of machine drives. We have first individually studied Fuzzy and ANFIS (Adaptive Neuro-Fuzzy Interference System) controller in controlling speed for a separately excited DC motor. Afterwards, we have analyzed both results to conclude that which technique is better to be adopted for precisely controlling the speed of DC motor. Outcomes from MATLAB fuzzy logic toolbox for simulation of our schematic has been provided in this research work. Our study parameters include input voltage of DC motor, its speed, percentage overshoot and rising time of the output signal. Our proposed research has interpreted the outcomes that ANFIS controller is better than Fuzzy controller because it produces less percentage overshoot and causes less distortion of the output signal as the overshoot percentage of ANFIS controller is 8.2% while that of Fuzzy controller is 14.4%.

Keywords

ANFIS, DC Motor, Fuzzy Logic, Percentage Overshoot, Rising Time

1. Introduction

Motor drives have been in use for long time and are an efficient way of transferring mechanical energy into desirable output in industries. Although there are two types of motor drives currently being used in every industry but DC motors can be considered much better than AC motors especially when considering transportation equipment because of their maximum torque producing quality at stalls which is very poor in AC motors. Also energy recovery mechanism observed in DC motors is much better than in AC motors [1]. Moreover dc motors provide a low horsepower rating at a much cheaper rate than AC drives [2]. To achieve maximum productivity, every single thing of a machine should be taken into account and analyzed accordingly. In motor control systems, hundreds of problems are faced such as change in load dynamics. The most important affecting factor will be noise parameter which is too much various and unpredictable affecting the functioning of the machine [3]. Similarly, another main factor is speed which should be monitored constantly according to the requirement for a desirable and reliable output.

A DC motor as the name indicates is a motor initiated usually by direct current and is converted into mechanical energy according to the requirement. DC motors are ruling the world due to their extensive use in modern technologies and in almost every industry such as to operate steel rolling mills, electric screw drivers, sewing machines, hard disk drives, air compressors, reciprocating machine etc. [4].

DC motors are generally classified into two types:
- Self-excited DC motor;
- Separately excited DC motor.

The basic difference between the two types is that self-excited DC motor is initiated by its own field circuit while separately excited DC motor is initiated by an external input supply.

Our research work aims at speed controlling of separately excited DC motor rather than self-excited DC motor. We need variable speed drives in our everyday industries such as automotive, petrochemical, food and beverage etc. However, position control of machine drive is also important but once a position is adjusted by some mechanism then its need not to be changed accordingly again and again. However, speed of an object needs to be changed as required such as of motor used in blender. Sometimes it is required to blend the mixture at high speed and sometimes at medium or low speed. Therefore, a technique should be devised for variable speed control rather than variable position control. The major reason of working on separately excited DC motor is that initiation of the motor is independent of internal circuitry of the machine. This gives us an advantage of generating output as desired by varying input supplied voltage with accurate and better speed control as compared to self-excited DC motors. Separately excited DC motors are now extensively ruling the industries due to their marvelous inventions such as OEM battery-powered applications separately excited electric golf car dc motor etc.

When considering hardware of a separately excited DC motor, speed can be controlled by following methods:
- By controlling Armature voltage of the machine [5];
- By adding variable resistance to armature circuit resistance [6].

Research studies have been done on using different controllers to control speed of separately excited DC motor. Several mathematical models have been used to control speed of drive as discussed in [7] [8]. Different types of controllers used are Proportional Integral Controller (PI), Proportional Integral Derivative Controller (PID) etc.

The speed of motor is usually dependent on the type of motor used. Usually when a motor starts, it draws a higher current than an expected value due to a static friction associated with the motor. The higher current will always remain proportional to the input voltage. The input voltage verses speed of a specified machine drive has been analyzed carefully and corresponding outputs have been studied. Results are then analyzed to conclude that which controller is better. Fuzzy and ANFIS controller are monitored so as to give less percentage overshoot and less rising time to make an efficient system and to minimize distortion.

Our research work has aimed on achieving precise and accurate speed control of separately excited DC motor by using Fuzzy Controller and ANFIS Controller. The purpose is to provide a better speed control method by comparative study of two controllers.

2. Background

Machines have successfully replaced uncountable human efforts into efficient and reliable output. Both DC and AC machines are equally important suiting to the required application. Several researches have been done on improving reliability and efficiency in machines. They have not discussed that despite of so many advantages of AC machines for why only DC motors speed should be precisely and accurately controlled and how are they better than AC machines [9]-[11]. Similarly, no comparison of separately excited DC motor and self-excited DC motor is shown [12]-[15]. All parameters of DC motors are correlated such as load dynamics, angular machines, speed of drive etc. Angular position can be affected by changes in load and speed until and unless ideal case is assumed. Publication such as [16] has not focused on how angular position will be affected with variable speed of the drive. DC motors speed can be controlled by various methods of which most commonly used is fuzzy controller based on Mamdani and Sugeno systems. [17] [18] have not discussed the reason of using Mamdani system rather than preferring Sugeno system.

A feasible, proficient, workable and ultra-efficient system should always be designed for negligible percentage overshoot and minimum rising time. However, lowering of percentage overshoot and minimizing rise time often contradict and is not possible at one time. Publications like [19]-[21] have not shown any conclusion for controlling speed of DC drives by considering percentage overshoot and rising time.

To incorporate any method in a practical and systematic way for efficacious, operative and dynamic performance, results should be analyzed on the basis of comparison to conclude which method is better to be implemented. [22] does not contain any comparative technique.

3. Methodology

In this research work, MATLAB Simulink has been used to implement working of controllers. Speed is monitored by using following two types of controllers and then their comparison on various factors is taken into account.

- Fuzzy Controller;
- ANFIS Controller.

3.1. Implementation of Fuzzy Controller

A Fuzzy Logic Controller (FLC) is formed by interpreting the analog or continuous values of 0 and 1 despite analyzing the digital values. FLC basically controls a process by assimilation of expert human knowledge into a pattern containing a relationship between inputs and outputs. Fuzzy control rules (mostly conditional rules) are then applied on the pattern of input and output.

MATLAB Fuzzy Controller Implementation Process
First of all type "fuzzy" in command window or have to click on "start" menu and open the "fuzzy editor" from "toolboxes". Now click "Edit" and add an input from "Add variable". Now you can select the "sugeno" FIS from "File". There are two types of Fuzzy logic systems which can be used in control systems:

- Mamdani;
- Sugeno.

Here, we have used Sugeno system in our research work because output membership functions of sugeno shows us either constant or linear result, which is not possible when using mamdani systems.

As shown in **Figure 1**, Fuzzy controller has four main processes to operate:

1. Fuzzification;
2. Fuzzy base rules;
3. Interference engine;
4. Defuzzification.

Fuzzification converts our measured data (e.g. speed of car is 15 mph) into rhetorical data (car is moving too slow). Fuzzy base rules define some rules on the information provided by fuzzification. The interference engine provides us appropriate coherence and analysis for an output simulation. Defuzzification gives us an output on the basis of defined set of membership function and rules. The complete process is shown in **Figure 1**.

Following graphical tools are used to create, analyze and view output of Fuzzy logic Controller:

- **Fuzzy Inference System (FIS) Editor**: It has a command on handling basic issues of the control system such as defining of input and output variables. Fuzzy logic toolbox can hold unlimited amount of inputs but corresponding there will be a huge number of membership functions which will become difficult for us to handle.
- **Membership Function Editor**: It defines the appearance and shape of membership function as per input. This is shown in **Figure 2** and **Figure 3**.
- **Rule Editor**: It builds and edits the set of rules which is associated with the behavior of the system.

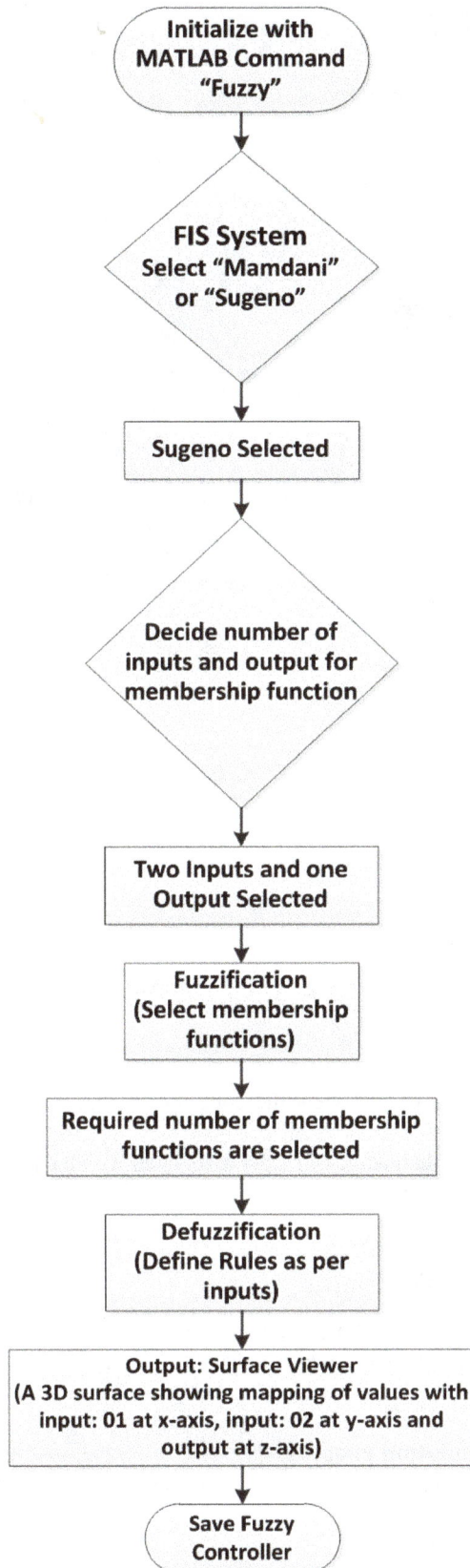

Figure 1. Flow chart of fuzzy controller simulation process.

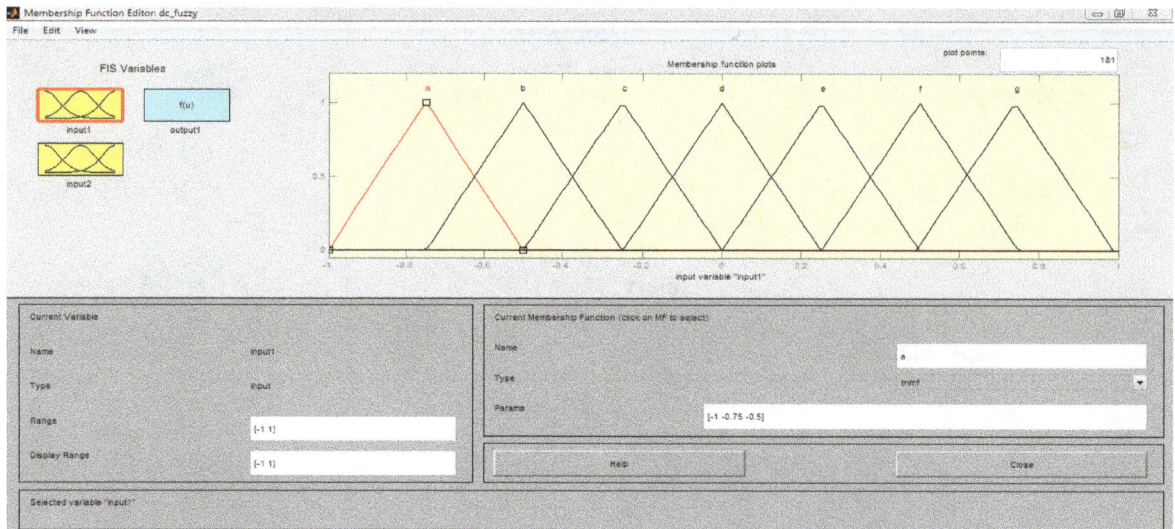

Figure 2. Overlapping of Input: 1 of DC motor.

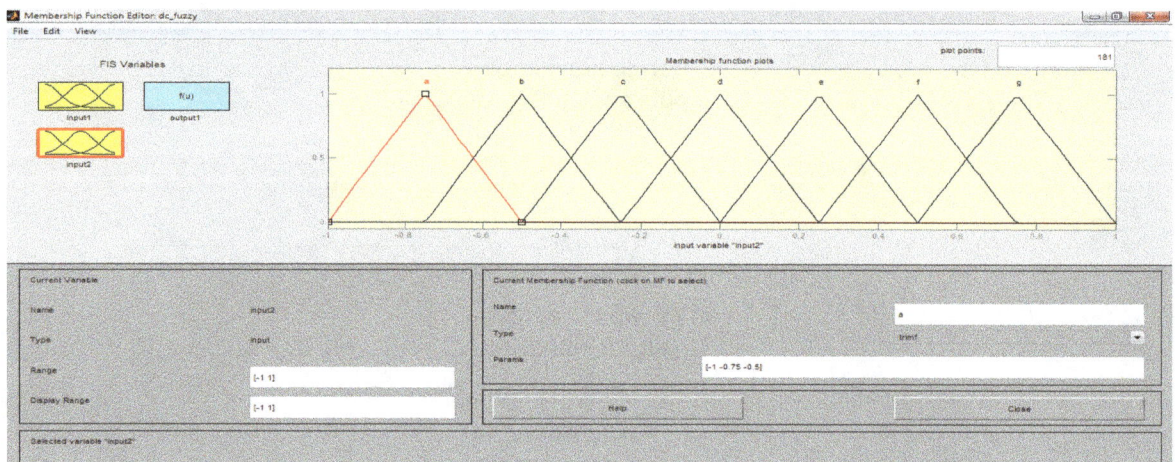

Figure 3. Overlapping of Input: 2 of DC motor.

- **Rule Viewer**: It is used to examine and view the controller output on the basis of defined rules. This is shown in **Figure 4**.
- **Surface Viewer**: It generates a 3-D linkage of output associated with the particular number of inputs.

   Afterwards, a rule viewer for DC motor can be seen in Fuzzy editor, which defines set of rules for Input: 1 ANDed with Input: 2.

3.2. Implementation of ANFIS Controller

ANFIS is a hybrid network which consists of a combination of two controllers; Fuzzy logic and neural network. These both controllers result in a single entity which enhances the features of controlling machine than using a single controller alone.

MATLAB ANFIS Controller Simulation Process

ANFIS editor window opens by typing "anfisedit" in MATLAB command window. The complete process is shown in **Figure 5**.

   ANFIS GUI involves the following steps:
- **Load Data**: This will load our previously saved data from .dat extension file. After loading the data, ANFIS editor will be displayed as shown in **Figure 6**.

Figure 4. Rule viewer for DC motor.

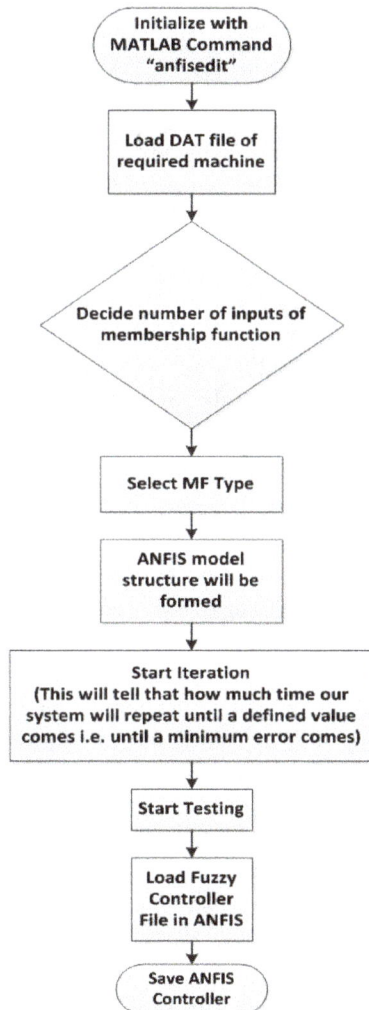

Figure 5. Flow chart of ANFIS controller simulation process.

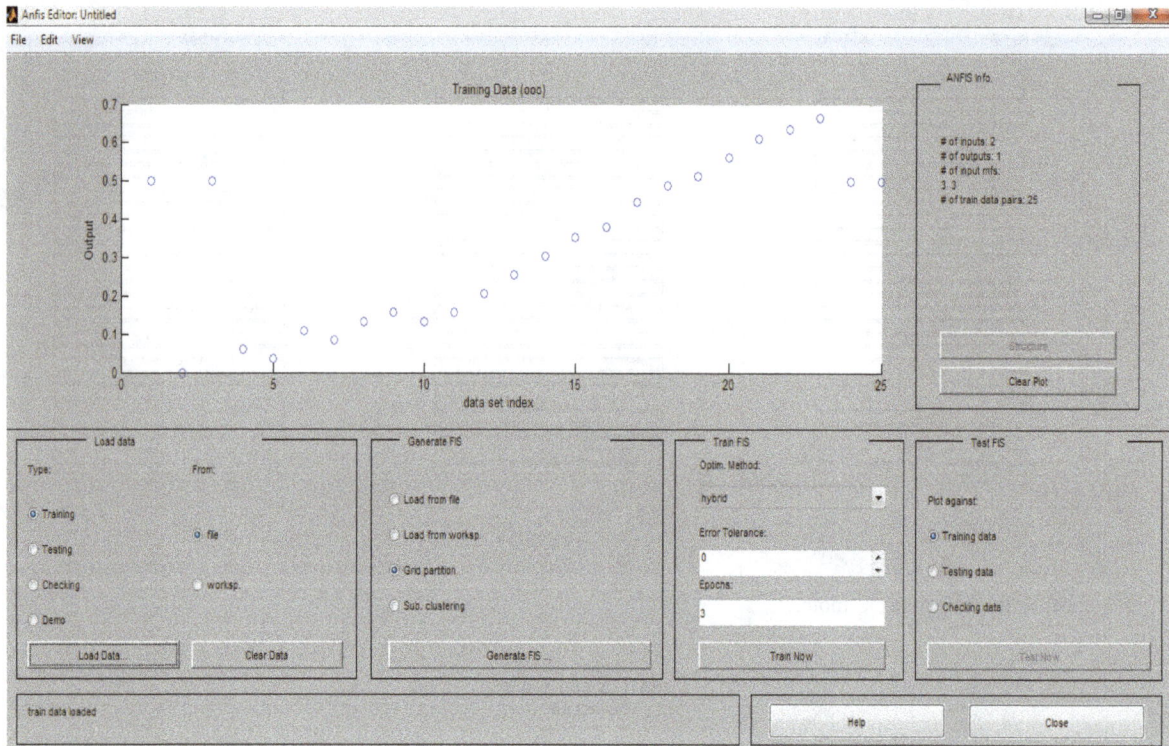

Figure 6. ANFIS Editor for DC motor.

- **Generate FIS**: FIS model can be generated by using any one of the following techniques.
1) **Grid partition**: It generates data via grid portioning.
2) **Sub clustering**: It generates data by analyzing the number of clusters in the given set of data. ANFIS structure can be observed clicking option of "Structure" as shown in **Figure 7**.
- **Training and validation of FIS:** This process trains the FIS model generated, repeats itself until and unless required number of epoch is reached and goal of training error is attained. Put epochs = 25 as given in load data, then "Train now". "Train now" shows the value of "Epochs error". This is shown in **Figure 8**.

Testing of the FIS is carried out by clicking "Test now". "Test now" shows the value of "Average testing error". This is shown in **Figure 9**.

3.3. MATLAB Simulink

Separate responses of DC motor for Fuzzy and ANFIS are then analyzed on MATLAB Simulink as shown in **Figure 10** and **Figure 11** respectively. Then their comparative interpretations are done.

In the circuit shown in **Figure 11** the ANFIS file is loaded in ANFIS block which was saved earlier. In ANFIS circuit, system is initialized by taking step input. The circuit has been made steady and stabilized through Gain6 and Gain7. Also specific oscillations during simulation have been adjusted successfully through these gains. The specified blocks of Tapped delay and Gain are employed as the differential circuit. Gain1 and Gain3 will amplify the signal. In Transfer Function block we can insert the transfer function of DC machine. Each Gain is calculated through specified transfer functions which are selected on the basis of technique used. That's why the value of certain gains is different in both the circuits.

The comparison of both of these circuits is shown in **Figure 12**.

Tabular analysis of rising time and percentage overshoot is calculated from graphical response of Fuzzy and ANFIS controller. This is shown in **Table 1**.

Fuzzy Controller output response has given us the percentage overshoot of 14.4% which is 6.2% greater than that of ANFIS controller. However, it gives a rising time of 0.072 sec less than that of ANFIS controller which can be counted as a drawback of ANFIS controller.

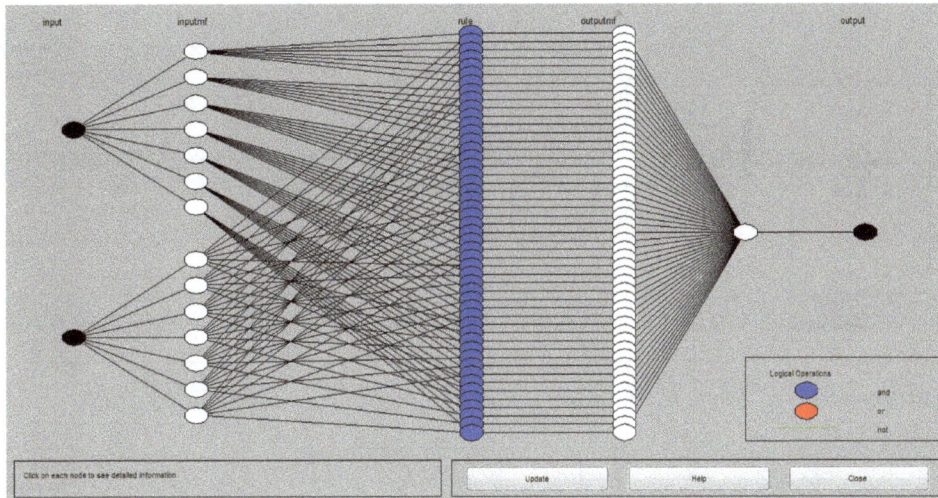

Figure 7. Structure of DC motor.

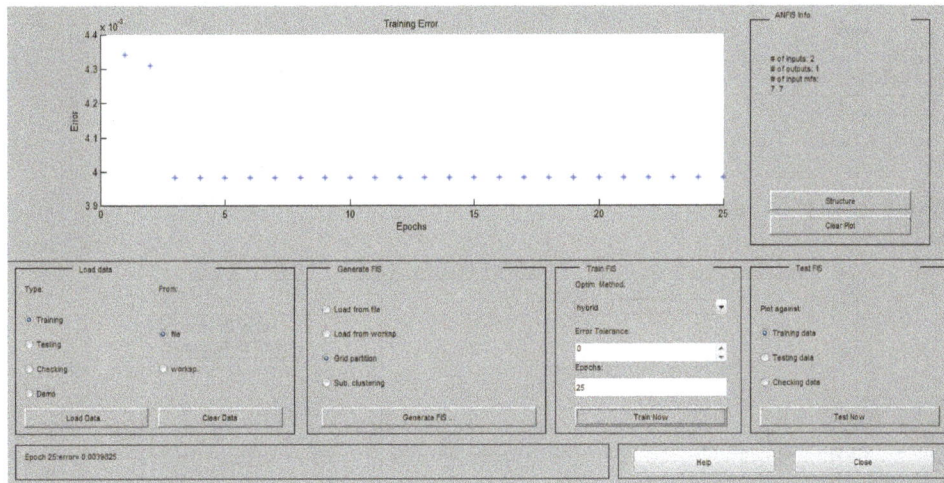

Figure 8. Train output for DC motor.

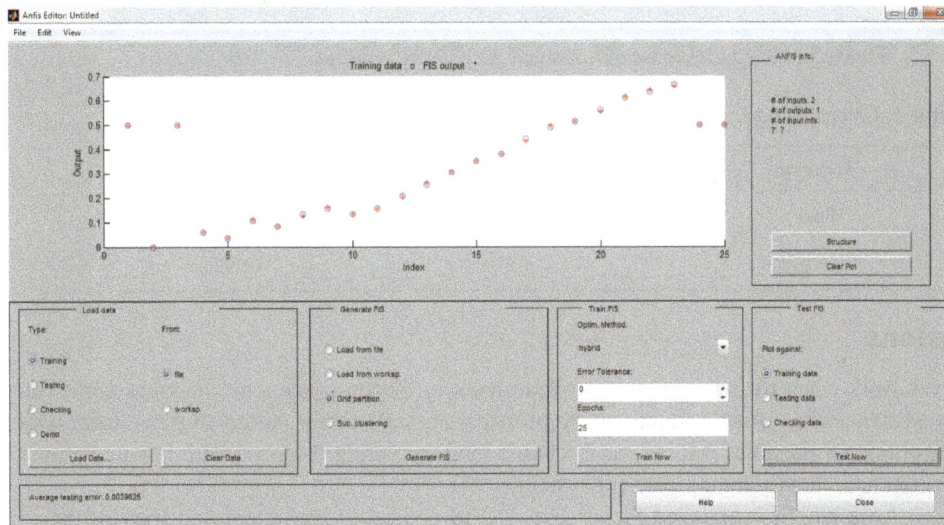

Figure 9. Test output for DC motor.

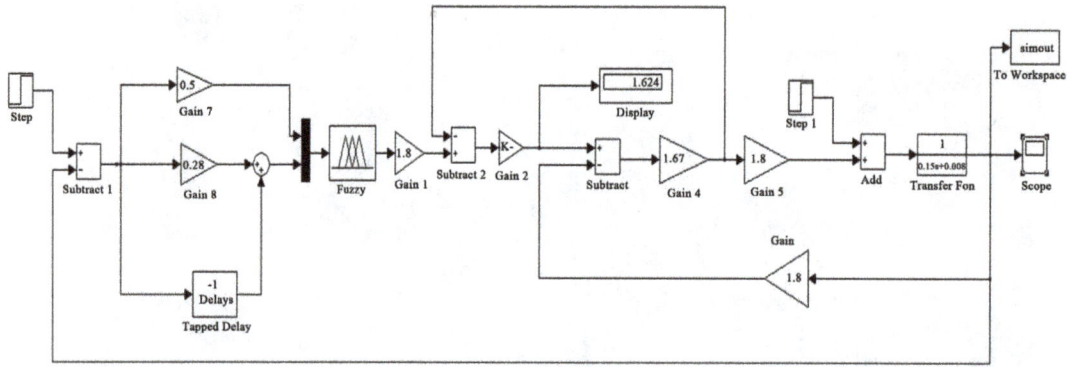

Figure 10. Fuzzy model of DC motor.

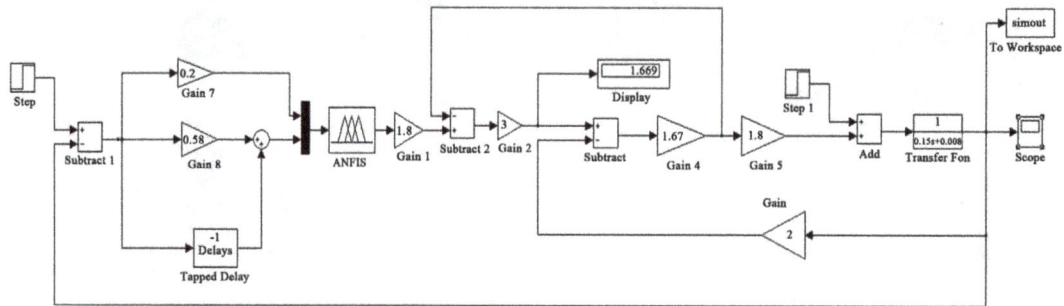

Figure 11. ANFIS model of DC motor.

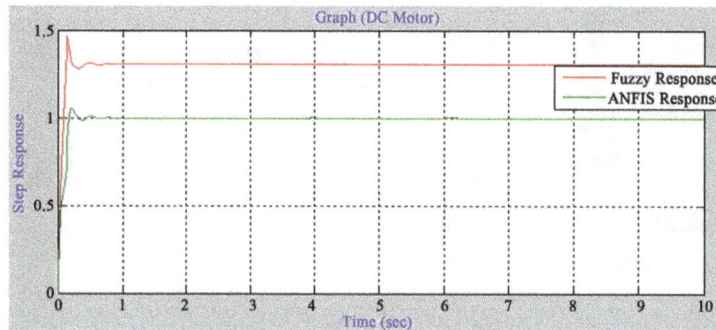

Figure 12. Output response of fuzzy and ANFIS performance.

Table 1. Comparison between Fuzzy and ANFIS performance.

Technique	Percentage overshoot	Rising time
Fuzzy	14.4%	0.720 sec
ANFIS	8.2%	0.792 sec

4. Conclusions

Machine drives work when their speed is controlled precisely. For analysis a lot of issues come across such as changes in load dynamics, variable inputs, noise propagation and certain unknown parameters which result in unpredictable output of machines. Moreover, a reliable load regulating response with low and almost negligible noise propagation is necessary for a proficient system.

Our research work has first analyzed Fuzzy and ANFIS controller separately and then we have interpreted both outcomes to show a comparison that which technique should be used in controlling speed of DC motor. A system should always be designed for less percentage overshoot and less rising time. Often there is a contradiction when adjusting percentage overshoot for minimum rising time. Different research work has been done for lessening percentage overshoot and minimizing rising time. [23] has controlled the speed of DC motor by giving a less percentage overshoot of 10.1% and 2.1% at two different reference speeds. The comparative study in this research paper has shown that ANFIS controller is much better than Fuzzy controller as it gives a percentage overshoot of 8.2% than that of fuzzy controller which is 14.4%. Percentage overshoot indicates an outcome when a signal surpasses its steady-state value. ANFIS technique gives a lower percentage overshoot because of phases such as epoch and training involved in its simulation. Training phase repeats itself until and unless minimum error is reached. This minimum error limit reached is synchronized with given value of epoch which gives a low percentage overshoot then Fuzzy technique. However, a minimum adjustment of 0.072 sec rising time has to be made with ANFIS controller but due to less percentage overshoot, it should be considered a more adoptable technique.

References

[1]   Bansal, U.K. and Narvey, R. (2003) Speed Control of DC Motor Using Fuzzy PID Controller. *Advance in Electronic and Electric Engineering*, **3**, 1209-1220.

[2]   Afrasiabi, N. and Yazdi, M.H. (2013) DC Motor Control Using Chopper. *Global Journal of Science, Engineering and Technology*, **8**, 67-73.

[3]   Atri, A. and Ilyas, Md. (2012) Speed Control of DC Motor Using Neural Network Configuration. *International Journal of Advanced Research in Computer Science and Software Engineering*, **2**, 209-212.

[4]   Saleem, F.A. (2013) Dynamic Modeling, Simulation and Control of Electric Machines for Mechatronics Applications. *International Journal of Control, Automation and Systems*, **1**, 30-42.

[5]   Kushwah, M. and Patra, A. (2014) Tuning PID Controller for Speed Control of DC Motor Using Soft Computing Techniques—A Review. *Advance in Electronic and Electric Engineering*, **4**, 141-148.

[6]   Waghmare, R.J., Patil, S.B., Shid, U.S. and Siddha, U.Y. (2013) Need of Electronic Starter for DC Motor. *International Journal of Science, Engineering and Technology Research*, **2**, 1203-1206.

[7]   Addasi, E.S. (2013) Modelling and Simulation of DC-Motor Electric Drive Control System with Variable Moment of Inertia. *ACEEE International Journal on Electrical and Power Engineering*, **4**, 52-57.

[8]   Gupta, R. and Ruchika (2013) Thyristor Based DC Motor Control with Improved PF & THD. *International Journal on Electrical Engineering and Informatics*, **5**, 519-536.

[9]   Suradkar, R.P. and Thosar, A.G. (2012) Enhancing the Performance of DC Motor Speed Control Using Fuzzy Logic. *IJERT*, **1**, 103-110.

[10] Munde, V. and Jape, V.S. (2013) Fuzzy Logic for Controlling Speed of DC Motor. *IJES*, **2**, 33-39.

[11] Salim, Ohri, J. and Naveen (2013) Speed Control of DC Motor using Fuzzy Logic based on LabView. *International Journal of Scientific & Research Publications*, **3**, 1-5.

[12] Dewangan, A.K., Shukla, S. and Yadu, V. (2012) Speed Control of a Separately Excited DC Motor Using Fuzzy Logic Control Based on Matlab Simulation Program. *International Journal of Scientific & Technology Research*, **1**, 52-54.

[13] Omar, B.A., Haikal, A.Y. and Areed, F.F. (2012) An Adaptive Neuro-Fuzzy Speed Controller for a Separately Excited DC Motor. *International Journal of Computer Applications*, **39**, 29-37.

[14] Saini, M. and Sharma, N. (2012) Speed Control of Separately Excited DC Motor Using Computational Method. *International Journal of Engineering Research & Technology (IJERT)*, **1**, 86-93.

[15] Kushwah, R. and Wadhwani, S. (2013) Speed Control of Separately Excited DC Motor Using Fuzzy Logic Controller. *International Journal of Engineering Trends and Technology (IJETT)*, **4**, 2518-2523.

[16] Sailan, K. and Kuhnert, K-D. (2013) DC Motor Angular Position Control Using PID Controller for the Purpose of Controlling the Hydraulic Pump. *International Conference on Control, Engineering & Information Technology (CEIT'13), Proceedings Engineering & Technology*, **1**, 22-26.

[17] Thorat, A.A., Yadav, S. and Patil, S.S. (2013) Implementation of Fuzzy Logic System for DC Motor Speed Control Using Microcontroller. *International Journal of Engineering Research and Applications (IJERA)*, **3**, 950-956.

[18] Ahmed, H., Singh, G., Bhardwaj, V., Saurav, S. and Agarwal, S. (2013) Controlling of DC Motor Using Fuzzy Logic Controller. *Conference on Advances in Communication and Control Systems* 2013 (CAC2S 2013), DIT University, 6-8 April 2013, 666-670.

[19] Dechrit, M., Benchalak, M. and Petrus, S. (2011) Wheelchair Stabilizing by Controlling the Speed Control of Its DC

Motor. *World Academy of Science, Engineering and Technology*, 310-314.

[20] Gupta, R., Lamba, R. and Padhee, S. (2012) Thyristor BasedSpeed Control Techniques of DC Motor: A Comparative Analysis. *International Journal of Scientific and Research Publications*, **2**, 14-18.

[21] Meha, S.A., Haziri, B., Gashi, L.N. and Fejzullahu, B. (2011) Controlling DC Motor Speed Using PWM from C# Windows Application. *15th International Research/Expert Conference Trends in the Development of Machinery and Associated Technology TMT* 2011, Prague, 12-18 September 2011.

[22] Singhal, R., Padhee, S. and Kaur, G. (2012) Design of Fractional Order PID Controller for Speed Control of DC Motor. *International Journal of Scientific and Research Publications*, **2**, 1-8.

[23] Sadiq, A., Mamman, H.B. and Ahmed, M. (2013) Field Current Speed Control of Direct Current Motor Using Fuzzy Logic Technique. *International Journal of Information and Computation Technology*, **3**, 751-756.

12

An Extended Sliding Mode Observer for Speed, Position and Torque Sensorless Control for PMSM Drive Based Stator Resistance Estimator

Pierre Tety[1], Adama Konaté[2], Olivier Asseu[2*], Etienne Soro[2], Pamela Yoboué[2]

[1]Institut National Polytechnique Houphouët Boigny (INPHB), Yamoussoukro, Côte d'Ivoire
[2]Ecole Supérieure Africaine des Technologies de l'Information et de la Communication (ESATIC), Abidjan, Côte d'Ivoire
Email: *oasseu@yahoo.fr

Abstract

This paper presents a robust sixth-order Discrete-time Extended Sliding Mode Observer (DESMO) for sensorless control of PMSM in order to estimate the currents, speed, rotor position, load torque and stator resistance. The satisfying simulation results on Simulink/Matlab environment for a 1.6 kW PMSM demonstrate the good performance and stability of the proposed ESMO algorithm against parameter variation, modeling uncertainty, measurement and system noises.

Keywords

PMSM, Extended Sliding Mode Observer, Feedback Control

1. Introduction

Drive applications with PMSM are receiving more and more interest because of their better performance in dynamic and steady state responses, from their greater power density, larger torque/ampere, best efficiency, lower cost and easier maintenance [1] [2]. To achieve high-performance field oriented control, accurate rotor position information, which is usually measured by rotary encoders or resolvers, is indispensable. However, the use of these sensors increases the cost, size, weight, and wiring complexity and reduces the mechanical robustness and the reliability of the overall PMSM drive systems.

---

*Corresponding author.

The goal of the research for this dissertation was to develop a rotor position/speed/load torque sensorless control system with performance comparable to the sensor-based control systems for PMSMs over their entire operating range.

The naturally structure of non-linear multivariable state of PMSM models induces the use of robust feedback linearization method [3] [4] in order to permit a decoupling and good dynamic stability of the PMSM variables in a field-oriented (d, q) coordinate so that stator currents can be separately and independently controlled.

However, this feedback control technique requires the knowledge of the instantaneous speed which is often difficult to access or not usually measurable in practice. Also parameters variations (more specifically the stator resistance and load torque variation) and noises injected by the inverter in the PMSM can induce a lack of field orientation and a state-space "coupling", which can involve a performance degradation of the system.

In order to achieve better system dynamic performance, the approach proposed in this paper consists to design extended observers allowing an on-line estimation of speed, position, load torque and stator resistance.

The Extended Kalman Filter (EKF) presented in [5] [6] can be used for the problem of states estimations for PMSM sensorless control. Unfortunately the initialization and the optimal choice of covariance and gain matrix are delicate. These matrixes play a critical role in robustness of the EFK.

Another approach proposed in [7]-[10] to estimate the state variables in a PMSM is the use of Sliding Mode Observer (SMO). This nonlinear estimator, based on the variable structure system theory, has been chosen to be of type "Sliding Mode" for having many advantages like: robustness to disturbances, low sensitivity to the system parameters vibrations, some gains easily adjusted compared with the EKF.

Thus, this paper proposes a sixth-order Discrete-time Extended Sliding Mode Observer (DESMO) to provide not only Speed/rotor position estimation but also the load torque and stator resistance reconstruction for the PMSM

After a brief review of the PMSM model, the simulation results for a 1.6 kW PMSM drive system are presented to validate the high robustness of the proposed DESMO approach against parameter variations, measurement and system noises.

2. Model of PMSM

By assuming that the saturation of the magnetic parts and the hysteresis phenomenon are neglected; by considering the case of a smooth-air-gap PMSM (where the inductances are equal: $L_d = L_q$) and according to the field oriented principle where the direct axis current (I_d) is always forced to be zero which simplifies the dynamics and achieve maximum electromagnetic torque per ampere, the PMSM model in the rotor reference (d, q) frame are as follows [2] [5]:

$$
\begin{cases}
\dot{X} = F(X) + G \cdot U \\
Y = H(X) = \left[h_1(X) \; h_2(X) \right]^{\mathrm{T}} = \left[I_d \; I_q \right]^{\mathrm{T}}
\end{cases}
\tag{1}
$$

with $X = \left[I_d \; I_q \; \Omega \; \theta \right]^{\mathrm{T}}$, $U = \left[V_d \; V_q \right]^{\mathrm{T}}$

$$
F(X) = \begin{bmatrix} f_1(X) \\ f_2(X) \\ f_3(X) \\ f_4(X) \end{bmatrix} = \begin{bmatrix} -\dfrac{R_s}{L_d} I_d + \dfrac{L_q}{L_d} p \cdot I_q \cdot \Omega \\[2mm] -\dfrac{R_s}{L_q} I_q - \dfrac{L_d}{L_q} p \cdot I_d \cdot \Omega - \dfrac{p \cdot \Phi_f}{L_q} \Omega \\[2mm] -\dfrac{f}{J} \cdot \Omega + \dfrac{p \cdot \Phi_f}{J} I_q - \dfrac{T_L}{J} \\[2mm] p \cdot \Omega \end{bmatrix}; \quad
G = \begin{bmatrix} \dfrac{1}{L_d} & 0 \\[2mm] 0 & \dfrac{1}{L_q} \\[2mm] 0 & 0 \\[2mm] 0 & 0 \end{bmatrix} = \begin{bmatrix} g_1 & 0 \\ 0 & g_2 \\ 0 & 0 \\ 0 & 0 \end{bmatrix}
$$

This relation (1) shows that the PMSM dynamic model can be represented as a non-linear function of speed and stator resistance which varies with temperature. A variation of this parameter can induce, for the PMSM, a lack of field orientation, performance and stability. Thus, to preserve the reliability and robustness stability under the stator resistance variation, a robust input-output linearization via feedback control, proposed by [3] [4], is used to provide a good regulation and convergence of the currents for the PMSM drive. However, the resolution of the feedback control for the PMSM requires an on-line estimation of the speed value that is not measurable.

Thus, in order to take into account the load torque and stator resistance variations, this work uses a full sixth-order Discrete-time ESMO method to provide an on-line estimation of currents, speed, rotor position, load torque and stator resistance in a PMSM.

3. Discrete-Time ESMO Model

Let us consider the dynamic model of the PMSM given by the system (1). Assume that among the state variable, only the currents $\left(I_d, I_q\right) = \left(z_1, z_2\right)$ are measurable. Consider that $\left(\hat{z}_1, \hat{z}_2\right)$ are the estimates of the currents and denote $\left(\hat{x}_1, \hat{x}_2\right)$ the estimates of the speed (Ω) and position (θ). Thus, In order to solve at the same time the problem of the load torque and stator resistance estimations in a PMSM, a six-dimensional extended state vector defined by $X_e = \begin{bmatrix} I_d & I_q & \Omega & \theta & T_r & R_s \end{bmatrix}^t = \begin{bmatrix} z_1 & z_2 & x_1 & x_2 & x_3 & x_4 \end{bmatrix}^t$ has been introduced.

Thus the proposed ESMO structure is a copy of the model (1), extended to the load torque and stator resistance equation, and by adding corrector gains with switching terms [8]:

$$\dot{\hat{X}}_e = Q\left(\hat{X}_e, U\right) + K \cdot J_S \tag{2}$$

with $\hat{X}_e = \begin{bmatrix} \hat{I}_d & \hat{I}_q & \hat{\Omega}_r & \hat{\theta} & \hat{T}_r & \hat{R}_s \end{bmatrix}^T = \begin{bmatrix} \hat{z}_1 & \hat{z}_2 & \hat{x}_1 & \hat{x}_2 & \hat{x}_3 & \hat{x}_4 \end{bmatrix}^T$

$$Q\left(\hat{X}_e, U\right) = \begin{pmatrix} -\dfrac{\hat{x}_4}{L_d} \cdot z_1 + p \cdot \hat{x}_1 \cdot z_2 + \dfrac{V_d}{L_d} \\[2mm] -p \cdot \hat{x}_1 \cdot z_1 - \dfrac{\hat{x}_4}{L_q} \cdot z_2 - \dfrac{p \cdot \Phi_f}{L_q} \cdot \hat{x}_1 + \dfrac{V_q}{L_q} \\[2mm] \dfrac{p.\Phi_f}{J} z_2 - \dfrac{f}{J} \cdot \hat{x}_1 - \dfrac{\hat{x}_3}{J} \\[2mm] p \cdot \hat{x}_1 \\[2mm] \tau \\[2mm] \varepsilon \end{pmatrix}$$

where the parameters (τ, ε) present the slow variation of (T_r, R_s); K is the observer gain matrices and the switching "J_s" that depends on the estimated currents, is given by:

$$J_s = \begin{bmatrix} sign\left(S_1\right) \\ sign\left(S_2\right) \end{bmatrix} \quad \text{with } S = \begin{bmatrix} S_1 \\ S_2 \end{bmatrix} = M^{-1} \cdot \begin{bmatrix} z_1 - \hat{z}_1 \\ z_2 - \hat{z}_2 \end{bmatrix} = \begin{bmatrix} p \cdot z_2 & -\dfrac{z_1}{L_d} \\[2mm] -\left(p \cdot z_1 + \dfrac{p \cdot \Phi_r}{L_q}\right) & -\dfrac{z_2}{L_q} \end{bmatrix}^{-1} \begin{bmatrix} z_1 - \hat{z}_1 \\ z_2 - \hat{z}_2 \end{bmatrix} \tag{3}$$

Setting

$$\begin{bmatrix} \tilde{z}_1 \\ \tilde{z}_2 \end{bmatrix} = \begin{bmatrix} z_1 - \hat{z}_1 \\ z_2 - \hat{z}_2 \end{bmatrix} \quad \text{and} \quad \begin{bmatrix} \tilde{x}_1 \\ \tilde{x}_2 \\ \tilde{x}_3 \\ \tilde{x}_4 \end{bmatrix} = \begin{bmatrix} x_1 - \hat{x}_1 \\ x_2 - \hat{x}_2 \\ x_3 - \hat{x}_3 \\ x_4 - \hat{x}_4 \end{bmatrix}$$

the estimation error dynamics is given by:

$$\begin{bmatrix} \dot{\tilde{z}}_1 \\ \dot{\tilde{z}}_2 \\ \dot{\tilde{x}}_1 \\ \dot{\tilde{x}}_2 \\ \dot{\tilde{x}}_3 \\ \dot{\tilde{x}}_4 \end{bmatrix} = \begin{bmatrix} p \cdot z_2 \cdot \tilde{x}_1 - \dfrac{Z_1}{L_d} \tilde{x}_4 - K_1 J_s \\[2mm] -\left(p \cdot z_1 + \dfrac{p \cdot \Phi_r}{L_q}\right) \cdot \tilde{x}_1 - \dfrac{Z_2}{L_q} \tilde{x}_4 - K_2 \cdot J_s \\[2mm] \dfrac{-f}{J} \cdot \tilde{x}_1 - \dfrac{1}{J} \cdot \tilde{x}_3 - K_3 \cdot J_s \\[2mm] p \cdot \tilde{x}_1 - K_4 \cdot J_s \\[2mm] -K_5 \cdot J_s \\[2mm] -K_6 \cdot J_s \end{bmatrix}$$

The condition for convergence is verified by chosen the following observer gain matrices K_1, K_2, K_3, K_4, K_5 and K_6:

$$K = \begin{bmatrix} K_1 \\ K_2 \\ K_3 \\ K_4 \\ K_5 \\ K_6 \end{bmatrix} = \begin{bmatrix} p \cdot z_2 & -\dfrac{z_1}{L_d} \\ -\left(p \cdot z_1 + \dfrac{p \cdot \Phi_r}{L_q}\right) & -\dfrac{z_2}{L_q} \\ \alpha - \dfrac{f}{J} & 0 \\ p & 0 \\ n & n \\ 0 & \alpha \end{bmatrix} \cdot \Gamma ; \quad \Gamma = \begin{bmatrix} \beta & 0 \\ 0 & \beta \end{bmatrix} \tag{4}$$

From the expression of K, it can be seen that there are three adjusting gains: (α, β and n) > 0, which play a critical role in the potential stability of the scheme with respect to stator resistance, speed and load torque estimation. These three adjusting gains must be chosen so that the estimator satisfies robustness properties, global or local stability, good accuracy and considerable rapidity.

In order to implement the ESMO algorithm in a DSP for real-time applications, the proposed extended sliding mode observer must be discretized using Euler approximation (1st order) proposed in [11]. The Discrete-time Extended Sliding Mode Observer (DESMO) should be written as:

$$\begin{cases} \hat{X}_e(k+1) = \hat{X}_e(k) + T_e \cdot Q\left(\hat{X}_e(k), U(k)\right) + K(k) \cdot J_s(k) \\ \hat{Y}_e(k) = H\left(\hat{X}_e(k)\right) \end{cases} \tag{5}$$

with

$$\hat{X}_e = \begin{bmatrix} \hat{I}_d(k) & \hat{I}_q(k) & \hat{\Omega}_r(k) & \hat{\theta}(k) & \hat{T}_r(k) & \hat{R}_s(k) \end{bmatrix}^T, \quad U(k) = \begin{bmatrix} V_d(k) & V_q(k) \end{bmatrix}^T$$

$$K = \begin{bmatrix} K_1(k) \\ K_2(k) \\ K_3(k) \\ K_4(k) \\ K_5(k) \\ K_6(k) \end{bmatrix} = \begin{bmatrix} T_e \cdot p \cdot z_2(k) & -T_e \cdot \dfrac{z_1(k)}{L_d} \\ -T_e \cdot \left(p \cdot z_1(k) + \dfrac{p \cdot \Phi_r}{L_q}\right) & -T_e \cdot \dfrac{z_2(k)}{L_q} \\ \alpha - T_e \cdot \dfrac{f}{J} & 0 \\ T_e \cdot p & 0 \\ n & n \\ 0 & \alpha \end{bmatrix} \cdot \Gamma ; \quad \Gamma = \begin{bmatrix} \beta & 0 \\ 0 & \beta \end{bmatrix}$$

$$J_s(k) = \begin{bmatrix} sign(S_1(k)) \\ sign(S_2(k)) \end{bmatrix} \quad \text{avec} \quad S(k) = \begin{bmatrix} S_1(k) \\ S_2(k) \end{bmatrix} = T_e \cdot \begin{bmatrix} p \cdot z_2(k) & -\dfrac{z_1(k)}{L_d} \\ -\left(p \cdot z_1(k) + \dfrac{p \cdot \Phi_r}{L_q}\right) & -\dfrac{z_2(k)}{L_q} \end{bmatrix}^{-1} \begin{bmatrix} z_1(k) - \hat{z}_1(k) \\ z_2(k) - \hat{z}_2(k) \end{bmatrix}$$

$$Q\left(\hat{X}_e(k),U(k)\right)=\begin{pmatrix} -\dfrac{\hat{x}_4(k)}{L_d}\cdot z_1(k)+p\cdot\hat{x}_1(k)\cdot z_2(k)+\dfrac{V_d(k)}{L_d} \\[2mm] -p\cdot\hat{x}_1(k)\cdot z_1(k)-\dfrac{\hat{x}_4(k)}{L_q}\cdot z_2(k)-\dfrac{p.\Phi_r}{L_q}\cdot\hat{x}_1(k)+\dfrac{V_q(k)}{L_q} \\[2mm] \dfrac{p\cdot\Phi_r}{J}z_2(k)-\dfrac{f}{J}\cdot\hat{x}_1(k)-\dfrac{\hat{x}_3(k)}{J} \\[2mm] p\cdot\hat{x}_1(k) \\[2mm] \tau \\[2mm] \varepsilon \end{pmatrix}$$

$$\hat{Y}_e(k)=\begin{pmatrix} I_d \\ I_q \end{pmatrix}=H\left(\hat{X}_e(k)\right)=\begin{bmatrix} 1 & 0 & 0 & 0 & 0 & 0 \\ 0 & 1 & 0 & 0 & 0 & 0 \end{bmatrix}\cdot\hat{X}_e(k)$$

where k means the kth sampling time, *i.e.* $t=k\cdot T_e$ with T_e the adequate sampling period chosen without failing the stability and the accuracy of the discrete-time model.

4. Simulation Results and Discussion

Finally, the proposed scheme (**Figure 1**), a combination nonlinear feedback control and DESMO approach, is carried out for a 1.6 kW PMSM by the simulation on SIMULINK /MATLAB in order to evaluate its robustness and effectiveness in the presence of measurement noise and parameter variations.

The nominal parameters of the PMSM are given in the **Table 1**. The sampling period is $T_e=1$ ms.

Two kinds of tests have been performed (with nominal and non-nominal parameters) in order to compare the behaviour of the DESMO algorithm with respect to parameter variation and the presence of about 20% noise on the simulated currents:

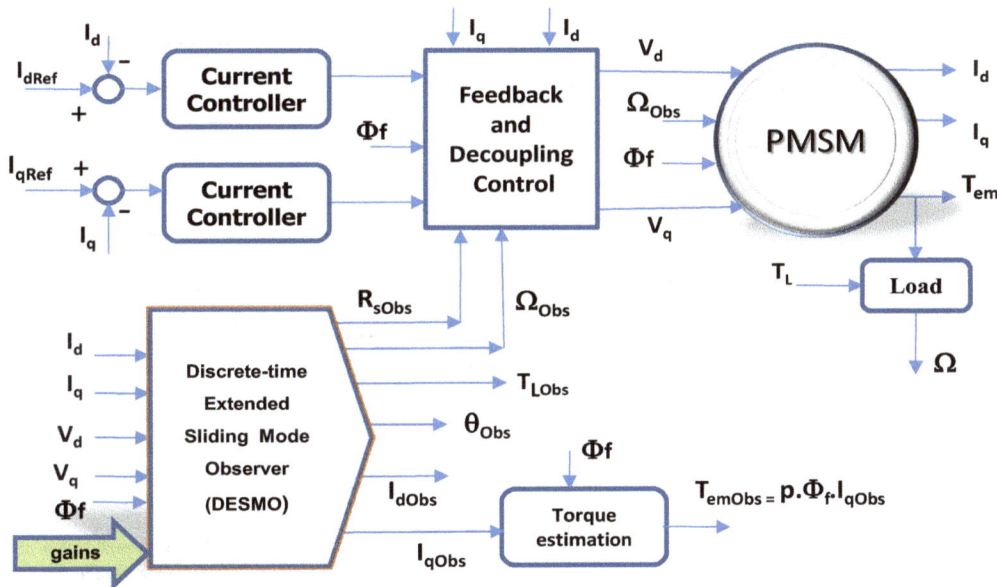

Figure 1. Simulation scheme.

Table 1. Nominal parameters of the PMSM.

$P_{mn}=1.6$ kW	$U_n=220/380$V	$f_n=0.0162$ N.m.sec.rad^{-1}
$p=3$	$\Omega_n=1000$ rpm	$J_n=0.0049$ kg·m^2
$R_{sn}=2.06\ \Omega$	Phif$_n=0.29$ Wb	$L_{qn}=L_{dn}=9.15$ mH

- **Figure 2** shows the simulation results with nominal parameters for a load torque ($T_L = 2$ N.m);
- **Figure 3** illustrates the results where the stator resistance varies ($R_s = 2.R_{sn}$) with a load torque $T_L = 3$ N.m and a step variation in current I_d (4 to 3 A).

For each test, the comparative simulation and estimated results are presented. Better estimation performance yielded by the proposed DESMO is obvious from the observation results. Thus it can be seen that the estimation waves are quite similar to the simulation ones. The observed speed, position and load torque indicate the good

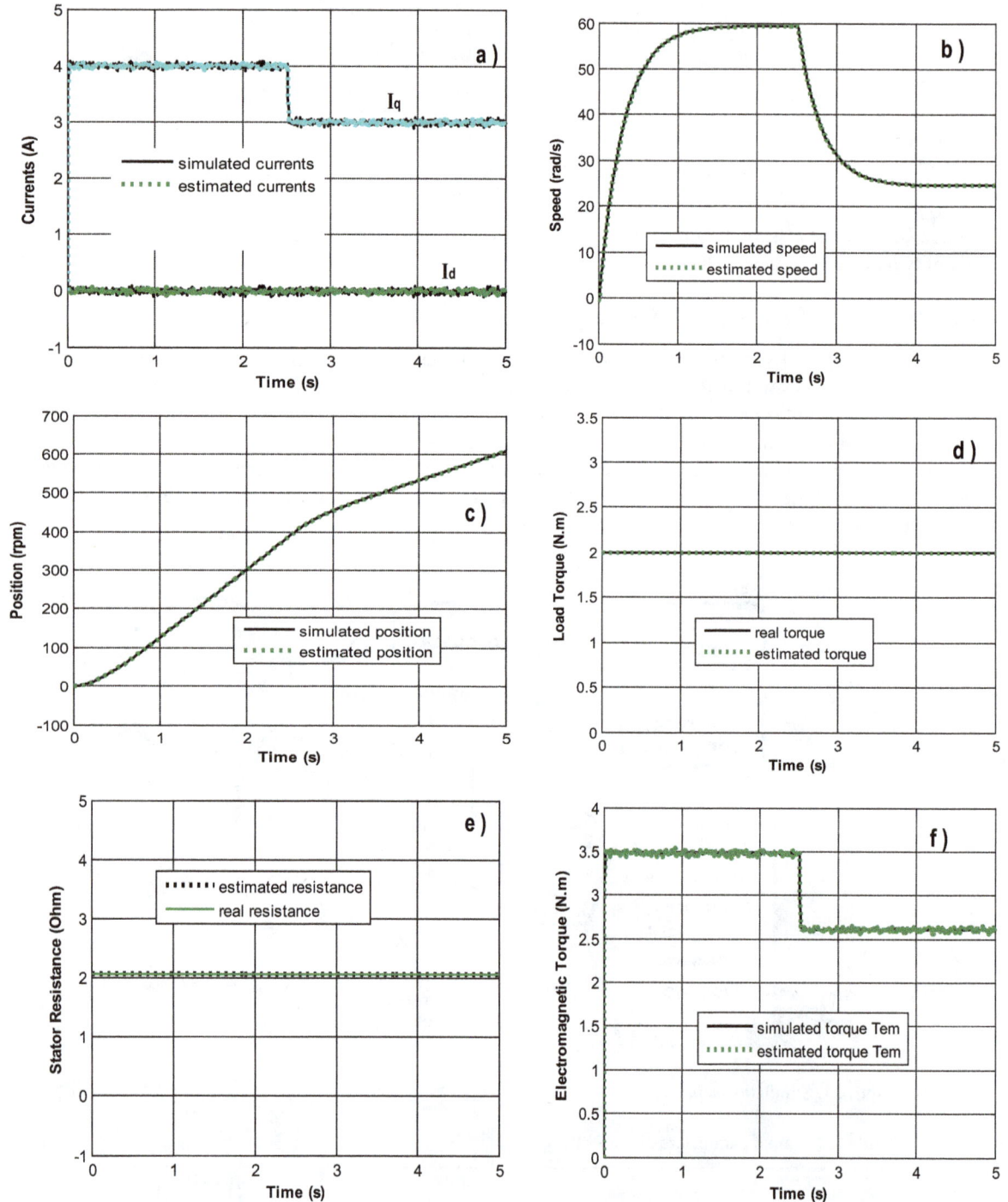

Figure 2. Nominal case ($R_s = R_{sn}$): Comparison between estimated and simulated values for $T_L = 2$ N.m in the presence of measurement noise.

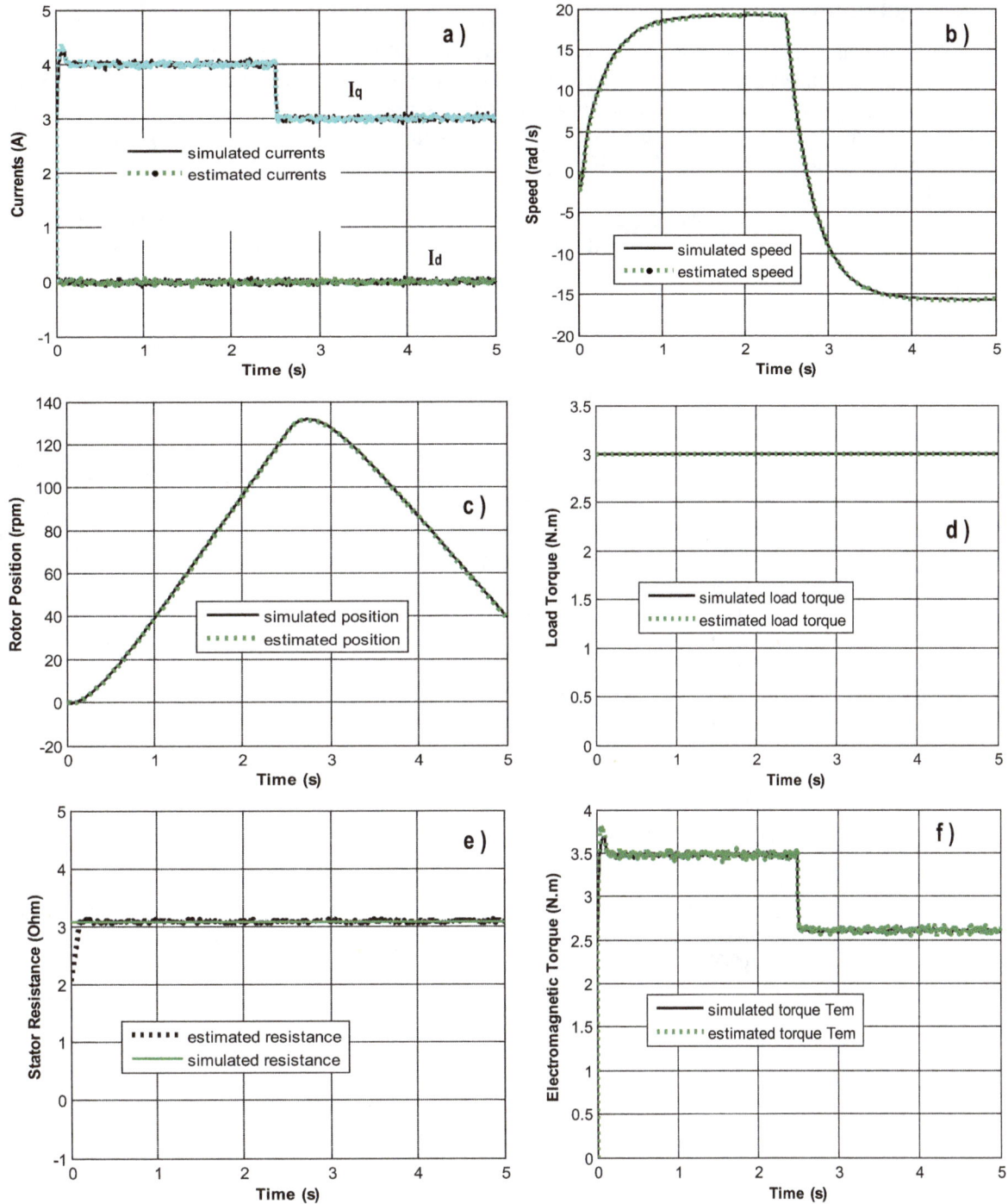

Figure 3. Non Nominal case ($R_r = 1.5 \times R_{rn}$): Comparison between estimated and simulated values for $T_L = 3$ N.m in the presence of measurement noises.

orientation (the current I_q converges very well to zero) which is due to a favourable stator resistance estimation. Also we can see an absence or a rejection of noises on the speed, position and load torque values in the both figure cases. Furthermore A variation in load torque cannot influence on the speed/position response that remains acceptable.

All those good waveforms show that the agreement between the observation dynamic performance and the simulated ones is demonstrated.

5. Conclusions

In this research, a robust feedback linearization strategy and a DESMO algorithm are used not only to decouple and then control independently the currents of the PMSM in a field-oriented (d, q) coordinate but also to provide the unmeasurable state variable estimation (speed, position, stator resistance and load torque). A series of simulations tests have been achieved on the PMSM. The results obtained have demonstrated a good performance of this robust decoupling controland DESMO algorithm against stator resistance variations, measured noise and load torque.

Thus, in the industrial applications, one will appreciate very well the experimental implement of this robust estimator for the reconstitution of the speed, position and the torque as well as the stator resistance.

References

[1] Dehkordi, A., Gole, A.M. and Maguire, T.L. (2005) PM Synchronous Machine Model for Real-Time Simulation. *International Conference on Power Systems Transients*, Montreal, Paper No. IPST05-159.

[2] Pillay, P. and Krishnan, R. (1988) Modeling of Permanent Magnet Motor Drives. *IEEE Trans. Industrial Electronics*, **35**, 537-541. http://dx.doi.org/10.1109/41.9176

[3] Jun, L. and Yuzhou, L. (2006) Speed Sensorless Nonlinear Control for PM Synchronous Motor Fed by Three-Level Inverter. *IEEE International Conference on Industrial Technology*, Mumbai, 15-17 December 2006, 446-451.

[4] Marino, R., Tomei P. and Verrelli, C.M. (2006) Nonlinear Adaptive Output Feedback Control of Synchronous Motors with Damping Windings. *32nd Annual Conference on IEEE Industrial Electronics*, Paris, 6-10 November 2006, 1131-1136. http://dx.doi.org/10.1109/IECON.2006.347363

[5] Taibi, D., Titaouine, A., Benchabane, F. and Bennis, O. (2015) Stability Analysis of the Extended Kalman Filter for Permanent Magnet Synchronous Motor. *Journal of Applied Engineering Science & Technology (JAEST)*, **1**, 51-60.

[6] Aissa, A. and Mokhtari, B. (2012) Extended Kalman Filter for Speed Sensorless Direct Torque Control of a PMSM Drive Based Stator Resistance Estimator. *Journal of Electrical and Control Engineering*, **2**, 33-39.

[7] Kazraji, S.M., Soflayi, R.B. and Sharifian, M.B. (2014) Sliding-Mode Observer for Speed and Position Sensorless Control of Linear-PMSM. *Electrical Control and Communication Engineering*, **5**, 20-26. http://dx.doi.org/10.2478/ecce-2014-0003

[8] Lindita, D. and Aida, S. (2013) An Improved Performance of Sensorless PMSM Drive Control with Sliding Mode Observer in Low Speed Operation. *International Journal of Engineering Trends and Technology (IJETT)*, **4**, 2205-2211.

[9] Yue, Z., Wei, Q. and Long, W. (2012) Compensation Algorithms for Sliding Mode Observers in Sensorless Control of IPMSMs. *IEEE International Electric Vehicle Conference (IEVC)*, Greenville, 4-8 March 2012, 1-7. http://dx.doi.org/10.1109/IEVC.2012.6183241

[10] Foo, G. and Rahman, M.F. (2010) Sensorless Sliding-Mode MTPA Control of an IPM Synchronous Motor Drive Using a Sliding-Mode Observer and HF Signal Injection. *IEEE Transactions on Industrial Electronics*, **57**, 1270-1278. http://dx.doi.org/10.1109/TIE.2009.2030820

[11] Lewis, F. (1992) Applied Optimal Control Estimation-Digital Design and Implementation. Prentice Hall, New York, 448 p.

Nomenclature

T_{em}, T_l: Electromagnetic and load torques (N.m).

I_d, I_q: (d, q)-axis stator currents (A).

p, J, f: p: pole number; J: inertia (kg·m^2); f: Damping coefficient (Nm.s/rad).

L_d, L_q: (d, q)-axis inductances (H).

R_s, T_e: Stator resistance (W) and Sampling period (s).

V_d, V_q: D-axis and q-axis stator voltage (V).

Φ_f, θ: Rotor magnet flux linkage (Wb); θ: Rotor position at electrical angle (rpm).

ω_r, Ω, ω_r: Rotor electrical radian speed; Ω: Mechanical rotor speed (rad/s).

13

Static Simulation of a 12/8 Switched Reluctance Machine (Application: Starter-Generator)

Sihem Saidani, Moez Ghariani

Laboratory of Electronics and Information Technology (LETI), Electric Vehicle and Power Electronics Group (VEEP), National School of Engineers of Sfax, University of Sfax, Sfax, Tunisia
Email: Sihem.saidani@yahoo.fr, moez.ghariani@isecs.rnu.tn

Abstract

Because its high efficiency, its simple stator and rotor structures, the low cost and high reliability, speed operation combined with robust and low cost construction, the switched reluctance machines have represented. In recent years, an interesting alternative to other machine types has been chosen for traction applications especially starter-generator. Their rotors do not generate significant heat, resulting in easy cooling. Their unidirectional flux and current may generate lower core losses and require a simple converter design. Moreover, the switched reluctance machines are known for their high reliability and capability of operating in four quadrants for a variable speed drive. Despite those merits, switched reluctance machine has not been extensively used until recently because of its problems of torque ripples and noise. Additionally, researchers have faced many difficulties to build a SRM model because it is inherently multivariable. It has strong coupling and especially a high nonlinearity. In this paper, we deal with many modeling methods. Numerical, analytical and intelligent approaches are studied. The important aim in this research is to use static results from FEMM simulation as flux-linkage, co-energy, static torque to form a dynamic model of a switched reluctance machine used next as a starter-generator of a hybrid vehicle.

Keywords

Hybrid Vehicle, Switched Reluctance Machine, Finite Element Analysis, Look up Table, FEMM, MATLAB-Simulink

1. Introduction

The design and the materials used in production preciseness plus the time are coming the major factors taken in motor manufacturing. As a matter of fact, interior permanent magnet machines (IPM) seem to be the best candidate in hybrid vehicles. However, and due to many problems as the earth magnet materials and increasing prices especially the Dysprosium and Neodymium, the auto makers have looked for new alternatives. Thus, the switched reluctance machines (SRM), having a near performances especially high efficiency, is one of the best replacing candidates. This machine is characterized by a high starting torque and efficiency similar to a high-efficiency IM. As its rotor is so simple, it is usually made by steel laminations in order to minimize the core losses without any windings or magnets. The SRM is the machine suited for harsh environments. Thus, the torque ripple, vibration, and acoustic noise are the most disadvantages of the SRM. The acoustic noise in SRMs is caused by the radial forces that inversely increase with the air gap and especially when the force frequency is near the stator resonant frequency. Many issues aimed to analyze SRM design and predict its performance in order to attend the best performances.

Obviously, the static characteristics exactly the flux linkage and the torque versus current profiles play an important role in the prediction of the dynamic behavior of switched reluctance motors and then as their functions as starters and generators in micro-hybrid vehicle.

2. SRM as a Starter Alternator

The electric motor of a starter alternator system must have a high starting torque for initial acceleration and then and high efficiency to extend the battery range and a wide operating speed range as generator. Besides, the SRM offers a wide choice of pole configurations and phase number. Consequently, a lower number of phases can reduce the converter cost, but it increases the torque ripple for example as three phases 12/8 poles is better than 3 phases 6/4 poles as we will explain later. Different design found with the technology of the start-stop system has been described in [1] are used as a European markets for an example: Bosh, Valeo, INA (Schaeffler Technologies), Benso represented in different configurations as:

- Belt Driven starter generator.
- Enhanced starter.
- Direct Starter.
- Integrated Starter Generator.

3. Modeling Methods of Switched Reluctance Machine

Due to saturation effects in the SRM, establishing the accurate nonlinear mapping relationship of the flux linkage with respect to phase current and rotor position is the basis of performance calculation. For many years, different modeling methods have been proposed by worldwide researchers such as O. Ichinokura in 2003, S. Song also Z. Lin. Generally, all these methodologies can be divided into three categories: The numerical, the analytical and the intelligent methods.

3.1. Intelligent Method

There are many intelligent methods that can build a model of SRM, such as artificial neural network (ANN), fuzzy inference system (FIS), B-spline neural network (BSNN) and others. Based on optimization algorithm of Levenberg-Marquardt (LM) in 2005-2006 the BPNN seemed to execute with perfect result the nonlinear relationship between the flux linkage and the electromagnetic torque. This approach had given through its results a strong robustness and a great ability especially for a non user or professional to model easily the SRM. Many researchers had been investigated in the dynamic behavior of (SRM) [2] by monitoring its dynamic response (torque and speed), and then they worked n minimizing the torque ripples and building different types of sensors in order to reduce the cost first and second to improve the SRM and reliability and performance. Many researches have proved the powerful problem-solving methodology of the Fuzzy logic with the control and the information process. Moreover the Fuzzy logic control (FLZ) seemed to be extremely a simple way to find precise solutions.

3.2. Analytical Methods

In the analytical approach, the nonlinear relationships of the phase inductance and torque to phase current and rotor position can be represented using different analytical expressions: [3]

$$L(\theta,i) = L(i)_0 + \sum_{k=1}^{n} \left[L(i)_k \cos(nN_r\theta) \right] \tag{1}$$

where $L(\theta,i)$ is the inductance of an independent phase, L_0 and L_k are a constant coefficients, N_r is the number of rotor's pole and θ is the rotor position and n is the maximum harmonic order. Generally, the inductance values at two different rotor positions L_{min} and L_{max}. For the first harmonic ($n = 1$), we can write:

$$L(\theta,i) = L(i)_0 + L(i)_1 \cos(N_r\theta) \tag{2}$$

where:

$$\begin{cases} L(i)_0 = \dfrac{L_{min} + L_{max}}{2} \tag{3} \\[2mm] L(i)_1 = \dfrac{L_{min} - L_{max}}{2} \tag{4} \end{cases}$$

The current pulses are applied on precise rotor position and the motor creates torque in the direction of increasing inductance. In **Figure 1**, we represent:

- The torque is positive when $dL/d\theta > 0$ from θ_1 to θ_2, and the SRM so it is the motoring mode.
- (see **Figure 13**).
- The torque is negative when $dL/d\theta < 0$ from θ_4 to θ_5, and the SRM so it is the generating mode.

The inductances L_{min} and L_{max} can be seen as a constant. Thus, these inductances can be calculated analytically by polynomial fitting using MATLAB for example, as L_{max} and L_{min} given by this relation:

$$L_{max/min} = \sum_{k=0}^{n} a_k i^k \tag{5}$$

where the a_k are a constant coefficients, Using the relation:

$$\Psi = L(\theta,i) \times i \tag{6}$$

The curve of flux linkage can be deduced and so the static torque deduced in (8):

$$L(\theta,i) = \dfrac{L_{min} + \sum_{k=0}^{n} a_k i^k}{\dfrac{2 + \left(L_{min} - \sum_{k=0}^{n} a_k i^k\right)\cos(N_r\theta)}{2}} \tag{7}$$

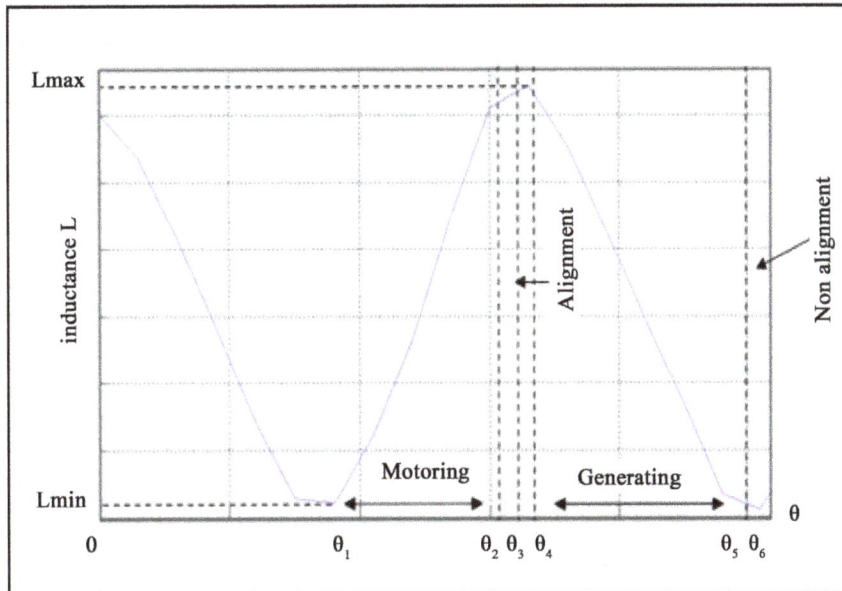

Figure 1. Inductance vs. rotor position angle.

$$T = -\frac{N_r}{4}i^2\left(L_{\min} - \sum_{k=0}^{n}\frac{2a_k}{k+2}i^k \sin(N_r\theta)\right) \tag{8}$$

3.3. Numerical Method

Lately and due to the development of power electronics, it becomes possible to control the SRM. There for, even with acoustic noise, torque ripple, special converter technology control etc., many researchers have developed new methods of design throw geometric modeling, finite analysis (FEA, BEA..) in order to optimize its performance and make the more and more suitable for the drive-train application especially in the hybrid vehicle.

1) *FEA*

Recently, the finite element analysis has become useful in the machine design stage. Where, the flux distribution may be computed in all the regions of the machine, with no geometric limitations or boundary conditions of several material properties. Thus, much useful detailed information has facilitated the range of analysis like noise and vibration issues.

2) *The BEA*

The nonlinear boundary element method is also another alternative way to modulate a machine design. Its principles and equation formulations have been well documented in [2]. The BEA, easier of processing so suitable for both design optimization and dynamic simulation, allows a medium movement without meshing boundaries and a very smooth integral operations without even the unpredictable differentiation used in FEA [3] [4].

3) *Finite Element Method (FEM)*

The finite element method can solve the governing differential equations in order to obtain the magnetic potentials that are minimizing the parameters to have optimized energy. Many tools of finite elements have been developed last years in order to compute the accurate properties and magnetic characteristics of machine without constructing a real prototype. FEMM is a program that can solve electromagnetic problems on 2D (two-dimensional planar). Thus, we can resolve linear and nonlinear magneto static, harmonic magnetic, linear electrostatic and steady state heat problems. The determination of magnetic characteristics facilitates the optimization and control of switched reluctance machine.

In particular, the switched reluctance machine phase magnetization characteristics vary hardly as a function of both excitation current and rotor position. Thus, numerical methods have used in order to calculate the magnetic field and to predict the magnetization characteristics.

4. Finite Element of Switched Reluctance Machine

The Finite Element Method Magnetics (FEMM) as a numerical technique is used to find approximate solutions of partial differential and integral equations. Thus, the solution is approached basically either on eliminating the differential equation completely or rendering the partial differential equation (PDE) into an approximating system of ordinary differential equations. Then, these ODE are numerically integrated using standard techniques as Runge-Kutta, Euler's method etc. For example, FEMM version 4.2, used in our work, is using the finite element method.

The (FEM) takes into account the property of non-linear used magnetic plus the SRM geometry. The solutions of magnetic field distributions allow the torque, flux and flux linkage, losses, energy and co-energy, and inductance to be evaluated. Inductance is normally defined using flux linkage (A), or stored magnetic energy (W).

Using FEMM, we can generate several characteristics for example the inductance variation the magnetic field energy, Maxwell's static and instantaneous [5] torque profiles and output power) have been obtained by integral Equation 9 and Equation 10) as examples

4.1. Torque

Using the Maxwell Stress Tensor (T_{ST}) the force applied to one part of the magnetic circuit and that can be obtained by integrating the next equation:

$$T_{TS} = L\int_S \left\{ r \times \left[\left(\frac{1}{\mu_0}\right)(B\cdot n)B - \left(\frac{1}{2\mu_0}\right)B^2 n \right] \right\} dS \tag{9}$$

where: T_{TS} is the static torque, L is the length, B is the induction vector in the elements and r is the lever arm.

4.2. The Magnetic Energy

Noted in FEMM (W) as the magnetic field in the specified region and used as an alternate method of getting inductance for linear problems (at least not heavily saturated). In case of nonlinear materials, the energy is computed with Equation (10):

$$W' = \int \left(\int_0^B H(B') \, dB' \right) dV \tag{10}$$

where: B is the magnitude of flux density (Tesla), H (A/m) is the magnitude of field intensity and V is the volume of the region obtained from integration over the 2D region.

4.3. The Flux Linkage

In nonlinear characteristics of the SRM, the value of flux and torque depends on with the rotor position and phase current [5] [6]. Therefore, the values of flux and torque must be known for better control of the motor. Thus, the flux or and torque curves can either be obtained using numerical experimentally or computation or by the Finite Element Analysis (FEA). As the matter of fact, the vector potential A witch determines the magnetic field inside the motor through the next expression called Poisson's equation:

$$\frac{\partial}{\partial x}\left(\gamma \frac{\partial A}{\partial x} \right) + \frac{\partial}{\partial y}\left(\gamma \frac{\partial A}{\partial y} \right) = -J \tag{11}$$

where γ the magnetic reluctivity and J is the current density vector. Using the 2D static analysis of FEMM, we can have when one phase is excited the aligned and unaligned inductance with a stator rotor poles position between 0° and 22.5° and it occurs at every 22.5° interval. For the machine J, at a fixed rotor position, we can calculate the co-energy, the static torque and the flux linkage by exciting separately one phase through injecting different values of winding current.

4.4. The Co-Energy

The switched reluctance machine is characterized with independent phases. Thus, to obtain the torque T we can also calculate the energy called also work $W'(i, \theta)$ given in the next Equation (12) (**Figure 2**):

$$\begin{cases} T = \left[\dfrac{\partial W'}{\partial \theta} \right]_{i=\text{constant}} & (12) \\[2mm] W' = \displaystyle\int_0^{i1} \Psi \, di & (13) \end{cases}$$

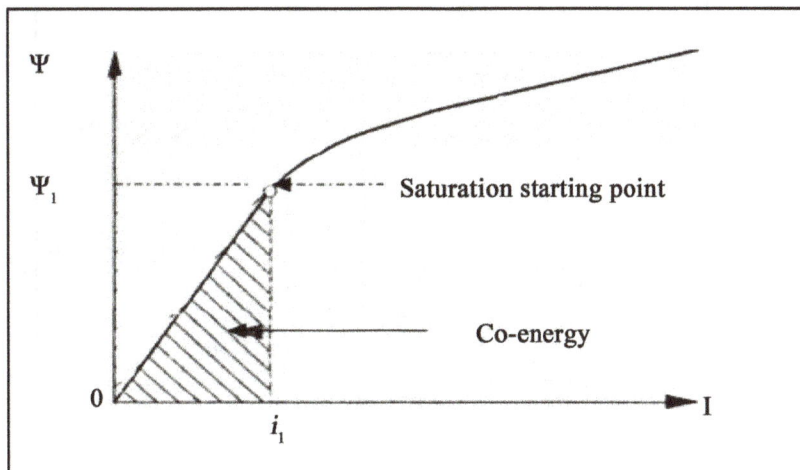

Figure 2. Magnetization curve.

The co-energy can also be used to calculate the inductance phase through the Equation (14):

$$W' = \frac{1}{2} L(\theta) i_1^2 \qquad (14)$$

Regarding to our machine and before getting the static characteristics and next dynamic we need to fulfill such important steps:

- Build the model of the motor according to its physical dimension and material.
- Build the model of the power converter according to its topology.
- Set the terminal and the boundary conditions.
- Add driving sources.
- Generate the finite element mesh.

For our example, **Figure 3** shows a 12/8 model of a switched reluctance machine build using FEMM.

Evidently, the nonlinearity of switched machine makes very difficult the modeling of the flux linkage or the inductance. Many researchers as have been made to resolve the problem of their calculation from varying the rotor position and the phase. The first way to plot the phase flux linkage is the variations of rotor position and phase current. The second way, at different phase currents we plot the phase inductance variation in function of rotor position. All these static characteristics are highly nonlinear. **Figure 4** shows different modeling techniques of switched reluctance machine.

5. Influence of Geometry on Static Behavior

The choice of machine's dimensions is considered essential to provide optimum performance of machine. There are four major factors:

- The air-gap length.
- The ratio of rotor diameter to machine outside diameter.
- The tooth width.
- The core back depth cover explanation of choosing the above aspects are explained below.

5.1. Air-Gap Length

In order to have the maximum torque, the air-gap length has to be made the smallest possible typically equal to 0.25 mm (in our case 0.32 mm). It is important element because it will fix the area of the co-energy curve W' (θ, i) witch as large as possible consequently the values of aligned inductance became so high.

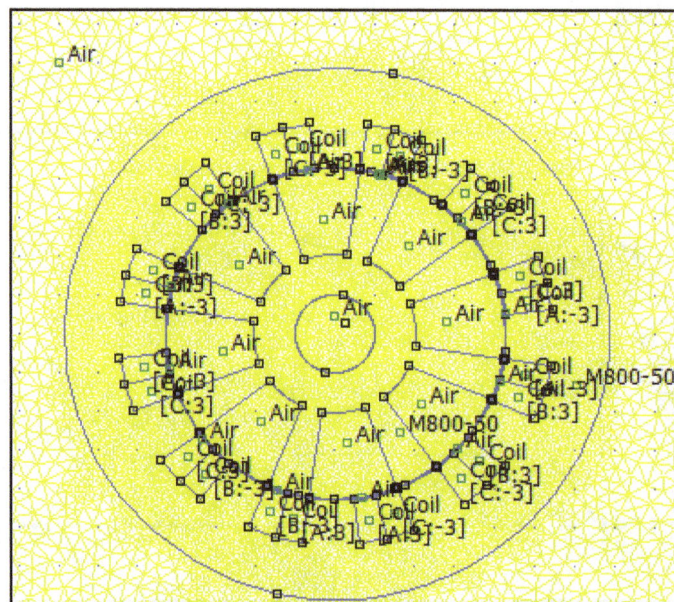

Figure 3. Triangular meshing of 12/8 SRM using FEMM (35,294 nodes, 70,361 elements).

5.2. Ratio of Rotor Diameter to Outside Diameter

The ratio of 0.5 to 0.6 is usually used for most machines. The factor of 1/2 comes from the fact that the stator phase sees two identical air gaps in series. The factors k_1 and k_2 respectively the ratio of rotor interpolar arc to stator pole arc and ratio of rotor pole length to rotor interpolar arc length. A study had been explained carefully to represent the relation between this ratio and the unaligned inductance (see **Table 1**):

where: $k_1 = \dfrac{l_{dr}}{l_{ds}}$ and $k_2 = \dfrac{h_{dr}}{l_{ds}}$ (see **Figure 5**).

5.3. Tooth Width

Different studies have been made as using diverse stator and rotor teeth widths were used to compare the torque ripples. The stator teeth have to be designed to satisfy two conditions:
- The narrow teeth could cause saturation and then the maximum flux which limits the SRM performances.
- The wide teeth may cause insufficient room for the winding. Therefore, an equal approximately widths between the stator teeth and slots.

Then, usually in order to estimate tooth width is as a ratio of about 0.3 to 0.5 can be fixed.

5.4. Core Back Depth

It is usually fixed as at least 0.6 times the width of the stator teeth. Despite, the core back is occasionally can be made thicker than 0.6 to increase mechanical stiffness and consequently may reduce mechanical noise.

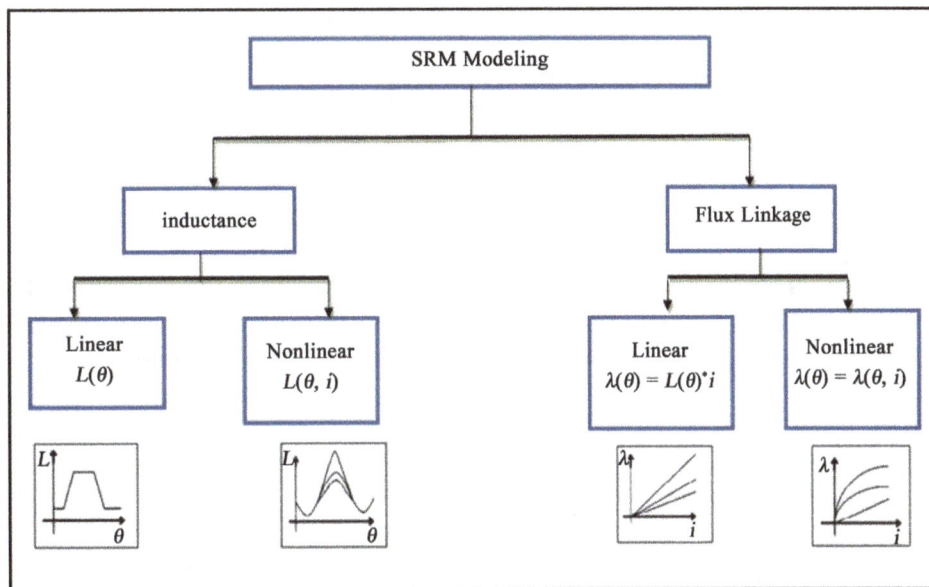

Figure 4. Classification of linear switched reluctance machine modeling.

Table 1. The unaligned inductance in function of ratio k_1/k_2.

k_2/k_1	1.05	1.20	1.5	2	2.5
0.125	17.09	12.98	10.22	8.14	7
0.1875	14.48	10.4	7.99	6.25	5.44
0.25	13.22	9.27	6.63	5.42	4.68
0.375	12.33	8.3	6.03	4.65	4.01
0.5	11.86	7.91	5.69	4.36	3.75
0.75	11.56	7.7	5.48	4.18	3.6
1	11.54	7.62	5.43	4.14	3.75

5.5. Other Factors

1) *Effect of soft magnetic materials on losses and torque ripple*

There are several factors either than geometric ones especially the material made of stator. As the matter of fact, the influence of different soft magnetic materials (SMC) with non-oriented grain also on the core losses and efficiency of a 6/4 SRM has been treated by H. Toda and K. Senda [7]. Besides, the impact of grain oriented (GO) and non-oriented (NO) steel on the electromagnetic characteristics of a 12/8 switched reluctance machine has been analyzed by Y. Sugawara and K. Akatsu. K. Vijayakumar, R. Karthikeyan and R. Arumugam have analyzed the influence of composite materials on the characteristics of electromagnetic torque, the average torque and torque ripple. Additionally, these researchers have studied the contribution of composite materials in high speed applications (relatively high frequency is considered), and made a comparison with the M19 material. To improve the performance of a 6/4 SRM, A. Chiba and H. Hayashi used amorphous materials.

It's well acknowledged that one of SRM's inherent problems in is the torque ripple due to switched nature of the torque production. The Equation (15) explain how to determined torque ripple. The torque ripple (T_{ripp}) represents the distance between the peak value and the common point of overlap in the static torque angle characteristics of two consecutive excited SRM phases (see **Figure 6**). Thus, we call the maximum value, in next **Figure 7**, the static torque "peak static torque" as a T_{max} and the minimum of the intersection point T_{int}:

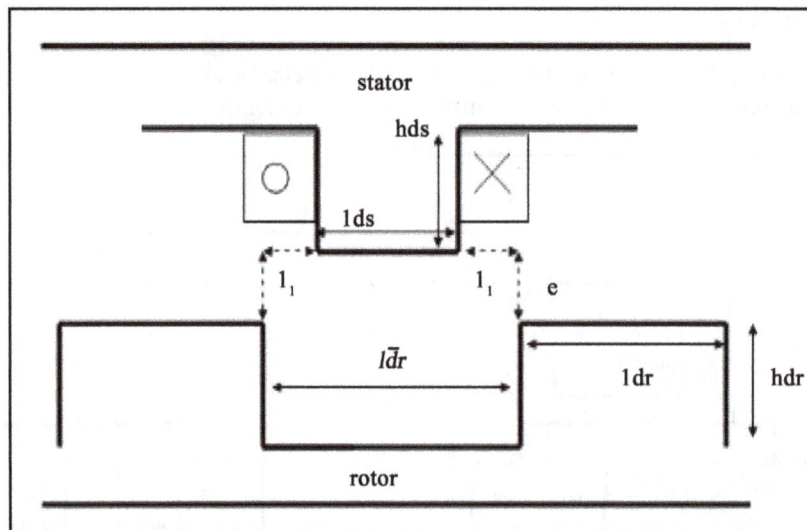

Figure 5. A plane projection of stator and rotor.

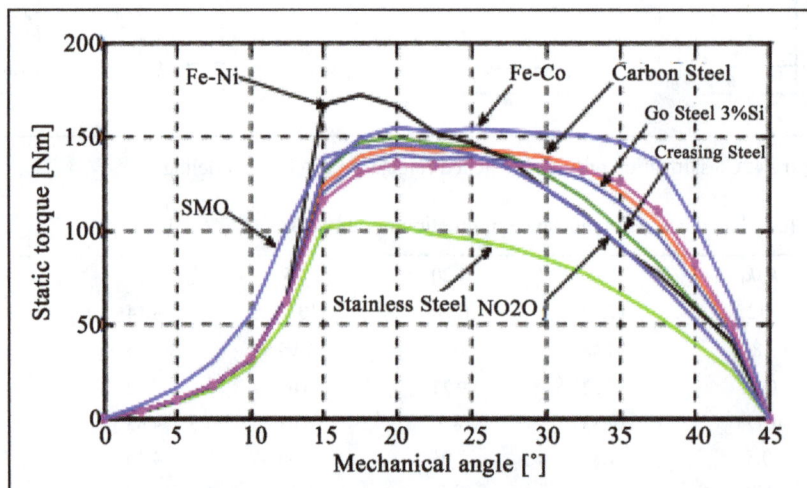

Figure 6. Static torque vs. a rotor position curves of soft materials.

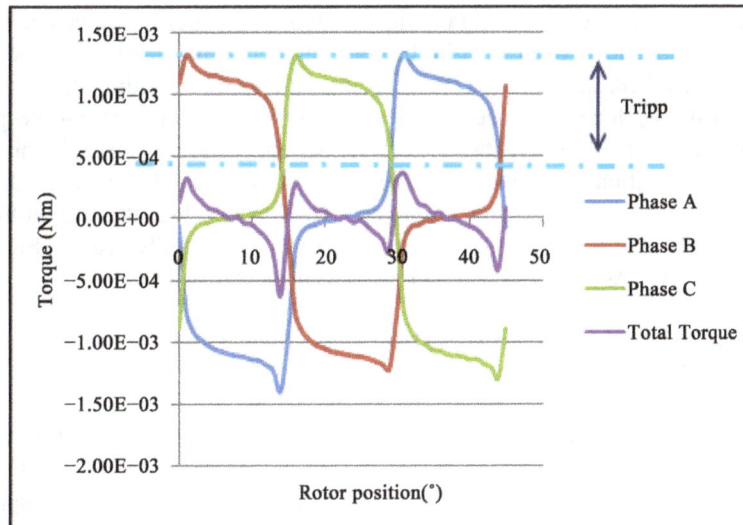

Figure 7. Torque ripple for 2 A phase current.

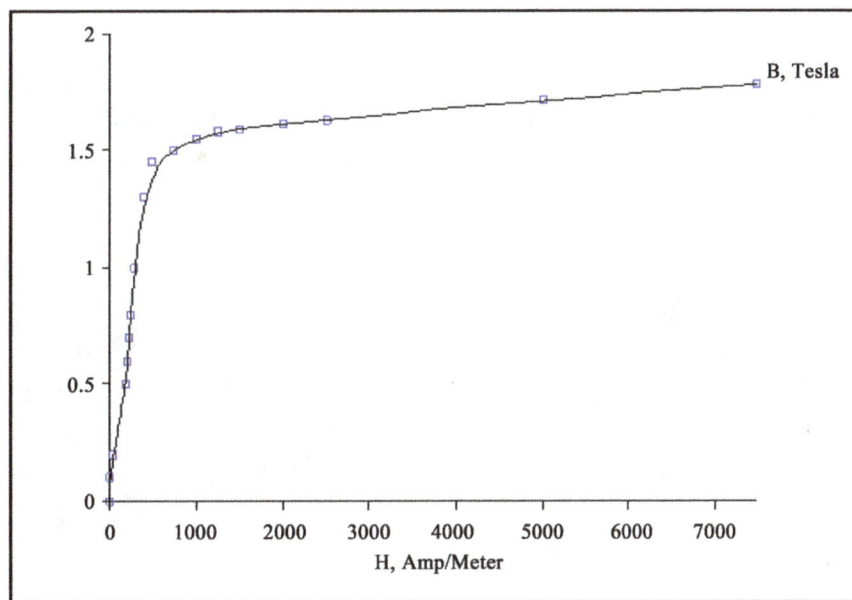

Figure 8. Magnetization curve of M800-50 (FeSi 3.5%).

$$\%\text{Torque Ripple} = \frac{T_{\max} - T_{\text{int}}}{T_{\max}} \tag{15}$$

The choice of magnetic material in function of the average torque given by the Equation (16):

$$T_{av} = \left(\frac{q \times N_r}{2\pi}\right) \times (W_a - W_u) \tag{16}$$

where, $(W_a - W_u)$ is the difference of co-energies at aligned and unaligned positions, N_r is the number of rotor poles and q number of machine's phase.

In our case of study we have chosen the M800-50 (FeSi 3.5%) with a static torque less than 150 nm as shown in **Figure 8**:

2) Impact of winding configuration

Additionally to the effect of soft magnetic materials on SRM configuration, different researches have been developed to improve the impact of winding configuration on the switched reluctance performances. Recently, new winding configurations have been used in the coupling between phases to produce higher torque density for 3-phase SRMs are studied. Barrie C. Mecrow has presented a new winding configuration that called Fully Pitched Winding SRMs witch can produce high density output torque. In order to keep their high output torque density and overcome their disadvantages as its longer end-winding and high copper losses performance [8], a new current distribution has been developed with short pitched windings like conventional SRMs. Called Mutually Coupled Switched Reluctance Motors (MCSRMs). Thus, the current distributions of 3 phases is A+_A-_B-_B+_C+_C-_A-_A+_B+_B-_C-_C+ in **Figure 9** as a mutually coupled SRMs, and A+_A-_B+_B-_C+_C-_A+_A-_B+_B-_C+_C- in **Figure 10** as a MCSRMs.

6. Static Design of 12/8 Switched Reluctance Machine

6.1. Finite Element Analysis of 12/8 SRM

The SRM is symmetric so only one phase is modeled. In our studies we have used for torque, flux linkage and inductance the phase A to be excited and to inform out the characteristics obtained. Using the finite elements method through FEMM package according to the steps schematized demands the respect of (see **Figure 11**). The coupled simulation program connecting FEMM and MATLAB software [9], as shown in flowchart of **Figure 12**:

6.2. Results

The SRM used for simulation is a 12/8, 3-phase-machine, the parameters of the machine are given in **Appendix A**. The finite element model was realized with the FEMM software (see **Appendix B**).

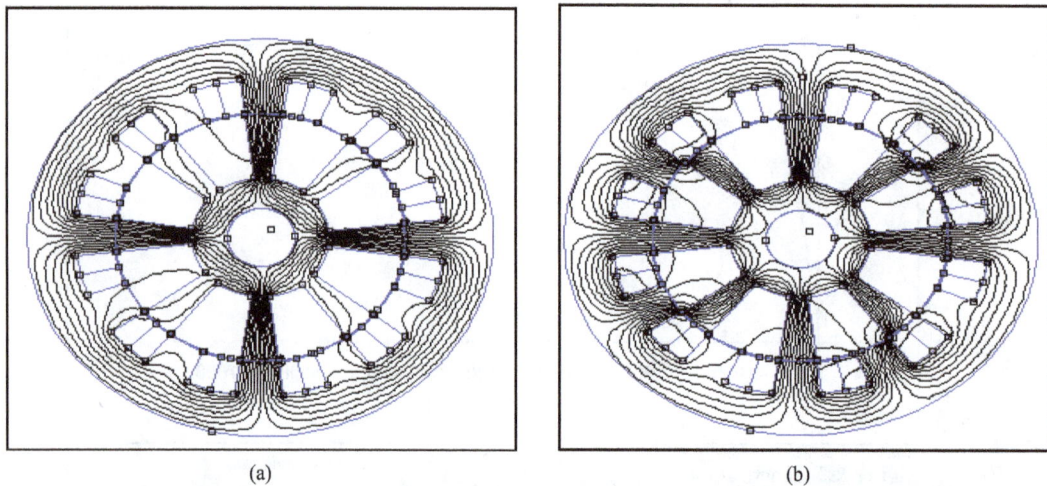

Figure 9. Flux distribution with the phase B excited in constant current (a) conventional SRM, (b) mutually coupled SRM (MCSRM).

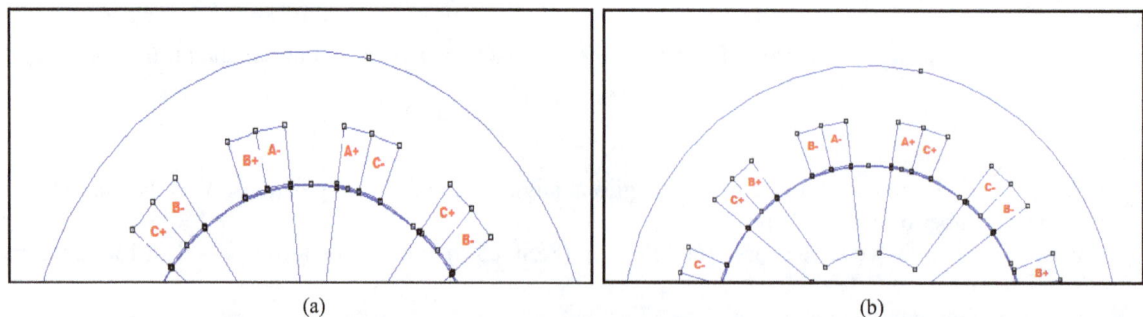

Figure 10. Winding configurations of SRM 12/8, (a) MCSRM; (b) conventional.

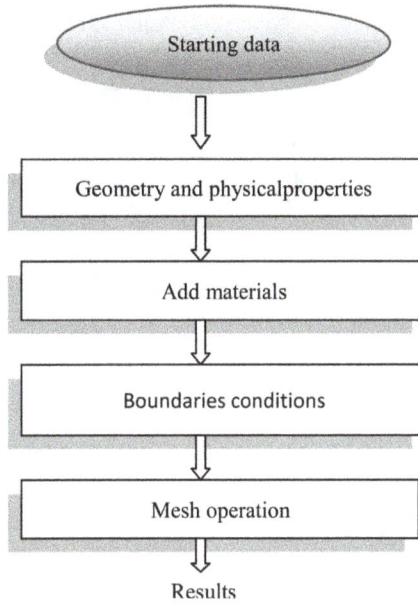

Figure 11. A flowchart of FEMM simulation.

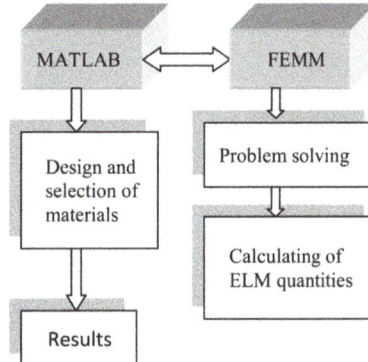

Figure 12. A flowchart of FEMM-MATLAB coupling.

Matlab/Simulink package is the software used in the simulation through the look up table of the torque $T(\theta, i)$ in **Figure 13** is and the self inductance $L(\theta, i)$ of one phase in **Figure 14**.

Thus, and as explained in, the look up table from Simulink library is represented as:

- The displacement angles are considered as a row parameters (horizontal parameters), that varied between (0° to 47°).
- The current is considered as column parameters (vertical parameters), which is varied between (2 to 30 A).

The look-up table described previously in **Figure 15** and **Figure 16** can be used to determinate the speed in function of time and the rotor position. The calculation of moment of inertia is essential then to calculate the torque using the relation:

$$T = T_e - T_1 = J\frac{d\omega}{dt}$$

(17)

where:

$$\begin{cases} T_i\text{: the load torque} \\ T\text{: the torque} \\ T_e\text{: the electromagnetic torque} \\ \omega\text{: the rotor speed} \end{cases}$$

Whereas the moment of inertia J can be calculated as explained in:

$$J = m \frac{D_1^2 + D_2^2}{2} \tag{18}$$

where:

D_1 and D_2 are respectively the external and the internal diameter of the machine and m is the rotor's weight. This relationship is resolved using the previous LUT in next **Figure 17**.

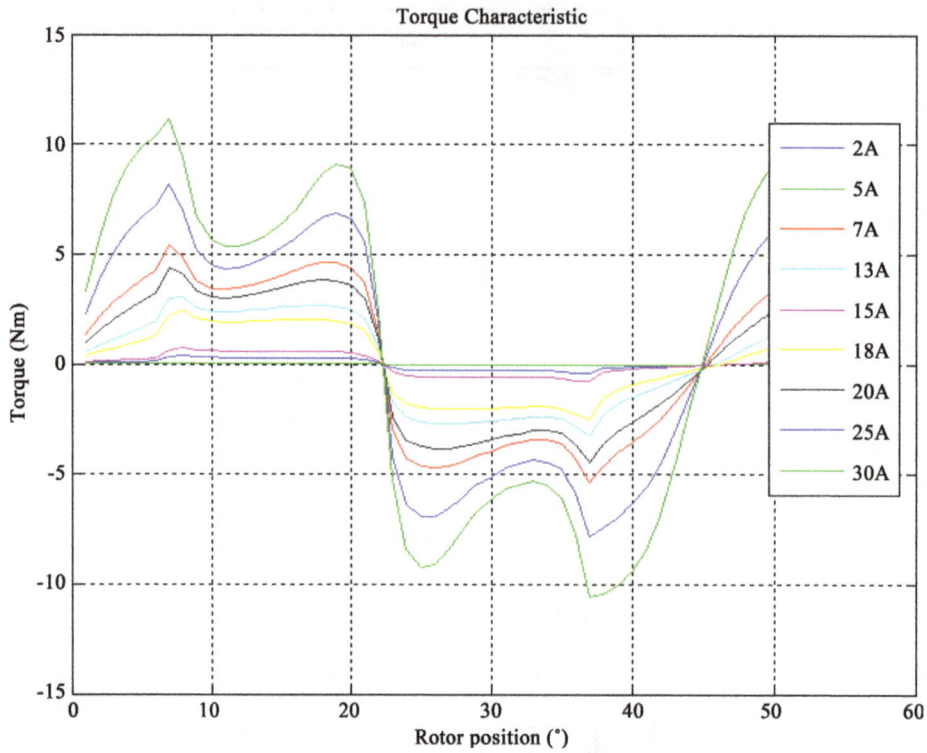

Figure 13. Static torque for different currents.

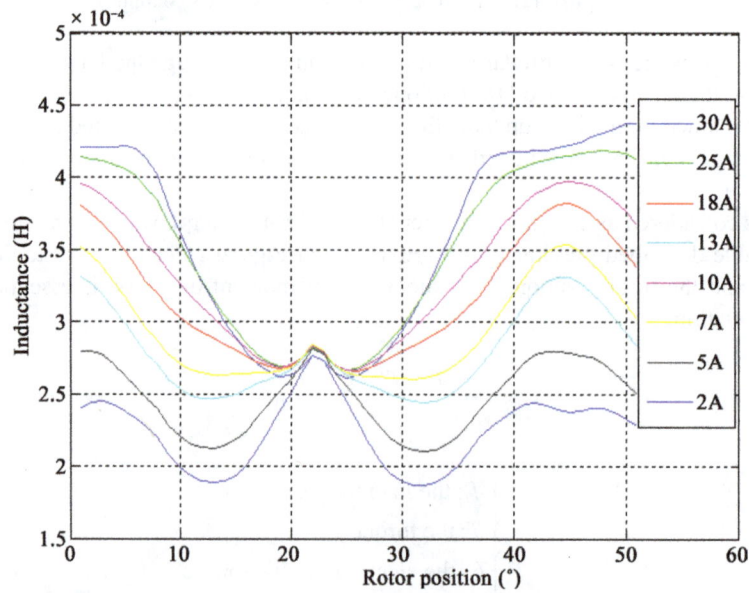

Figure 14. Self inductance for different rotor position ($\theta°$).

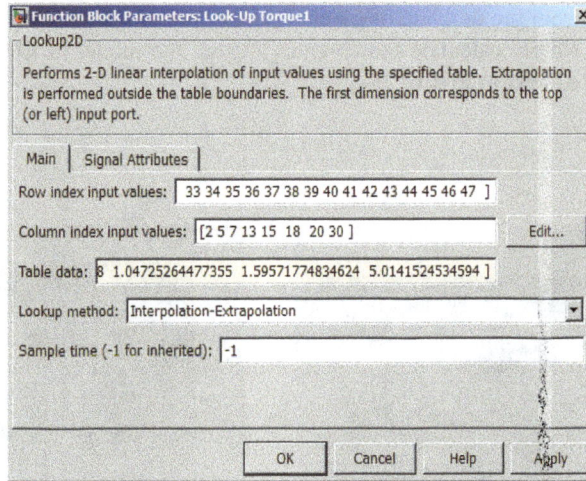

Figure 15. The look up table of the static torque.

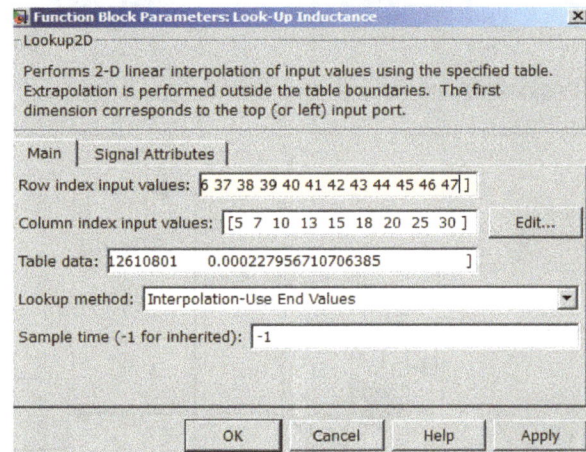

Figure 16. The look-up table of self-inductance.

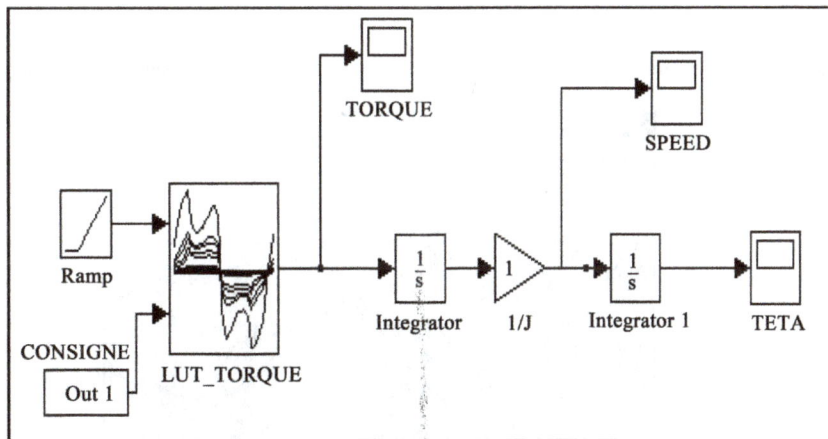

Figure 17. Speed calculation.

Figure18 next represents the scope 3 and 4 where the rotor position (θ) and the speed in function of time $\left(\dfrac{\mathrm{d}\omega}{\mathrm{d}t} \right)$:

The torque is obtained from the derivate of co-energy. Thus with a simple integration of the look-up table of $T(\theta, i)$ as in [2]. We can simply calculate the co-energy using the model in **Figure 19** and the result of scope "COENERGY" is presented next in **Figure 20**:

$$T = \frac{\mathrm{d}W}{\mathrm{d}t} \qquad (19)$$

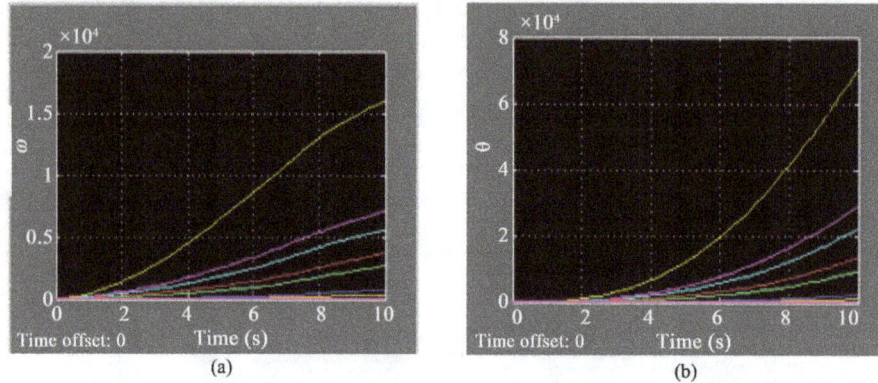

Figure 18. Speed (ω) versus time, (b) angle θ versus time.

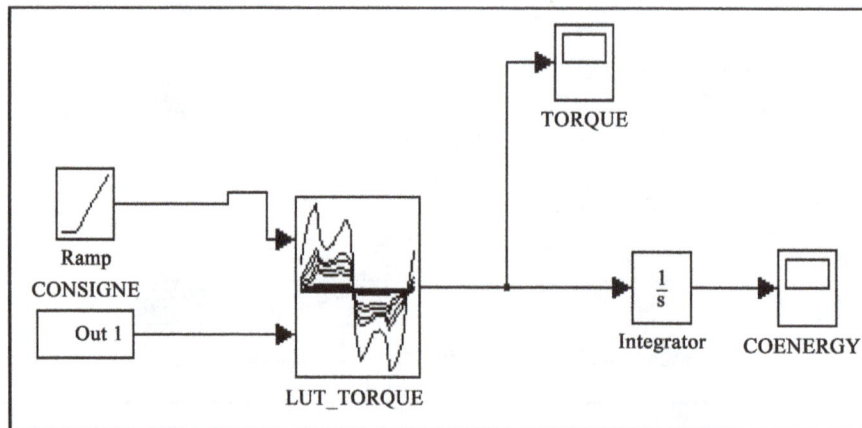

Figure 19. Calculation model using MATLAB/Simulink to determine co-energy from torque.

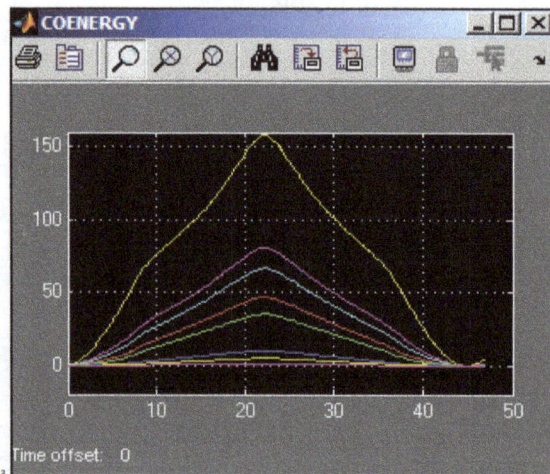

Figure 20. The co energy in function of rotor position (θ) of 12/8 SRM obtained from torque.

The flux in the air gap of the machine can be also calculated using the look-up table described in **Figure 15** and **Figure 16**. Using the expression of flux linkage (see **Figure 21**) relation:

$$\lambda(i,\theta) = L(\theta,i) \times i \tag{19}$$

Thus, in **Figure 22** we deal with the MATLAB/Simulink model used in calculation and in **Figure 22**. The result of the simulation: the flux for different current in function of rotor position:

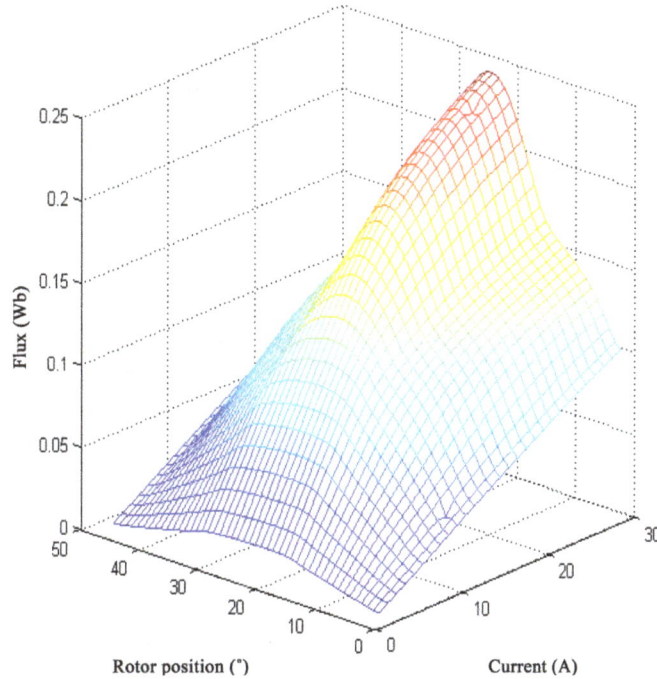

Figure 21. Flux linkage in 3-D.

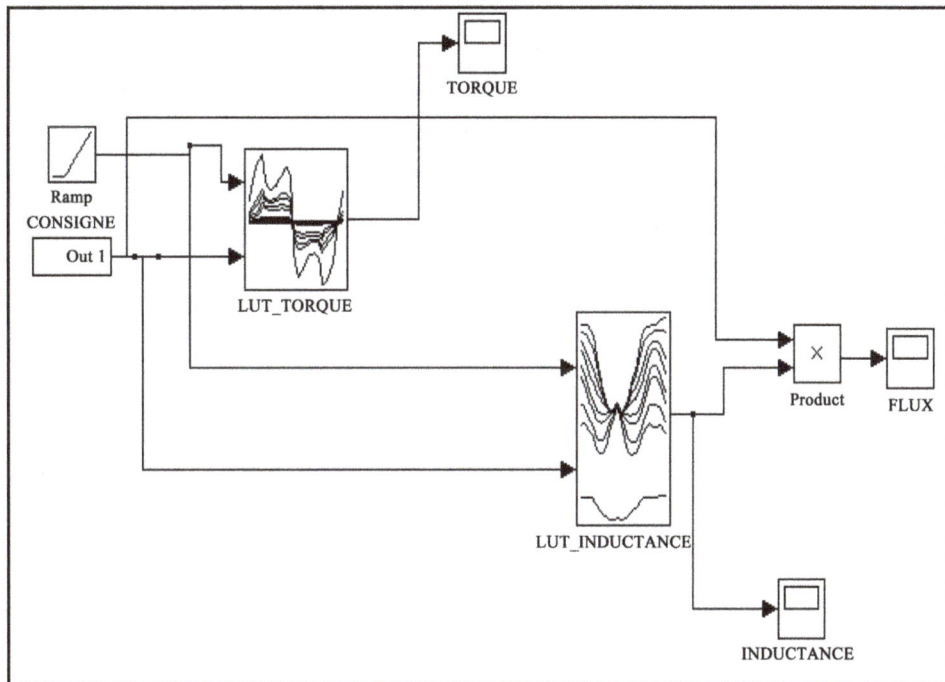

Figure 22. A one-phase inductance calculation.

The use of constant current in our simulation was a way to explain simply how to obtain the electromagnetic characteristics in general case. Although in our application we will use a 3 phase system of current or voltage to predict the dynamic behavior of the switched reluctance machine in generator and starter mode. Next, through an exciting of the three phases with a 3-system of current (see **Figure 23**) we recuperate the total torque in **Figure 24** which represent the sum of every phase torque excited separately, the flux linkage and the current with a maximum current equal to 300 A in **Figure 25**.

7. Conclusions

It is essential to have a suitable accurate model of the SRM that describes its static characteristics. In literature, several different methods of modeling the static are developed. All these models consider the variation of either the phase flux linkage or the phase inductance with rotor position accounting for magnetic saturation. The process of predicting static or even dynamic behavior of a machine needs first a structured methodology that begins with fixing the machine parameters through looking for the factors of impact (geometry, materials). Besides, the difficulty of studying a SRM is its highly nonlinear characteristics, which must be considered into modeling. However, to obtain a high quality control in either torque or speed control applications, it is necessary to have an accurate model of the machine that describes the static characteristics.

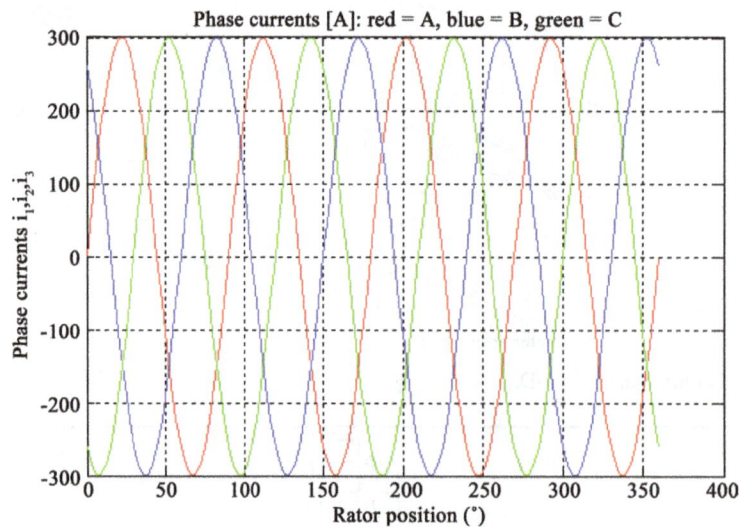

Figure 23. A 3-alimentation of the 12/8 SRM.

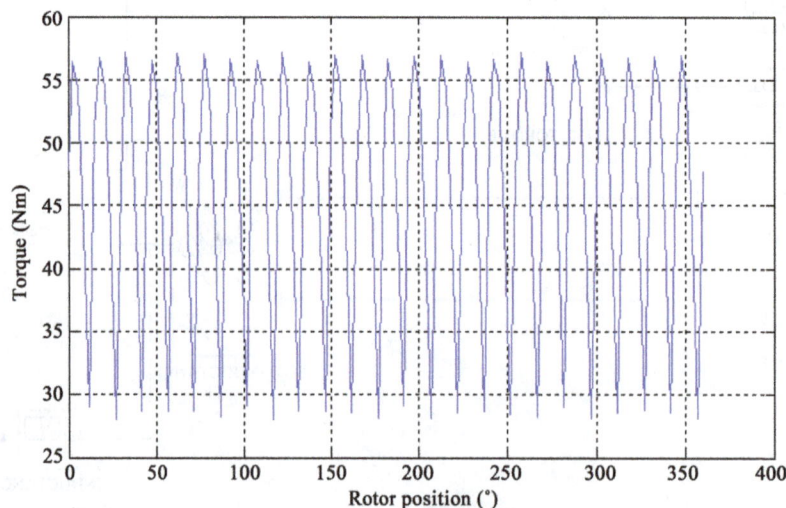

Figure 24. The total torque.

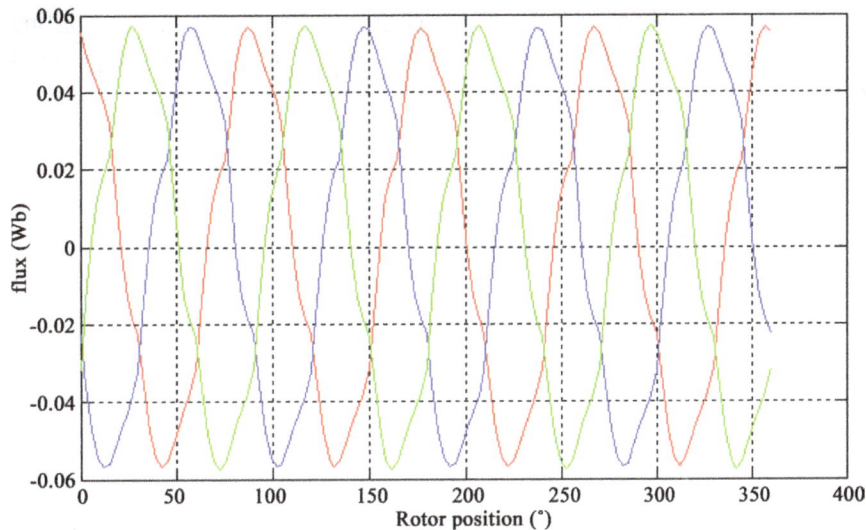

Figure 25. The flux of the tree phases.

Thus, in this paper, three principle approaches of modeling have been developed namely numerical, analytical and intelligent methods. The numerical one which has been chosen is developed in this paper. The FEMM, coupled to MATLAB for the conception, is a software used to draw a 2Da 12/8 3-phase SRM. The simulation for different rotor positions makes us prepare the flux-linkage curve and then is used to extract the other performance (static torque, inductance, co-energy). Then, MATLAB/Simulink is used as a package software to obtain the look-up table developed in form of block in Simulink models to calculate different SRM parameters as co-energy, inductance. The modeling technique LUT is a very simple and rapid approach in implementation and structure and establishes an excellent link between finite element method (represented here in FEMM) and dynamic simulation (with Simulink).

The dynamic model of one phase and typically the other stator phases can be prepared easily using static characteristics and enable us in the next work to develop a control strategy in a Simulink/MATLAB of the three phases of a SRM and then to optimize its characteristics in mode of the alternator and generator in order to avoid the acoustic noise produced by stator vibrations in hybrid vehicle operating mode.

References

[1]   Anderman, M. (2013) Assessing the Future of Hybrid and Electric Vehicles: Based on Private Onsite Interviews with Leading Technologists and Executives. The xEV Industry Insider Report.

[2]   Nagrial, M., Rizk, J. and Aljaism, W. (2010) Dynamic Simulation of Switched Reluctance Motor Using Matlab and Fuzzy Logic. *Proceedings of the* 14*th International Middle East Power Systems Conference* (*MEPCON'*10), Cairo, 19-21 December 2010, 819-824.

[3]   Stephen, J.E., Kumar, S.S. and Jayakumar, J. (2014) Nonlinear Modeling of a Switched Reluctance Motor Using LSSVM—ABC. *Acta Polytechnica Hungarica*, **11**, 143-158.

[4]   Tormey, D.P., Torrey, D.A. and Levin, P.L. (1990) Minimum Airgap-Permeance Data for the Doubly-Slotted Pole Structures Common in Variable-Reluctance-Motors. *Industry Applications Society Annual Meeting*, 1990, *Conference Record of the* 1990 *IEEE*, Seattle, 7-12 October 1990, 196-200.

[5]   Bal, G. and Uygun, D. (2010) An Approach to Obtain an Advisable Ratio between Stator and Rotor Tooth Widths in Switched Reluctance Motors for Higher Torque and Smoother Output Power Profile. *Gazi University Journal of Science* (*GU J SCI*), **23**, 457-463.

[6]   Vijayakumar, K., Karthikeyan, R. and Arumugam, R. (2008) Influence of Soft Magnetic Composite Material on the Electromagnetic Torque Characteristics of Switched Reluctance Motor. *IEEE Power India Conference on Power System Technology*, 12-15 October 2008.

[7]   Chi, H.P. (2005) Flux-Linkage Based Models for Switched-Reluctance Motors. PHD Philosophy, National Cheng Kung University, Tainan.

[8]   Rebahi, F., Bentounsi, A., Lebsir, A. and Benamimour, T. (2014) Soft Magnetic Materials for Switched Reluctance

Machine: Finite Element Analysis and Perspective. http://ieeexplore.ieee.org/stamp/stamp.jsp?arnumber=7076953

[9]   Sugawara, Y. and Akatsu, K. (2013) Characteristics of Switched Reluctance Motor Using Grain-Oriented Electric Steel Sheet. 2013 *IEEE ECCE Asia*, Melbourne, 3-6 June 2013, 1105-1110.

Appendix A

FEMM

Finite Element Method Magnetics (FEMM) is a finite element package for solving 2D planar and ax symmetric problems in low frequency magnetic and electrostatics. The program runs under runs under all versions Windows and XP. The Finite Element Method Magnetics (FEMM) is a program that can solve electromagnetic problems on 2D (two-dimensional planar). Thus, we can resolve linear and nonlinear magnetostatic, harmonic magnetic, linear electrostatic and steady state heat problems. The determination of magnetic characteristics facilitates the optimization and control of switched reluctance machine.

Appendix B

12/8 SRM Dimensions

The dimensions of the switched reluctance motor are next in **Table B1**:

Table B1. Geometry of the 12/8 SVRM.

	SR machine model
Stator outer diameter	137 (mm)
Stator inner diameter	85 (mm)
Stator teeth length	22 (mm)
Rotor outer diameter	42.48 (mm)
Air gap	0.325 (mm)
Number of turns	12
Stator pole arc	30(°)
Rotor pole arc	45(°)
Rotor teeth length	18.02 (mm)

14

Performance Analysis of Grid Connected and Islanded Modes of AC/DC Microgrid for Residential Home Cluster

Ahmed M. Othman[1,2], Hossam A. Gabbar[1,3*], Negar Honarmand[1]

[1]Faculty of Engineering and Applied Science, University of Ontario Institute of Technology, Oshawa, Canada
[2]Faculty of Engineering, Electrical Power & Machine Department, Zagazig University, Zagazig, Egypt
[3]Faculty of Energy Systems & Nuclear Science, University of Ontario Institute of Technology, Oshawa, Canada
Email: ahmed_othman80@yahoo.com, *hossam.gabbar@uoit.ca, negar.honarmand@uoit.ca

Abstract

This paper presents performance analysis on hybrid AC/DC microgrid networks for residential home cluster. The design of the proposed microgrid includes comprehensive types of Distributed Generators (DGs) as hybrid power sources (wind, Photovoltaic (PV) solar cell, battery, fuel cell). Details about each DG dynamic modeling are presented and discussed. The customers in home cluster can be connected in both of the operating modes: islanded to the microgrid or connected to utility grid. Each DG has appended control system with its modeling that will be discussed to control DG performance. The wind turbine will be controlled by AC control system within three sub-control systems: 1) speed regulator and pitch control, 2) rotor side converter control, and 3) grid side converter control. The AC control structure is based on PLL, current regulator and voltage booster converter with using of photovoltaic Voltage Source Converter (VSC) and inverters to connect to the grid. The DC control system is mainly based on Maximum Power Point Tracking (MPPT) controller and boost converter connected to the PV array block and in order to control the system. The case study is used to analyze the performance of the proposed microgrid. The buses voltages, active power and reactive power responses are presented in both of grid-connected and islanded modes. In addition, the power factor, Total Harmonic Distortion (THD) and modulation index are calculated.

Keywords

Microgrid, Photovoltaic Systems, Wind Power Generation, Hybrid AC/DC Networks

---

*Corresponding author.

1. Introduction

Smart Grid will be the future electricity distribution system. This intelligent system consists of advanced digital meters, distribution automation, communication systems and distributed energy resources [1]. Self-healing, high reliability and power quality, providing accommodations to a wide variety of distributed generation and storage options are some of the functionalities for a desired Smart Grid [2]. If photovoltaic generations, fuel cells, wind turbines and gas cogenerations are installed into utility grids directly then they can cause a variety of problems such as voltage rise and protection problem in the utility grid. In order to avoid these problems, the new concept in power system has been introduced and that is called microgrid [3] [4]. Renewable "Distributed Energy Resources" (DERs) can consist of small Photovoltaic (PV) generators and small wind turbines that can be installed anywhere such as customers' place. Microgrids consist of DER, including Distributed Generation (DG) and Distributed Storage (DS). In disasters, current distribution systems can face challenges to provide the required energy supply. Using the proposed microgrid in parallel with the grid, the distribution system can recover faster. Microgrids have the ability to be switched in and out of the transmission system. They can also operate independently from the system for a period of time. Therefore, microgrids can be either in grid-connected mode or islanding mode. Because of their ability to operate in islanding mode, main use of microgrid can be providing power in an emergency to the residential community. The use of microgrids can improve power delivery and it allows utilities grid to deliver power in urban areas. Microgrid can be connected to the main grid single Point of Common Coupling (PCC). Microgrids can work islanded or grid-connected. Islanding means that the microgrid continues to operate independently when disconnected from the grid. Microgrid is easily islanded by opening the circuit breaker at PCC. When islanded, microgrid should be able to supply the power to its loads without disruption. Microgrid should have the ability to resynchronize with grid when the condition caused islanding has been corrected [5]-[11].

There are many recent control methods for nonlinear-feedback control of power systems with application of power electronics. The dynamic model of voltage source converters (VSC) is a nonlinear one and VSCs enable connection of distributed power generation units to the grid. One of recent control methods in ref. [12] will depend on an H-infinity control problem for the voltage source converter that makes use of a locally linearized model of the converter. The H-infinity control will enable to compensate for the linearization errors, and also to eliminate the effects of external perturbations. The current paper will concern with PV controlling and Wind Turbine (WT) controlling. The proposed control action concerns with the PI portion to find optimal gain settings those dynamically minimize the error value between the reference value and the feedback one. More details will be shown in control design section.

The emphasis of the paper is to have a comprehensive modeling and analysis of the microgrid with both AC and DC operation. And in the same moment, the application of various control strategies for DC side (represented in Maximum Power Point Tracking (MPPT) for PV) and AC side (represented in Pitch control and rotor side converter for wind turbine).

2. Microgrid for Home Cluster

Figure 1 shows how a microgrid can be connected to a home cluster. Home cluster is the combination of dif-

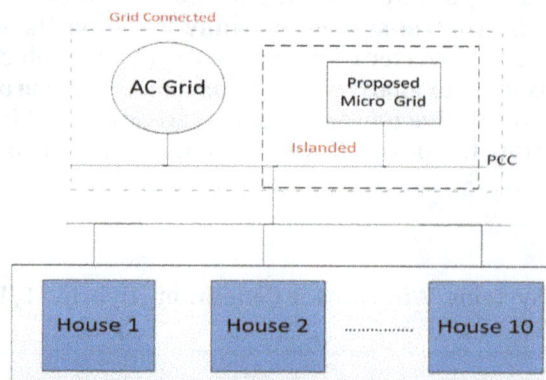

Figure 1. Microgrid for home cluster.

ferent houses together that can share the power between each other. The figure shows that the microgrid is connected to the AC line. The microgrid can be connected to the DC line as well but since the houses are connected to the AC line, for the simplicity this figure only shows the AC line. In the proposed home cluster, there are ten houses that are chosen in a small community for this project. These houses are located in Ontario, Canada. The average consumed energy for a household in Ontario is 11,221 KWh. These houses can be considered as one load. The microgrid can be connected at the main feeder from the utility and tied into the distribution system to the individual homes in a way that is approved by the utility. All microgrids have the ability to disconnect from the grid and they can operate on their own for a period of time. Being able to operate in the islanding mode is one of the key features of microgrids. The power loss for an extended period can have negative effects on the economy. Therefore, microgrids can have an important role in the power system. They can distribute power in consumption loads and they can improve power quality to the main grid. Because of their ability to operate in islanding mode, the main use of microgrid can be providing power to the residential community.

3. Description of Designed Microgrid for Home Cluster

The residential Microgrid (MG) system consists of Grid, protective relays, and control systems. The system is modeled using the MATLAB/Simulink SimPower Systems toolbox. Bus 1 is connected to the grid and Bus 2 is connected to AC distributed energy sources. Bus 3 is connected to DC distributed energy sources. The proposed hybrid MG consists of PV, wind turbine (WT), Fuel Cell (FC), Battery, Micro Gas Turbine (MGT), AC loads, AC distribution lines, DC distribution lines, DC loads and DC-AC-DC converters. The energy that is produced by DGs is stored in the battery. **Figure 2** shows the design of the microgrid. 10 houses are considered to be the load for this project. A single phase dynamic load is used as the load in the simulations.

The optimal mix and control strategies for operation of Wind and PV are considered for both DC-AC and AC-AC interface at different voltage levels with the Utility. The DC and AC key Interface buses are connected to different types of DC and AC loads like: resistance loads, DC motor load, AC loads, dynamic AC loads and three phase motor loads, the data of them appears in the appendix.

The proposed control design which operated with PV is installed and selected according to the connected bus. Proposed design will be connected to the AC side with Wind Turbine whereas another proposed design will be connected to the DC side with PV.

The rating for each proposed control design is according to the rated voltage and total current of the connected bus. Two proposed for both DC and AC sides is required for local controls for each DG unit. Each one can control its self DG, so it adapts the performance of one DG. The local control of PV, for example, may produce signal which is complementary to the signal of fuel cell local control.

4. Modeling of Microgrid Component

4.1. Wind Turbine

The mechanical power Pm captured by the blades of a wind turbine is defined in Equation (1).

$$P = 1/2 \cdot Cp\left(\beta, \xi\right) \rho \pi R_m^2 \cdot \left(V_{wind}\right)^3 \tag{1}$$

where Cp is a rotor power coefficient, β is a blade pitch angle, ξ is a tip-speed ratio (TSR), ρ is an air density, R_m is the radius of a wind turbine blade and V_{wind} is a wind speed [13].

The mathematical models of a Double-Feed Induction Generator (DFIG) are essential requirements for its control system. The voltage equations of an induction motor in a rotating-coordinate are shown in Equation (2) and Equation (3).

$$\begin{bmatrix} u_{ds} \\ u_{qs} \\ u_{dr} \\ u_{qr} \end{bmatrix} = \begin{bmatrix} -R_s & 0 & 0 & 0 \\ 0 & -R_s & 0 & 0 \\ 0 & 0 & R_r & 0 \\ 0 & 0 & 0 & R_r \end{bmatrix} \begin{bmatrix} i_{ds} \\ i_{qs} \\ i_{dr} \\ i_{qr} \end{bmatrix} + \rho \begin{bmatrix} \lambda_{ds} \\ \lambda_{qs} \\ \lambda_{dr} \\ \lambda_{qr} \end{bmatrix} + \begin{bmatrix} -\omega_1 \lambda_{qs} \\ \omega_1 \lambda_{ds} \\ -\omega_2 \lambda_{qr} \\ \omega_2 \lambda_{dr} \end{bmatrix} \tag{2}$$

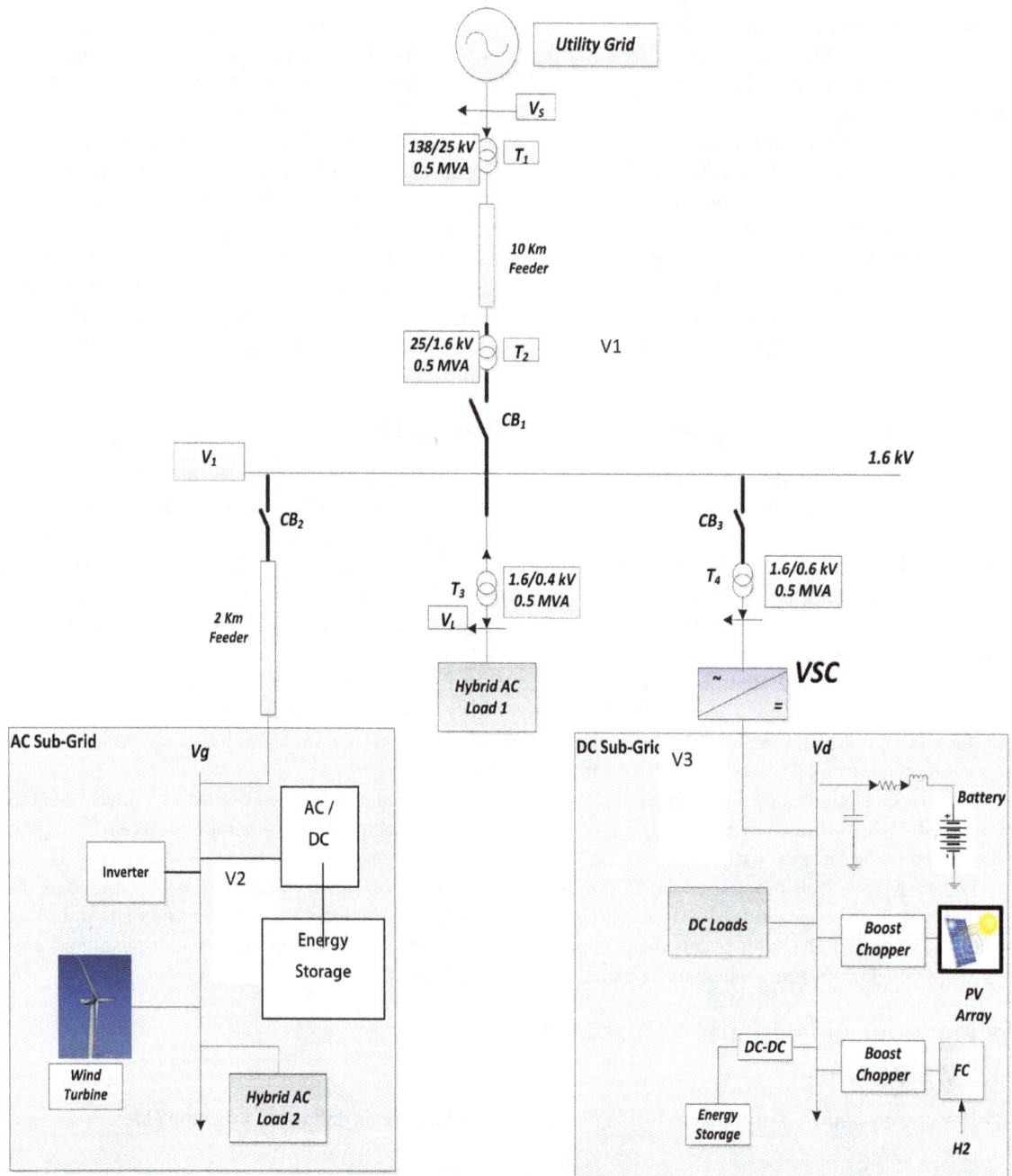

Figure 2. Proposed microgrid design.

$$\begin{bmatrix} \lambda_{ds} \\ \lambda_{qs} \\ \lambda_{dr} \\ \lambda_{qr} \end{bmatrix} = \begin{bmatrix} -L_s & 0 & L_m & 0 \\ 0 & -L_s & 0 & L_m \\ -L_m & 0 & L_r & 0 \\ 0 & -L_m & 0 & L_r \end{bmatrix} \begin{bmatrix} i_{ds} \\ i_{qs} \\ i_{dr} \\ i_{qr} \end{bmatrix} \qquad (3)$$

The dynamic equation of the DFIG is shown in Equation (4) and Equation (5).

$$\frac{J}{n_p} \frac{d\omega_r}{dt} = T_m - T_{em} \qquad (4)$$

$$T_{em} = n_p L_m \left(i_{qs} i_{dr} - i_{ds} i_{qr} \right) \tag{5}$$

The detailed modeling of the DFIG is presented by applying d-q reference frame along the reference frame is rotating with the same speed as the stator voltage. The stator and rotor voltages with the flux variables can be written as follows:

$$\frac{d}{dt} \lambda_{ds} = -R_s i_{ds} - \omega_1 \lambda_{qs} + u_{ds}, \quad \frac{d}{dt} \lambda_{qs} = -R_s i_{qs} + \omega_1 \lambda_{ds} + u_{qs} \tag{6}$$

$$\frac{d}{dt} \lambda_{dr} = -R_r i_{ds} - s\omega_1 \lambda_{qr} + u_{dr}, \quad \frac{d}{dt} \lambda_{qr} = -R_r i_{qs} + s\omega_1 \lambda_{dr} + u_{qr} \tag{7}$$

$$\frac{d}{dt} \omega_r = -\omega_s \cdot dS/dt \tag{8}$$

where the subscripts d, q, s and r represent d-axis, q-axis, stator, and rotor respectively. L is the inductance and λ is the flux linkage. u is the voltage and i is the current. ω_1 represents the angular synchronous speed and ω_2 is slip speed. T_m is the mechanical torque and T_{em} is the electromagnetic torque [14].

In order to have an effective control system of a wind turbine, the following criteria must be met:

1) The wind power must be captured as much as possible,
2) Power quality standards such as power factor and harmonics should be met and
3) Must be able to transfer the electrical power to the grid for different wind velocities [15].

Aerodynamic control, variable speed control, and grid connection control are the three subsystems of the control system. Pitch control is used to control the aerodynamics drive train. In addition, variable speed control is used to control the electromagnetic subsystem. Grid connection subsystem is controlled by output power conditioning [16] (**Figure 3**).

There are many recent references that are presenting comparative study of PI control action and feedback linearization control on DFIG. Those references confirm the performance of the PI controller at different operating conditions. PI control of DFIG-Based Wind Farm can take an active part to enhance the voltage control and other aspects in the system [17]-[19].

4.2. Wind Turbine Control

4.2.1. Rotor Side Controller Model

The objective of rotor-side converter controller is to govern the DFIG output real power, and to keep controlling the terminal voltage. The active power and voltage are controlled independently via u_{qr} and u_{dr}, respectively. **Figure 4** represents the blocks of the rotor side control.

The rotor side controller section has four states: $[x_1, x_2, x_3, x_4]$. x_1 represents the comparison between the stator supplied power and the target reference power, x_2 represents the comparison between the rotor current of q-axis and the related reference current. x_3 represents the comparison between the stator terminal voltage and the required reference voltage, and x_4 compares between the rotor current of d-axis and its reference current. The state equations can be represented as below:

$$\frac{d}{dt} x_1 = -\left(K_{i1}/K_{p1} \right) \cdot x_1 + \left(1/K_{p1} \right) \cdot i_{qr\ ref} \tag{9}$$

$$\frac{d}{dt} x_2 = K_{p1} \left(P_{ref} + P_s \right) + K_{i1} \cdot x_1 - i_{qr} \tag{10}$$

$$\frac{d}{dt} x_3 = \left(-K_{i3}/K_{p3} \right) \cdot x_3 + \left(1/K_{p3} \right) \cdot i_{dr\ ref} \tag{11}$$

$$\frac{d}{dt} x_4 = K_{p3} \left(v_{s_{ref}} - v_s \right) + K_{i3} \cdot x_3 - i_{dr} \tag{12}$$

where K_{p1} and K_{i1}: power regulator proportional and integrating gains; K_{p2} and K_{i2}: current regulator proportional and integrating gains; K_{p3} and K_{i3}: voltage regulator proportional and integrating gains; i_{dr_ref} and i_{qr_ref}: d and q axis references of current control; v_{s_ref}: reference of terminal voltage; and P_{ref}: reference of active power control.

4.2.2. Grid Side Controller Model

The objective of grid side controller is to keep controlling the DC coupled voltage, and the reactive power. **Figure 5** represents the model of grid side control.

The grid side controller section has three states: $[x_5, x_6, x_7]$. x_5 controls to the error between the DC voltage and its required reference. x_6 and x_7 controls to the error between the current in d- and q-axis and their reference values.

$$\frac{d}{dt}x_5 = v_{DC\_ref} - v_{DC} \tag{13}$$

$$\frac{d}{dt}x_6 = -K_{pdg} \cdot \Delta v_{DC} + K_{Idg} \cdot x_5 - i_{dg} \tag{14}$$

$$\frac{d}{dt}x_7 = i_{qg\_ref} - i_{qg} \tag{15}$$

where K_{pdg} and K_{idg}: DC-voltage regulator proportional and integrating gains, K_{pg} and K_{ig}: current regulator proportional and integrating gains, v_{DC_ref}: reference of DC coupled voltage, i_{qg_ref}: reference of q-axis current.

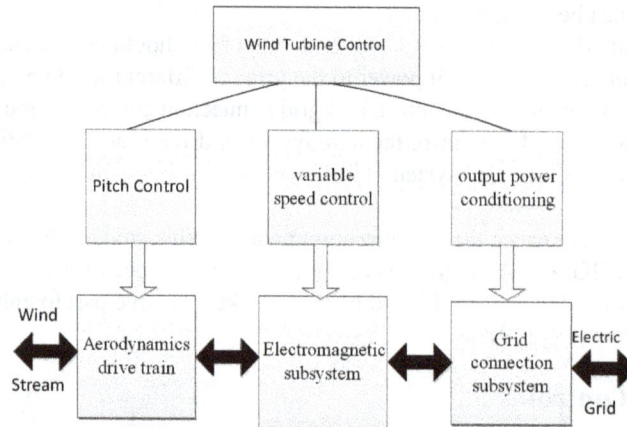

Figure 3. Wind turbine control architecture.

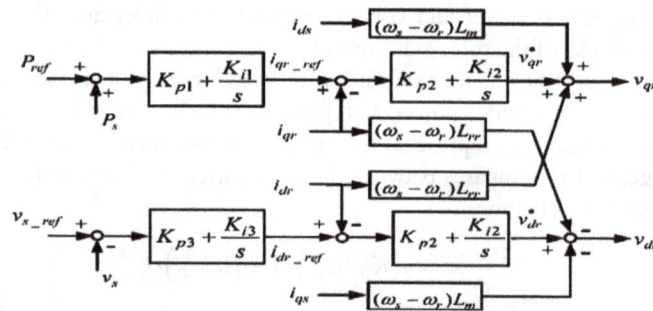

Figure 4. Rotor side controller model.

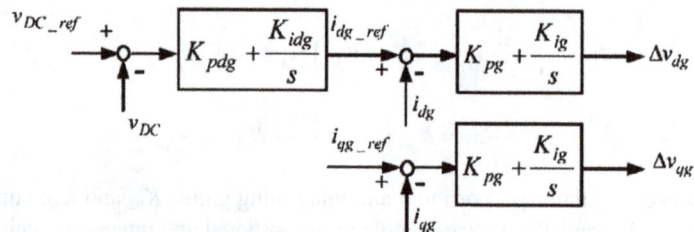

Figure 5. Grid side controller model.

4.2.3. Pitch Control Model

The objective of pitch control is to keep the wind turbine speed in the optimal zone; it is characterized as shown in **Figure 6**.

$$\frac{d}{dt}\beta = K_{i4} \cdot \Delta\omega_t - K_{p4} \cdot \frac{d}{dt}\Delta\omega_t \tag{16}$$

In order to control the wind turbine, the control system is designed. The control system has three sub control systems. The first one is side converter control. The next one is grid side converter control rotor and the last one is speed regulator and pitch control.

The control system consists of $V_{dc\text{-}ref}$, V_{dc}, I_d (the current in d-axis), $I_{d\text{-}ref}$, I_q (the current in q-axis), $I_{q\text{-}ref}$, $V_{d\text{-}ref}$, $V_{q\text{-}ref}$, I_{dr} (the rotor current in d-axis), $I_{dr\text{-}ref}$, I_{qr} (the rotor current in q-axis), $I_{qr\text{-}ref}$, dq to abc transformation block, PWM generator and PI controller. These control systems were designed in MATLAB in order to control the wind turbine in the microgrid.

From the rotor side of the DFIG, the turbine speed reference signal is taken while the active power P is taken from the stator side. Both of them will be shared in the optimization search to share the error value between the reference and the actual to activate the control action of the q axis reference currents control. The same will be done with regard to the system power factor control, the reactive power Q (both from the stator and the rotor sides) and the direct axis currents control.

The selection of the references values: turbine speed reference value w_ref and the active power reference value P_ref is set by estimating the required value to be stabilized in. It is adapted according to the rated value of the speed and the power.

The proposed controller is applied to find optimal gain settings those dynamically minimizes the error value between the reference value and the feedback one. There control strategy for each one will have four variables: VS and IS are the voltage and current of the input signals at the bus connected to the controller whereas VL and IL are the voltage and current of the output signals at the bus connected to the controller. The error signal will be sum of loops for: Load Voltage Stabilization, RMS-Current Minimization, and Dynamic Damping Loop for power oscillations and Load Ripple Current Damping.

$$e_{V_L} = V_{Lref_{p.u.}} - V_{L_{p.u.}}\left(\frac{1}{1+ST_1}\right), \; e_{I_L} = \left(I_{L_{p.u.}} - I_{L_{p.u.}}\left(\frac{1}{1+ST_2}\right)\right) \tag{17}$$

The pattern search optimization algorithm, in MATLAB platform, is implemented for tuning PID controller gains KP, KI and KD.

There are some regulators; those will adapt the operation of the loops. Voltage regulator is applied to regulate voltage by keeping and measuring the load voltage to near unity. Also, current regulator is applied to face any sudden current variation and to reduce oscillations.

4.3. PV Panel

The following Equations (6)-(9) are used in order to model the PV panel [14] (**Figure 7, Table 1**).

$$V_{OC} = \frac{nkT}{q} \times \ln\left(\frac{I_L}{I_O}+1\right) \tag{18}$$

$$I_{pv} = n_p I_{ph} - n_p I_{sat} \times \left[\exp\left(\left(\frac{q}{AKT}\right)\left(\frac{V_{pv}}{n_s}+I_{pv}R_s\right)\right)-1\right] \tag{19}$$

$$I_{ph} = \left(I_{sso} + k_i\left(T-T_r\right)\right) \cdot \frac{S}{1000} \tag{20}$$

$$I_{sat} = I_{rr}\left(\frac{T}{T_r}\right)^3 \exp\left(\left(\frac{qE_{gap}}{kA}\right)\cdot\left(\frac{1}{T_r}-\frac{1}{T}\right)\right) \tag{21}$$

A boost converter is applied to step up and convert the voltage of the PV module. The boost converter is shown in **Figure 8**, the output voltage is defined by the relation $V_o = V_{in}/(1-K)$, where K is duty cycle.

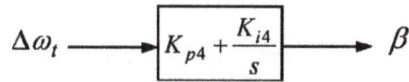

Figure 6. Pitch control model.

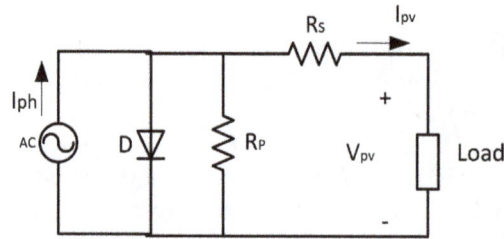

Figure 7. Equivalent circuit of a solar cell [13].

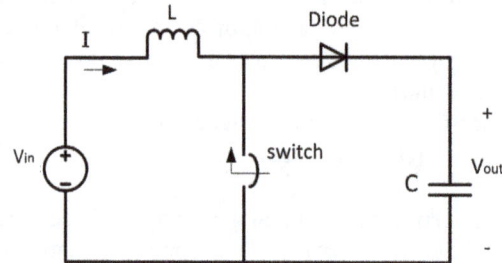

Figure 8. Boost converter equivalent circuit.

Table 1. Parameters for photovoltaic panel.

Symbol	Description
V_{oc}	Rated open circuit voltage
I_{ph}	Photocurrent
I_{sat}	Module reverse saturation current
q	Electron charge
A	Ideality factor
k	Boltzman constant
R_s	Series resistance of a PV cell
R_p	Parallel resistance of a PV cell
I_{sso}	Short circuit current
k_i	SC current temperature coefficient
T_r	Reference temperature
I_{rr}	Reverse saturation current at Tr
E_{gap}	Energy of the band gap for silicon
n_p	Number of cells in parallel
n_s	Number of cells in series
S	Solar radiation level
T	Surface temperature of the PV
I_O	Dark saturation current
I_L	Light generated current
n	Ideality factor

The state matrix have $x_1 = i_L$, $x_2 = V_c$ with $u = V_{in}$; and can be described by:

$$\frac{\mathrm{d}}{\mathrm{d}t}x_1 = \left[-(1+K)/L\right] \cdot x_2 + (1/L) \cdot u$$

$$\frac{\mathrm{d}}{\mathrm{d}t}x_2 = \left[-(1-K)/C\right] \cdot x_1 - (1/RC) \cdot x_2$$

(22)

PV Module: MPP Tracking

The output power of the PV is calculated by $P = VI$. The voltage of the PV and the current are represented by V and I respectively. In order to get the maximum output power point the conventional MPPT algorithms use dv/dp. The reference voltage is increased or decreased according to the operation region which is determined by measured ΔP and ΔV [20]. **Figure 9** shows the control system of the *PV* module. I_{pv} and V_{pv} are the inputs of the MPPT and the output is V_{ref}. The error will be the difference between V_{pv} and V_{ref} and that is the output of PWM.

There are many methods to get the MPP. These methods have various criteria like effectiveness, complexity, cost and others. The Perturb and Observe (P & O) is the most common method due to its ease of implementation. In the P & O algorithm, the operating voltage of the PV array is perturbed by a small increment, and the corresponding change in power, ΔP, is measured. If ΔP is positive, then the perturbation of the operating voltage transfer the PV array's operating point near to the MPP. Thus, further voltage perturbations in the same direction (that is, with the same algebraic sign) should transfer the operating point directed to the MPP.

An inverter is used on the AC side of the microgrid. In order to control the inverter, the control system is designed. **Figure 10** shows the control system of the inverter. The control system consists of PLL, Park Transformation block, DC voltage regulator, current regulator and a PWM generator. **Figure 3** shows the inverter control. In addition, AC/DC converter is used in order to connect the AC side to the DC side. A boost converter is used to connect to the PV array block and boost the voltage up to the VSC converter which then connects to the inverter. In addition, a MPPT controller is also designed to try and optimize the PV power output.

The control objective of the boost converter in grid connected mode is tracking the MPPT of the PV array. The boost converter regulates the PV terminal voltage. The control objective of back-to-back AC/DC/AC of the DFIG is regulating rotor side current. In order to achieve MPPT and to synchronize with AC grid, the rotor side needs to be regulated. In grid connected mode, the battery does not a significant role since the power is regulated and balanced by the utility grid.

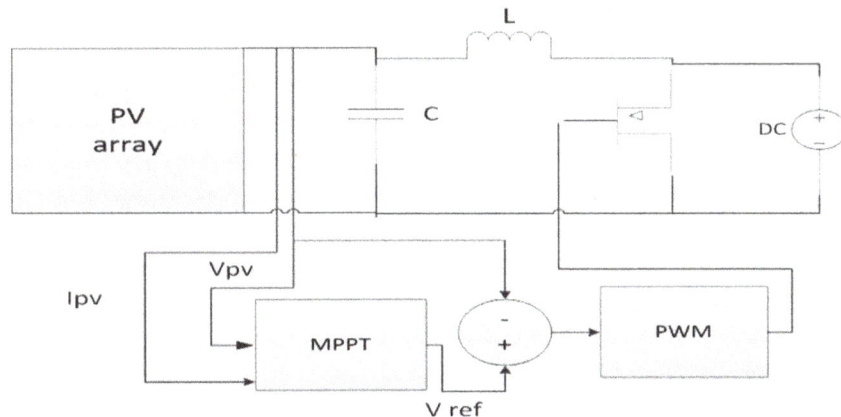

Figure 9. PV module: MPP tracking.

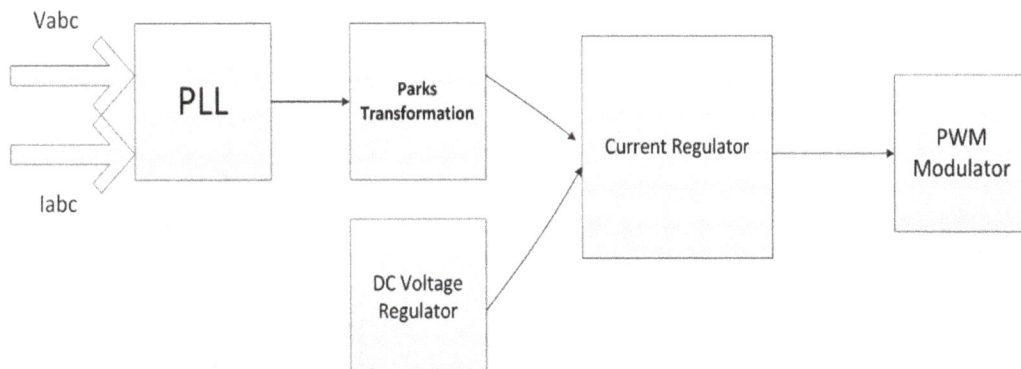

Figure 10. Inverter control.

There are some regulators; those will adapt the operation of the loops. Voltage regulator is applied to regulate voltage by keeping and measuring the load voltage to near unity. Also, current regulator is applied to face any sudden current variation and to reduce oscillations.

According to Xiong Liu, Peng Wang and Poh Chiang Loh when the microgrid is in islanding mode, based on system power balance, the boost converter and the back-to-back AC/DC/AC converter of the DFIG can operate either MPPT or off-MPPT. The main converter behaves as a voltage source in order to provide a stable voltage and frequency for the AC grid. It can operate either in inverter or converter mode. By acting as either an inverter or converter, the power exchange between the AC and DC link can be smoother. Based on power balance in the system, the battery converter can be in charging or discharging mode. System operating condition can determine the DC-link voltage and the voltage can be maintained by either the battery or the boost converter.

4.4. Battery

The state-space model of the battery can be represented based on the equivalent circuit that is shown in **Figure 11**.

C_{bk} is a bulk capacitor, which represents the element of charging energy storage. $C_{surface}$ represents the diffusion and surface capacitance, where R_s is the surface resistance. R_e is the end resistance and R_t is the terminal surface. V_{Cb} and V_{Cs} represents the voltages across the capacitors, respectively.

The state-space model of the battery is presented in [14], and can be summarized as follows:

$$\begin{bmatrix} \dfrac{\mathrm{d}}{\mathrm{d}t}V_{Cb} \\ \dfrac{\mathrm{d}}{\mathrm{d}t}V_{Cs} \\ \dfrac{\mathrm{d}}{\mathrm{d}t}V_o \\ \dfrac{\mathrm{d}}{\mathrm{d}t}V_{bk} \end{bmatrix} = \begin{bmatrix} -A & A & 0 & 0 \\ B & -B & 0 & 0 \\ (-A+B) & 0 & (A-B) & 0 \\ 0 & 0 & 0 & 0 \end{bmatrix} \begin{bmatrix} V_{Cb} \\ V_{Cs} \\ V_o \\ V_{bk} \end{bmatrix} + \begin{bmatrix} A \cdot R_s \\ B \cdot R_e \\ A(0.5R_s - R_t - D) + B(0.5R_e + R_t + D) \\ 0 \end{bmatrix} \tag{23}$$

where

$$A = 1/\left[C_{bk}\left(R_e + R_s \right) \right], \quad B = 1/\left[C_{surface}\left(R_e + R_s \right) \right], \quad D = R_e R_s / \left(R_e + R_s \right) \tag{24}$$

4.5. Fuel Cell

Fuel cells run on hydrogen, which can be derived from ethanol, methane, propane or natural gas. Industrially produced pure hydrogen can run a Fuel Cell. Another way to run a fuel cell is using the hydrogen that is generated from water. In fact, the electrolysis process can decompose it to oxygen and hydrogen gas. In order to do this electrolytic process, solar or wind energy can be used to generate the power for electrolyzing. Because of their cleanness, high efficiency, and high reliability, they are used in DG [18] [20]-[23].

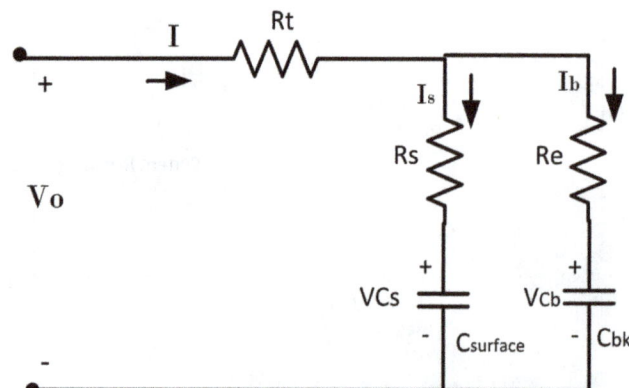

Figure 11. Equivalent circuit of a battery.

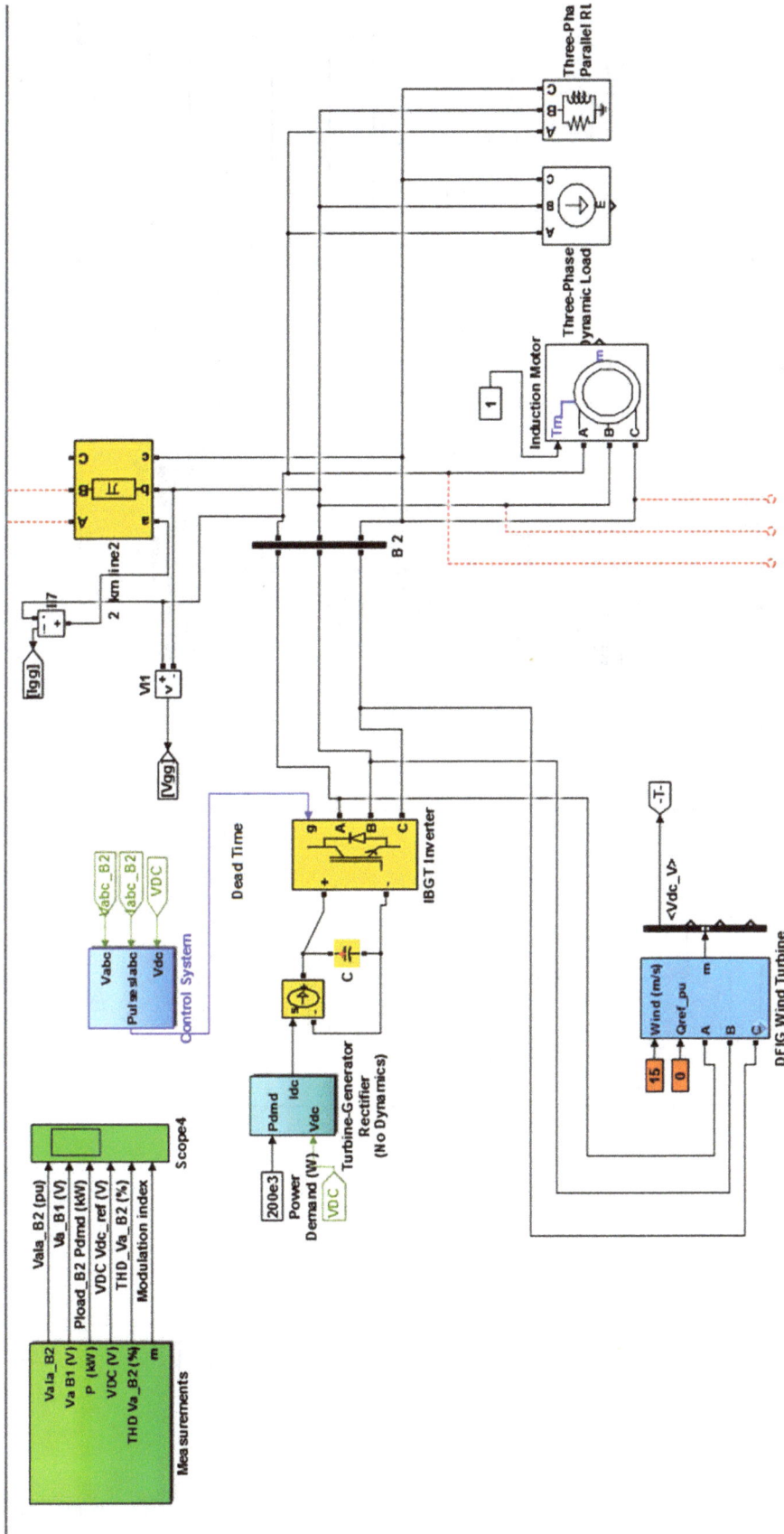

Figure 12. Simulink/ Simpower System (AC Portion).

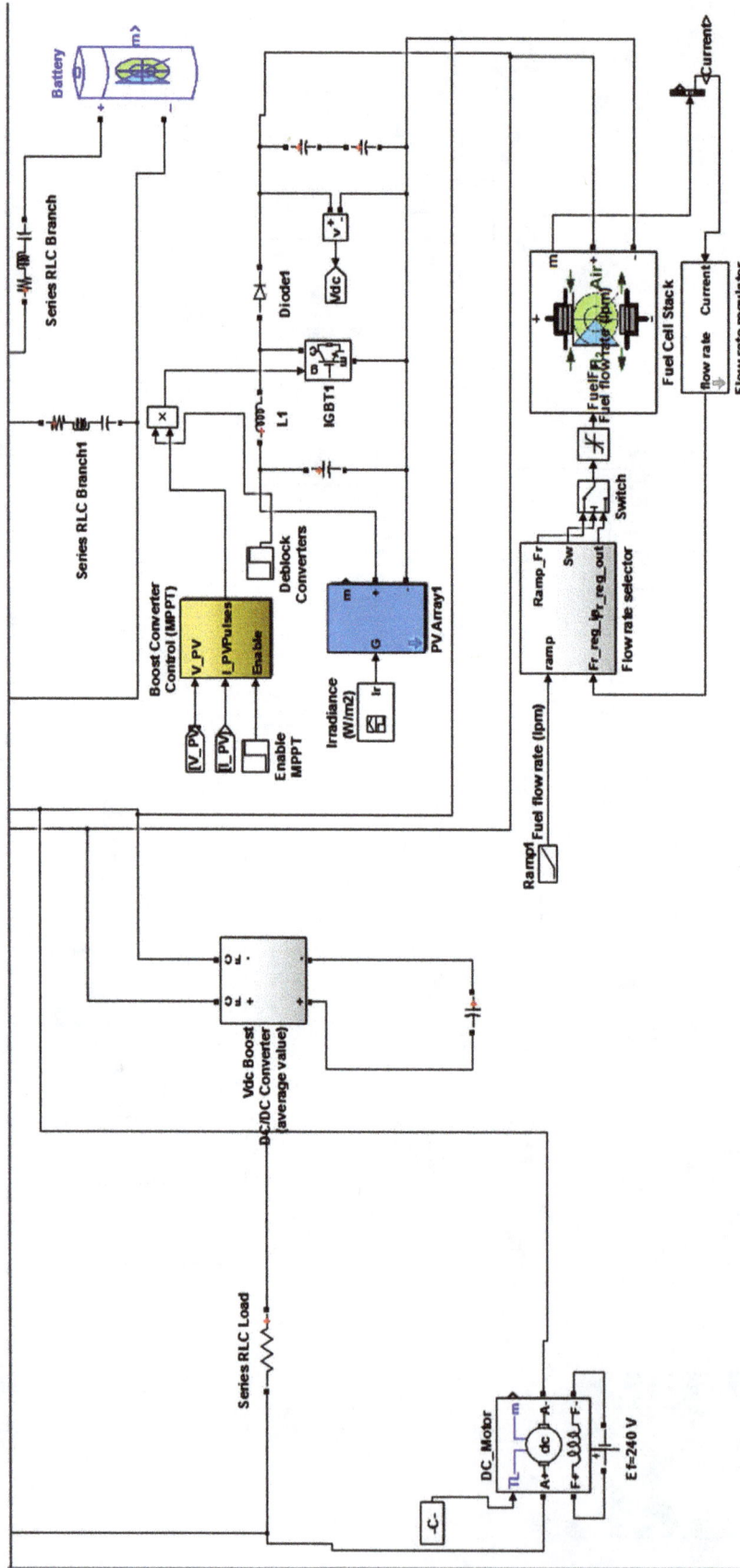

Figure 13. Simulink/ Simpower System (DC Portion).

The conditions for control method reliability are concerned with the microgrid size which can be considered as a function of reliability. Another condition is the ability of the system to supply the loading patterns without loss of its requirements as power and/or achieving certain level of the voltage profile. The dynamic response for the system output variables can be analyzed with respect to time specifications. The performance characteristics of a controlled system are specified in terms of the transient response. In specifying the transient response characteristic, it is common to specify the following terms: Delay time (td), Rise time (tr), Peak time (tp), Max over shoot (Mp) and Settling time (ts). Those parameters can be initial proof of the system stability. Another system plot can indicate the stability is the system Bode plot as frequency response for the system.

Many nonlinear control approaches consider the representation of the controller dynamics and voltage source converter dynamics in the dq reference frame and use PI compensators. Other control approaches work with input-output linearization, as well as input-state linearization to convert the nonlinear system to a decoupled linear one. Traditional control methods and adaptive AI control methods can be been presented and found in ref. [24] [25].

5. Microgrid Simulations and Analysis

The operations of this microgrid were investigated in two different modes. The first one is the Grid-Connected Mode and the second one is Islanded Mode. The simulation platform will be MATLAB/SIMULINK/SIMPOWER SYSTEM. Some figures are taken from the software used for the system modeling and simulation and are shown bellow in **Figures 11-13**.

5.1. Grid-Connected Mode (Case Study 1)

In this mode, the main converter operates in the PQ mode. The power is balanced by the utility grid. The battery is fully charged. AC bus voltage is maintained by the utility grid and DC bus voltage is maintained by the main converter. In this mode the voltage of the PV is 281.69 V and its power is 274.17 KW. **Figures 14-16** show the voltage in different buses when the microgrid is connected to the grid. **Figure 17** and **Figure 18** show the active and reactive power in the grid connected mode. The THD (%) at different buses is shown in **Table 3**. Power factor at different buses is shown in **Table 4**. The modulation index for this mode is 0.61.

To stress the effect of PI control, simulation figures, **Figure 19** and **Figure 20**, are used to compare between the dynamic response at different operating conditions with and without tuned-adapted PI control of DIFG local cascade control. It is clear that the change of the parameters of PI controller affects on the settlement and stability of the performance.

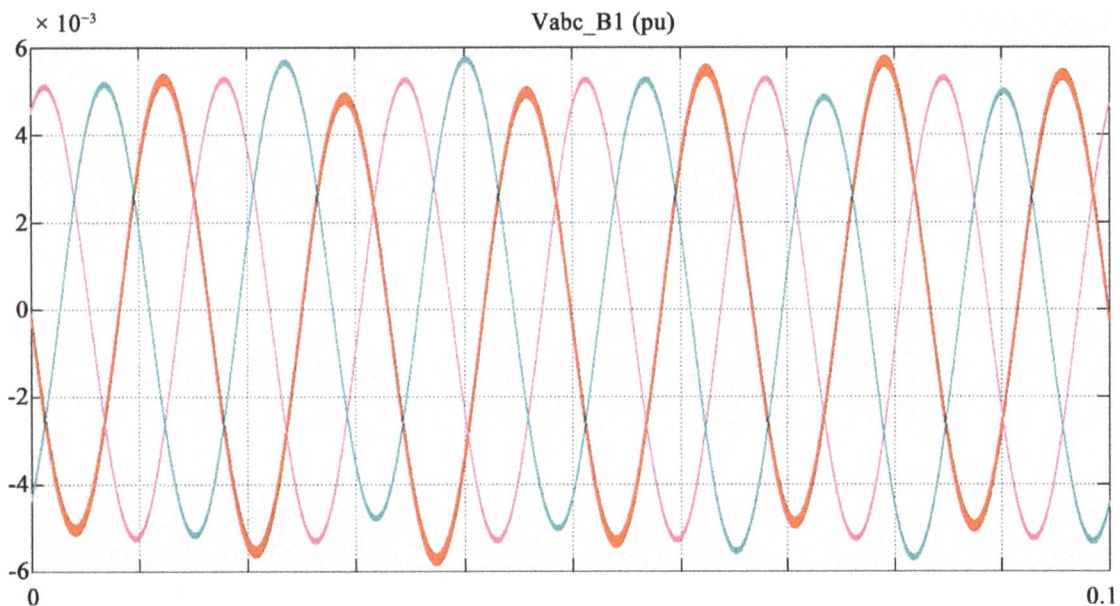

Figure 14. Voltage at bus 1 in grid-connected, with time.

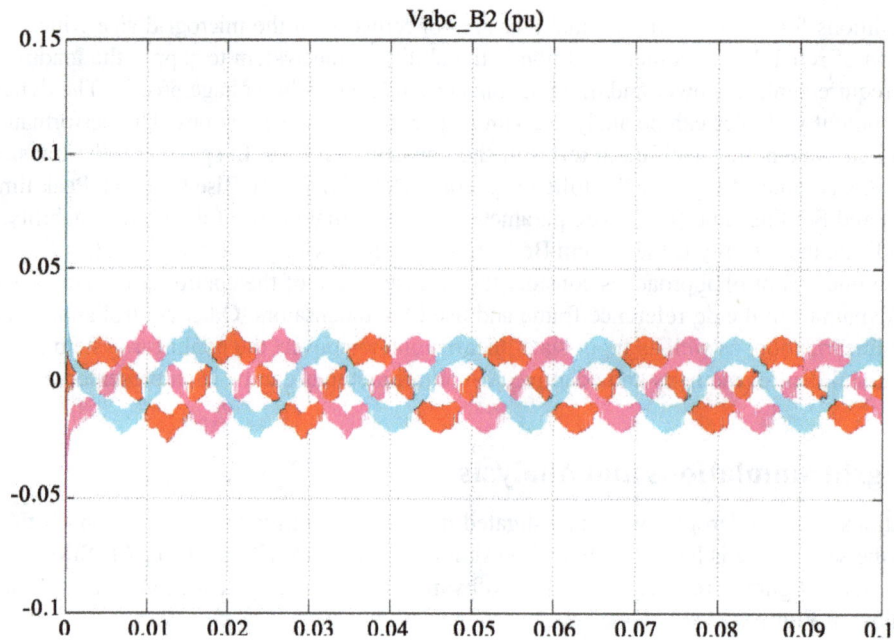

Figure 15. Voltage at bus 2 in grid-connected, with time.

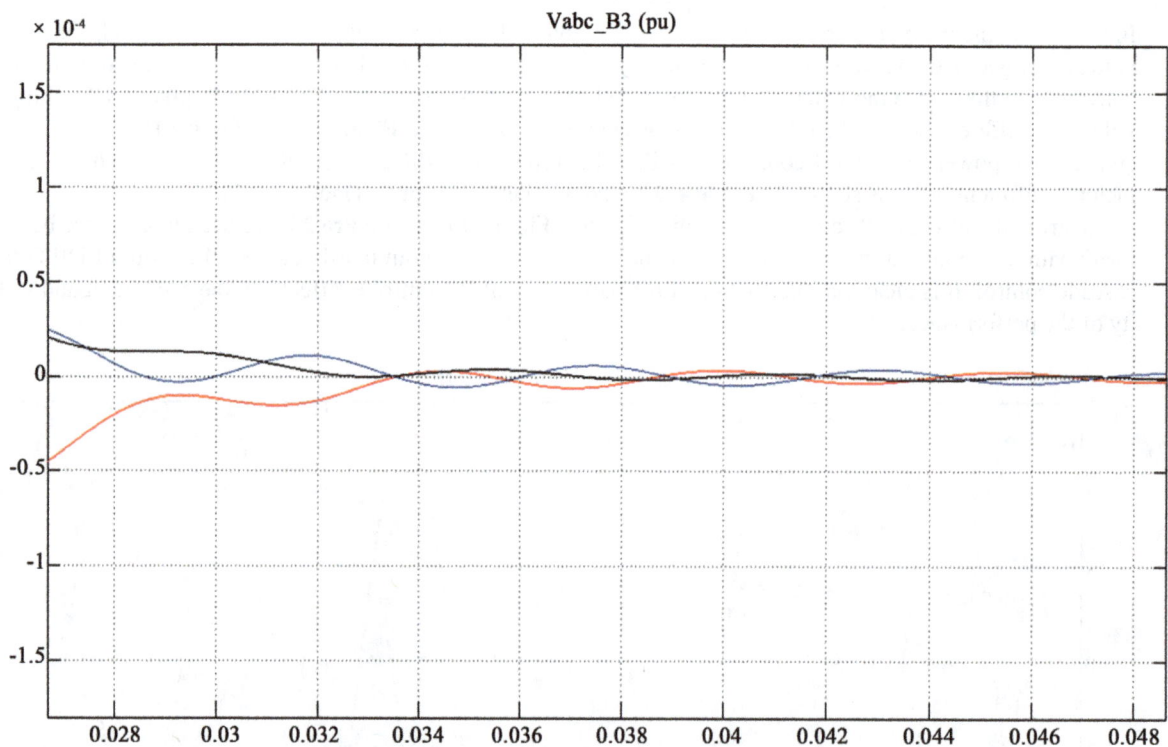

Figure 16. Voltage at bus 3 in grid-connected, with time.

MATLAB platform is applied for the stability analysis. The MATLAB power_analyze command will get the equivalent state-space of the Sim Power Systems model. It gets A, B, C, D matrices of the state-space described by the equations:

$$x' = A \cdot x + B \cdot u$$
$$y = C \cdot x + D \cdot y$$

where the state vector x represents the inductor currents and capacitor voltages, the input vector u represents the voltage and current sources, and the output vector y represents the voltage and current measurements of the model.

Nonlinear elements, like the switch devices, motors and machines, are acted by current sources driven by the voltages across the nonlinear element terminals. The nonlinear items generate additional current source inputs to the u vector, and additional voltage measurements outputs to the y vector. The above mentioned states are defining the A-matrix (elements and sizing). The dimension of matrix A is based on the above mentioned states. Using MATLAB code for eignvalues, the real parts of them are negative.

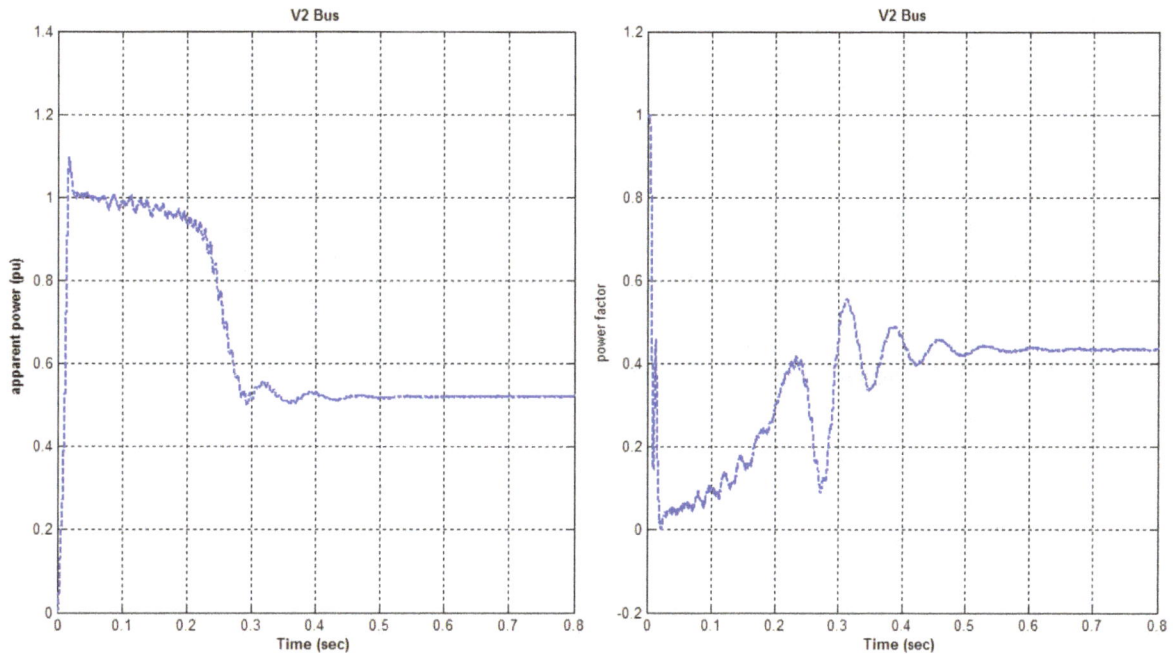

Figure 17. Apparent power and power factor at bus 2 in grid-connected.

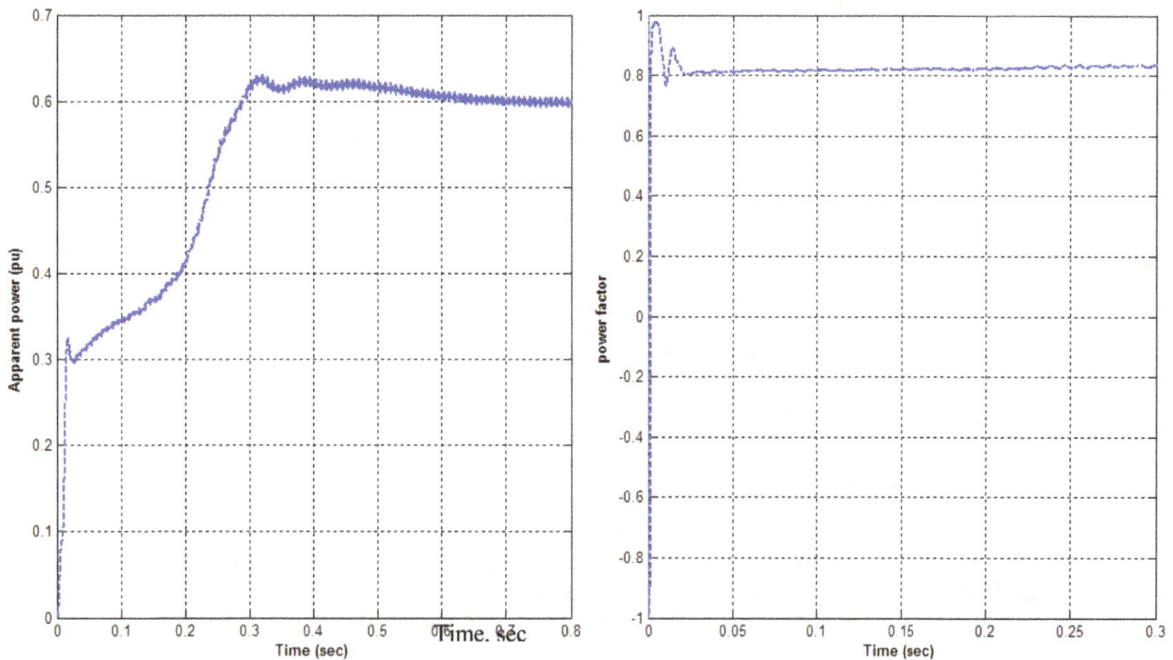

Figure 18. Active and reactive power at bus 2 in grid-connected.

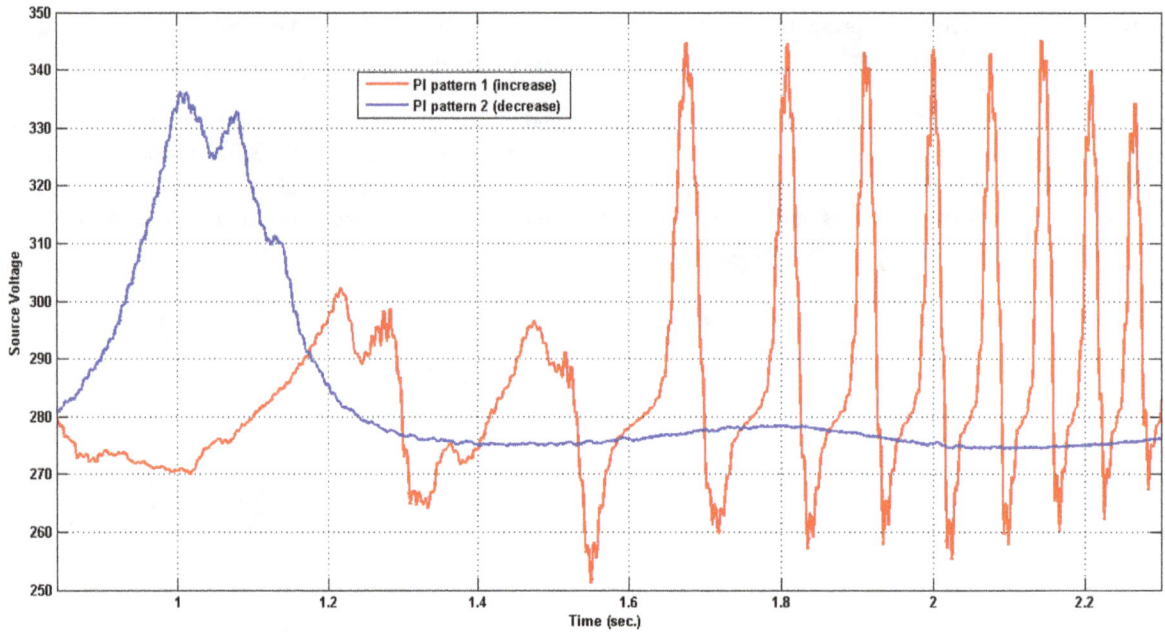

Figure 19. Effect 1 of changing PI parameters (increase and decrease).

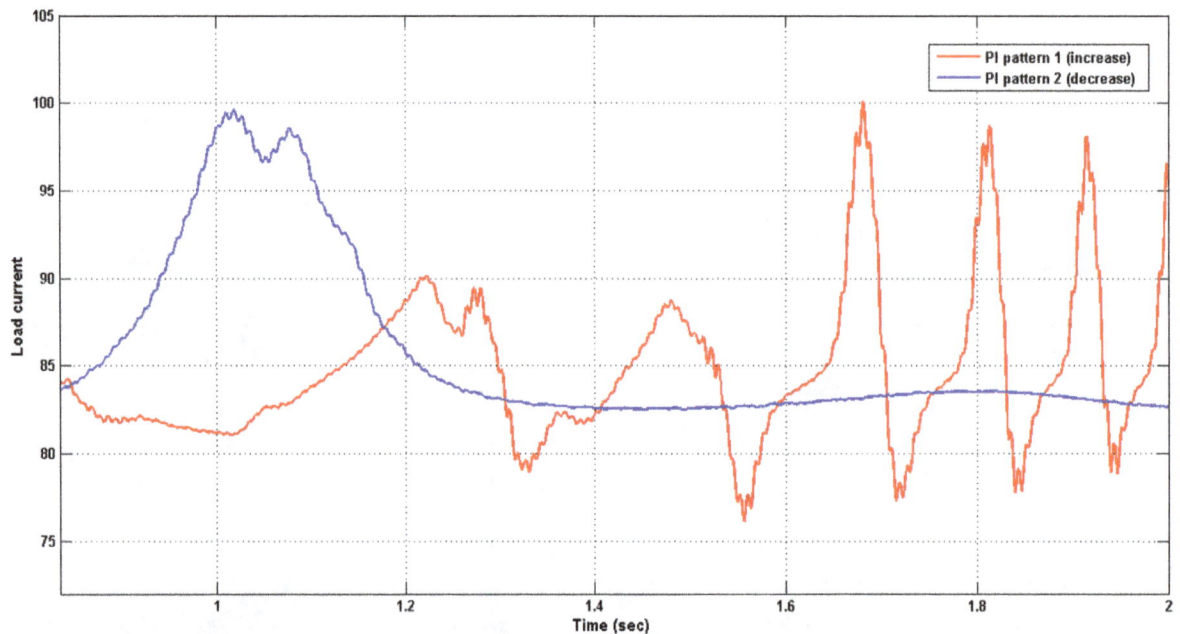

Figure 20. Effect 2 of changing PI parameters (increase and decrease).

Table 2 confirms that eignvalues has real parts with negative signs.

5.2. Islanded Mode (Case Study 2)

The DC bus voltage is maintained stable by the battery converter, PV and Fuel Cell. AC bus voltage is provided by the main converter and Wind Turbine. The nominal voltage and rated capacity of the battery are selected as 240 V and 65 Ah respectively. In this mode the voltage of the PV is 279.64 V and its power is 274.22 KW. **Figure 21** and **Figure 22** show the voltage in different buses when the microgrid is in islanding mode. **Figure 23** and **Figure 24** show the active and reactive power when the microgrid is in islanded mode. **Table 3** shows THD at different buses. **Table 4** shows power factor at different buses. Modulation index for this mode is 0.6113.

Table 2. The eignvalues of the system.

PART I of Eignvalues	PART II of Eignvalues
−2.10e+005	−8.55e+001 −4.20e+002i
−1.82e+009	−8.16e+001 +4.85e+002i
−1.10e+009	−8.16e+001 −4.85e+002i
−1.10e+009	−8.16e+001+4.85e+002i
−1.35e+008	−8.16e+001−4.85e+002i
−1.35e+008	−4.99e+000+4.07e+002i
−1.99e+007	−4.99e+000−4.07e+002i
−1.10e+007	−4.80e+001
−4.47e+006	−1.11e+002
−4.47e+006	−1.11e+002
−1.63e+006	−1.11e+002
−9.37e+006	−1.11e+002
−9.41e+006	−1.11e+002
−9.44e+006	−1.14e+001
−9.44e+006	−6.40e+000
−9.44e+006	−6.40e+000
−9.42e+006	−4.18e+000
−2.87e+004 +7.024e+005i	−9.88e−001
−2.87e+004 −7.024e+005i	−9.42e−001
−1.63e+006	−1.00e−002
−9.41e+006	−1.00e−001
−9.41e+006	−1.00e−001
−9.41e+006	−1.19e−002
−9.41e+006	−1.50e−007
−4.62+004 +4.07e+005i	−4.41e−009
−4.62e+004−4.07e+005i	−2.22e−009
−4.62e+004 +4.07e+005i	−3.93e−011−1.55e−010i
−4.62e+004 −4.07e+005i	−3.93e−011+1.55e−010i
−2.24e+004	−6.09e−011
−1.01e+002 +4.01e+004i	−4.08e−013
−1.01e+002 −4.01e+004i	−1.27e−014
−1.01e+002 +4.01e+004i	−4.79e−004
−1.01e+002 −4.01e+004i	−4.79e−004
−1.59e+004	−7.99e−005
−1.59e+004	−7.99e−005
−4.68e+001 +3.53e+004i	−7.50e−005
−4.68e+001 −3.53e+004i	−7.11e−005
−3.05e+002 +6.08e+003i	−7.11e−005
−3.05e+002 −6.08e+003i	−3.94e−005
−3.05e+002 +6.08e+003i	−3.75e−005
−3.05e+002 −6.08e+003i	−3.76e−005
−6.91e+003	−4.01e−005
−5.40e+000 +3.16e+003i	−4.01e−005
−5.40e+000 −3.16e+003i	−5.00e−002
−8.55e+001 +4.20e+002i	−2.00e+000

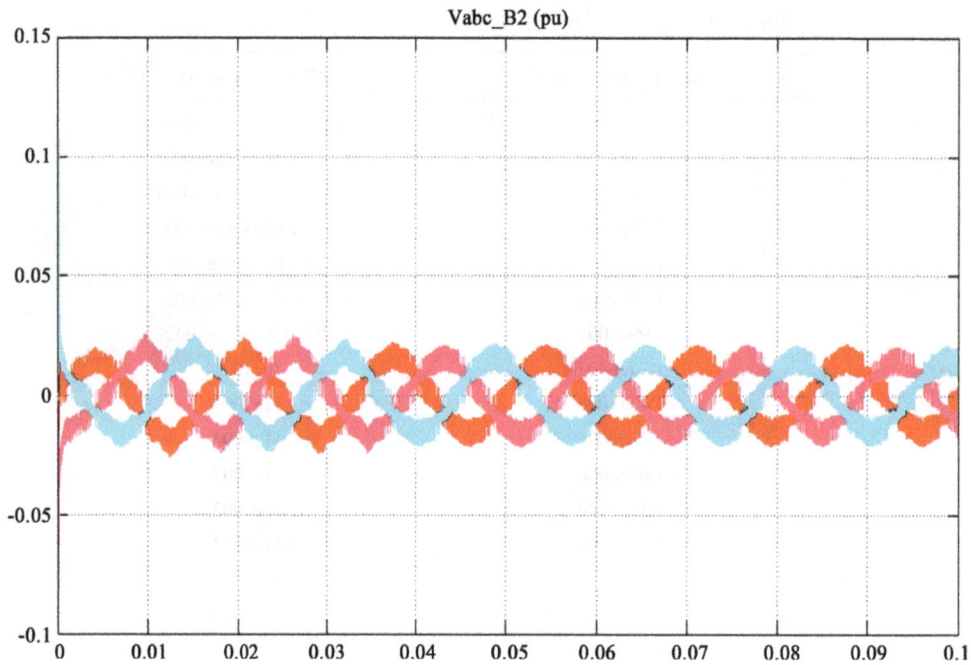

Figure 21. Voltage at bus 2 in islanding, with time.

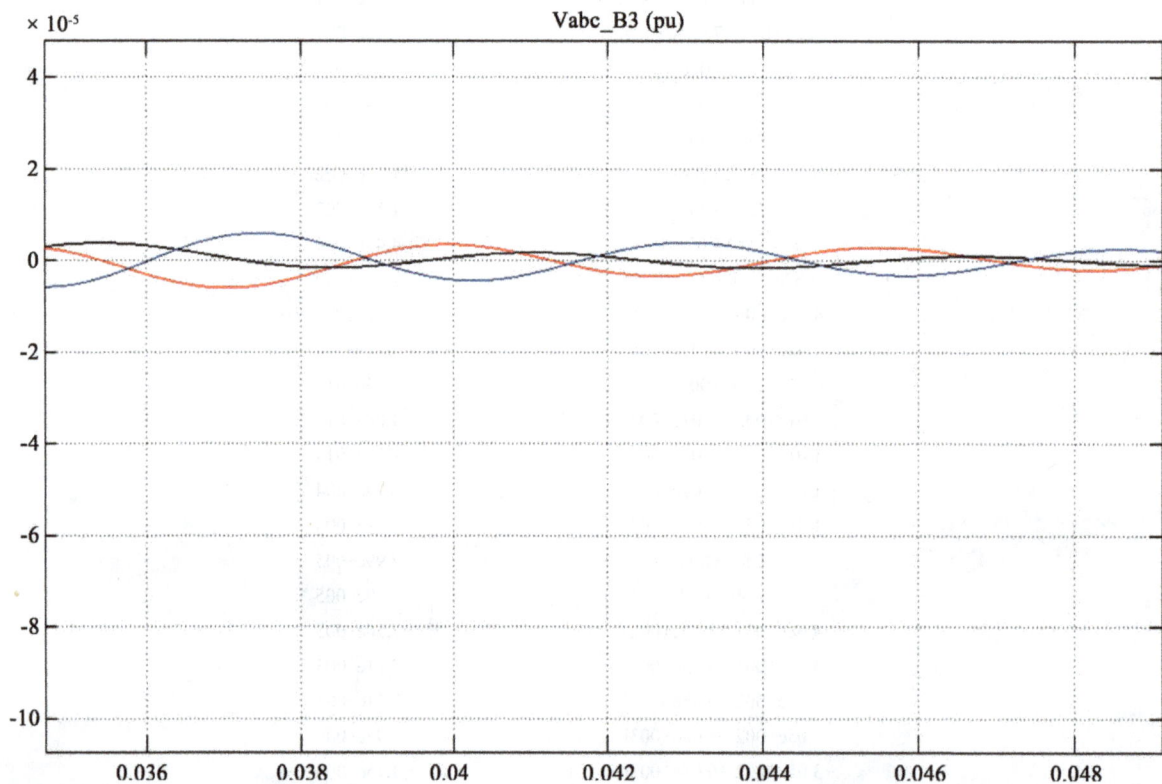

Figure 22. Voltage at bus 3 in islanding, with time.

Table 3 shows the THD % at different bus in different modes. All the THDs are less that 5% except at bus 2. One reason could be that there are a lot of AC loads which are connected at this bus and that cause a lot of noise. **Table 4** shows the power factor at different buses in different modes. The power factor is closer to one when the microgrid is operating in Grid-connected mode. **Table 5** shows the modulation index in different modes and it is

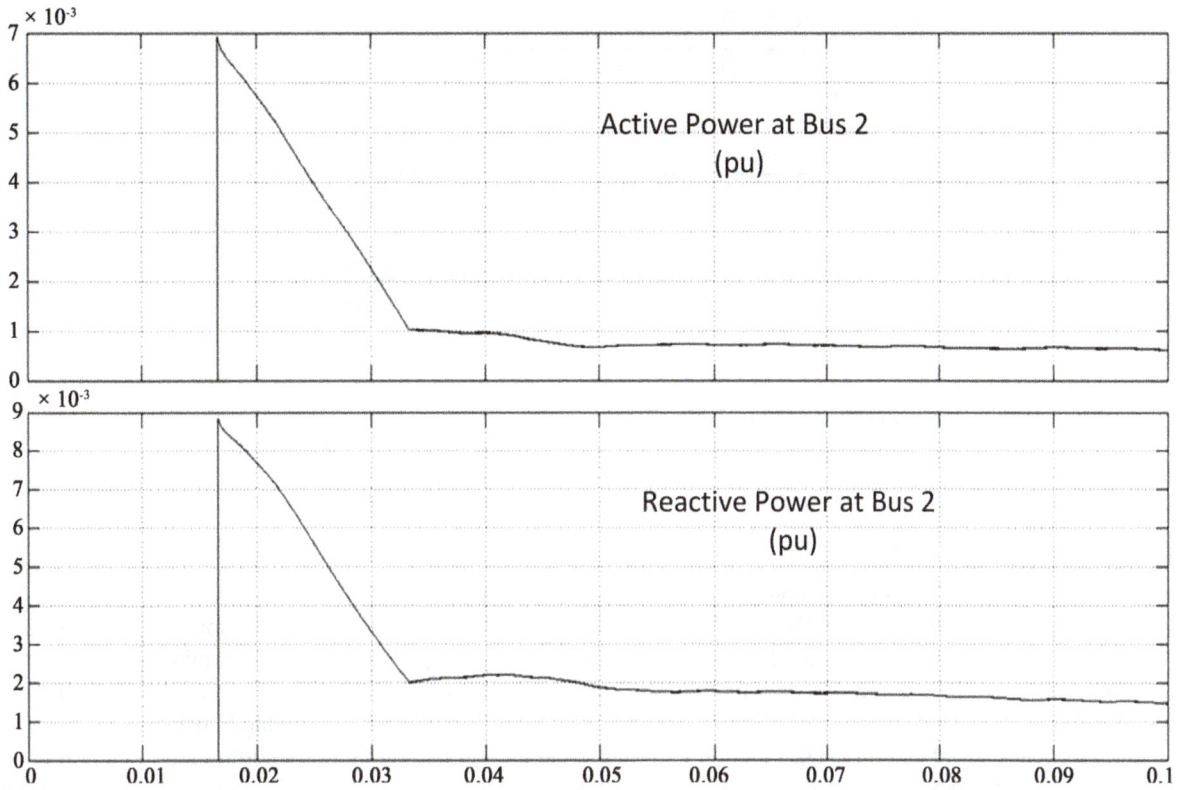

Figure 23. Active and reactive power at bus 2 in islanding, with time.

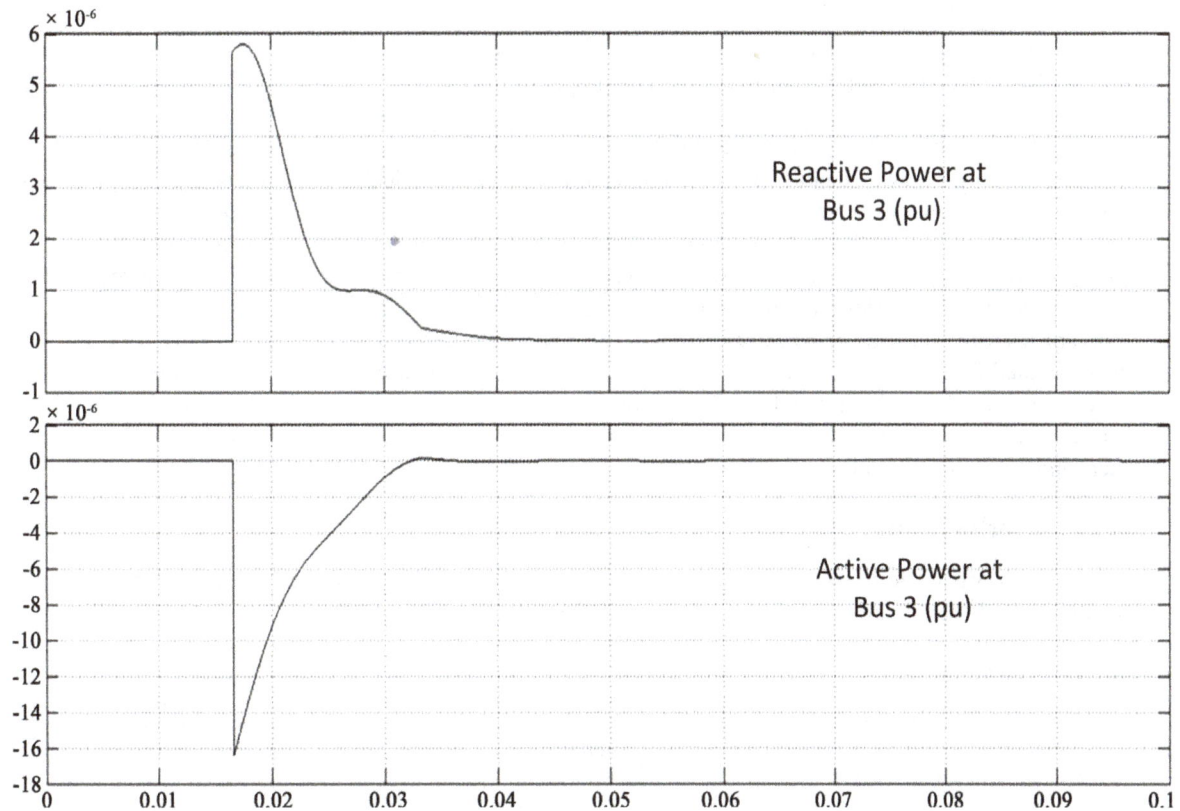

Figure 24. Active and reactive power at bus 3 in islanding, with time.

Table 3. THD at different buses.

THD %	Grid-connected	Islanding
Bus 1	1.898	Not applicable for islanding
Bus 2	33.85	34.73
Bus 3	2.66	3.359

Table 4. PF at different buses.

Power Factor	Grid-connected	Islanding
Bus 1	0.9915	Not applicable for islanding
Bus 2	1	1
Bus 3	0.9993	0.8283

Table 5. Modulation index in different modes.

Modes	Modulation Index
Grid-connected	0.61
Islanding	0.6113

almost the same in both modes. Voltages at Bus 1 to Bus 3 are in steady state. Active power is really high at Bus 1 since it is directly connected to the grid. On the other hand, the reactive power is really low. That means that the current waveform is in phase with the voltage waveform. The active power is still high in other two buses in both modes (Grid connected and islanding mode). The reactive power is high at Bus 2 in both modes. One of the reasons could be that the current waveform is out of phase with the voltage waveform due to using inductive and capacitive loads. Increased loses and extreme voltage sag can be the side effect of the current flow that is associated with the reactive power. As a result, having a minimum allowable power factor is necessary in distribution systems.

6. Conclusion

This paper presents dynamic modeling of the proposed microgrid that is powered with various DGs as wind turbine, PV solar PV cell, battery and fuel cell. The proposed microgrid can operate in grid-connected mode and also, can be used in an emergency situation to provide power to a residential community at the islanded mode. The response of buses voltages is presented in both of grid-connected and islanded modes. In addition to the voltage profile, the power factor, THD and modulation index are calculated. Furthermore, active power and reactive power responses were captured for different buses in the two modes. Each DG has appended control system with its modeling that is discussed to control DG performance. PI controller is important component inside the appended control system. For wind-power system, that control system includes speed regulator, pitch control, rotor-side converter control, and grid-side converter control. To stress the effect of PI control, simulations are used to compare between the dynamic response at different operating conditions with and without tuned-adapted PI control of DIFG local cascade control. It is clear that the change of the parameters of PI controller affects on the settlement and stability of the performance. A MPPT controller based on boost converter is discussed in order to control the PV system. The proposed microgrid can recover home cluster in terms of disasters and unexpected events. This makes the overall power system more resilient.

Acknowledgements

This research was made possible by a NPRP award NPRP 5-209-2-071 from the National Research Fund. The statements made herein are solely the responsibility of the authors.

References

[1]   Zhang, D., Papageorgiou, L.G., Samsatli, N.J. and Shah, N. (2011) Optimal Scheduling of Smart Homes Energy Consumption with Microgrid. *Proceedings of 1st International Conference on Smart Grids, Green Communications and IT Energy-Aware Technologies*, Venice/Mestre, 22-27 May 2011, 70-75.

[2]   Brown, R.E. (2008) Impact of Smart Grid on Distribution System Design. *Power and Energy Society General Meeting—Conversion and Delivery of Electrical Energy in the 21st Century*, Pittsburgh, 20-24 July 2008, 1-4.

[3] Kakigano, H., Miura, Y. and Ise, T. (2009) Configuration and Control of a DC Microgrid for Residential Houses. *Transmission & Distribution Conference & Exposition: Asia and Pacific*, Seoul, 26-30 October 2009, 1-4.

[4] Kakigano, H., Miura, Y., Ise, T., Momose, T. and Hayakawa, H. (2008) Fundamental Characteristics of DC Microgrid for Residential Houses with Cogeneration System in Each House. *Power and Energy Society General Meeting—Conversion and Delivery of Electrical Energy in the 21st Century*, Pittsburgh, 20-24 July 2008, 1-8.

[5] Pascual, J., San Martin, I., Ursua, A., Sanchis, P. and Marroyo, L. (2013) Implementation and Control of a Residential Microgrid Based on Renewable Energy Sources, Hybrid Storage Systems and Thermal Controllable Loads. *Energy Conversion Congress and Exposition (ECCE)*, Denver, 15-19 September 2013, 2304-2309.

[6] Kuo, Y.C., Liang, T.J. and Chen, J.F. (2001) Novel Maximum-Power-Point-Tracking Controller for Photovoltaic Energy Conversion System. *IEEE Transactions on Industrial Electronics*, **48**, 594-601.

[7] Kroposki, B., Lasseter, R., Ise, T., Morozumi, S., Papatlianassiou, S. and Hatziargyriou, N. (2008) Making Microgrids Work. *Power and Energy Magazine, IEEE*, **6**, 40-53.

[8] Katiraei, F., Iravani, R., Hatziargyriou, N. and Dimeas, A. (2008) Microgrids Management. *Power and Energy Magazine, IEEE*, **6**, 54-65.

[9] Surprenant, M., Hiskens, I. and Venkataramanan, G. (2011) Phase Locked Loop Control of Inverters in a Microgrid. *Energy Conversion Congress and Exposition (ECCE)*, Phoenix, 17-22 September 2011, 667-672.

[10] Jimeno, J., Anduaga, J., Oyarzabal, J. and Muro, A. (2011) Architecture of a Microgrid Energy Management System. *European Transactions on Electrical Power*, **21**, 1142-1158. http://dx.doi.org/10.1002/etep.443

[11] Gabbar, H.A., Honarmand, N. and Abdelsalam, A.A. (2014) Resilient Microgrids for Continuous Production in Oil & Gas Facilities. Saudi Arabia Smart Grid, Jeddah, 19-21 October 2014.

[12] Rigatos, G., Siano, P. and Cecati, C. (2014) An H-Infinity Feedback Control Approach for Three-Phase Voltage Source Converters. *IEEE IECON* 2014, Dallas, 29 October-1 November 2014, 1227-1232.

[13] Sungwoo, B. and Kwasinski, A. (2012) Dynamic Modeling and Operation Strategy for a Microgrid with Wind and Photovoltaic Resources. *IEEE Transactions on Smart Grid*, **3**, 1867-1876. http://dx.doi.org/10.1109/TSG.2012.2198498

[14] Xiong, L., Peng, W. and Poh, C. (2011) A Hybrid AC/DC Microgrid and Its Coordination Control. *IEEE Transactions on Smart Grid*, **2**, 278-286.

[15] Luis, A.S., Wen, Y. and Rubio, J. (2013) Modeling and Control of Wind Turbine. *Mathematical Problems in Engineering*, **2013**, Article ID: 982597.

[16] Pierce, K. and Jay, L. (1998) Wind Turbine Control System Modeling Capabilities. *Proceedings of the American Controls Conference*, Philadelphia, 26-27 June1998, 24-26.

[17] Sang, H., Bruey, S., Jatskevich, J. and Dumont, G. (2007) A PI Control of DFIG-Based Wind Farm for Voltage Regulation at Remote Location. *IEEE Power & Energy Society General Meeting*, Tampa, 24-28 June 2007, 1-6.

[18] Naguru, N., Karthikeyan, A. and Nagamani, C. (2012) Comparative Study of Power Control of DFIG Using PI Control and Feedback Linearization Control. *Advances in Power Conversion and Energy Technologies (APCET)*, Mylavaram, 2-4 August 2012, 1-6.

[19] Yu, C.Y. and Li, D.D. (2012) Fuzzy-PI and Feed forward Control Strategy of DFIG Wind Turbine. *IEEE Conference of Innovative Smart Grid Technologies (ISGT)*, Tianjin, 21-24 May 2012, 1-5.

[20] Yeong, K., Tsorng, J. and Chen, J. (2001) Novel Maximum-Power-Point-Tracking Controller for Photovoltaic Energy Conversion System. *IEEE Transactions on Industrial Electronics*, **48**, 594-601.

[21] Luna-Sandoval, G., Urriolagoitia-C, G., Hernández, L.H., Urriolagoitia-S, G. and Jiménez, E. (2011) Hydrogen Fuel Cell Design and Manufacturing Process Used for Public Transportation in Mexico City. *Proceedings of the World Congress on Engineering*, London, 6-8 July 2011, 2009-2014.

[22] http://www.canadiangeographic.ca/magazine/jun12/map/default.asp

[23] Gabbar, H. and Othman, A.M. (2015) Performance Optimisation for Novel Green Plug-Energy Economizer in Micro-Grids Based on Recent Heuristic Algorithm. IET Generation, Transmission & Distribution, 2, 1-10.

[24] Tsikalakis, A.G. and Hatziargyriou, N.D. (2011) Operation of Microgrids with Demand Side Bidding and Continuity of Supply for Critical Loads. *European Transactions on Electrical Power*, **21**, 1238-1254. http://dx.doi.org/10.1002/etep.441

[25] Buayai, K., Ongsakul, W. and Mithulananthan, N. (2012) Multi-Objective Micro-Grid Planning by NSGA-II in Primary Distribution System. *European Transactions on Electrical Power*, **22**, 170-187. http://dx.doi.org/10.1002/etep.553

Appendix

The proposed simulation is conducted using Matlab/Simulink/Sim Power software Environment, which is applied to the selected MG case study, as per the following specification:

- *Utility Grid*: 138 KV, 5GVA, X/R = 10.
- *Wind Turbine Generator*: V = 1.6 kV, P = 1 MW.
- *PV*: 240 V, 200 KW, Ns = 318, Np = 150, Tx = 293, Sx = 100, Iph = 5, Tc = 20, Sc = 205.
- *Fuel Cell*: 240 V, 200 KW, number of Cells = 220, nominal Efficiency, 55%.
- *Battery*: 240 V, Rated capacity: 300 Ah, Initial State-Of-Charge: 100%, discharge current: 10, 5A.
- *Hybrid AC Load* 1: linear load: 0.1 MVA, 0.8 lag pf, non-linear load: 0.2 MVA, Motorized load is an induction motor: 3phase, 0.3 MVA, 0.85 pf.
- Hybrid *AC Load* 2: linear load: 200 kVA, 0.8 lag pf., non-linear load: 200 kVA, Motorized load is an induction motor: 3 phase, 100 kVA, 0.8 pf.
- *DC Load*: resistive load: 100 kw, motorized load dc series motor: 100 kw.
- *VSC*: Fs/w = 1750 Hz Cs = 100 µF, Rs = 0.1 Ω, Ls = 10 mH.
- *PID*: Kp = 0 - 100, Ki = 0 - 30, Kd = 0 - 15, Ke = 0 - 10.

15

Interface for Intelligence Computing Design and Option of Technical Systems

Javanshir Mammadov, Tarana Tagiyeva, Akhmedova Sveta, Aliyeva Arzu

Department of "Information Technology and Programming" of Sumgait State University, Sumqayit, Azerbaijan
Email: cavan62@mail.ru

Abstract

As a result of analysis of the existent methods and tools of computer aided design of the technical systems of many industrial areas, the primary purpose of the article that consists in the decision of different project problems within the framework of one programmatic system with the use of comfortable programmatic interface is certain. Architecture of the program interface for computer designing and option of technical systems of different industrial areas on the basis of stage-by-stage automated designing principles with using programmatic and informative supports is worked out. In the article, the problem of computer designing, searching and option of elements of manipulator for flexible manufacture module is considered. As a method of logical simulation of the problem, production model of designing procedures of the program interface and intelligence option of a manipulator, its technical parameters are developed. On the basis of algorithmic scheme of searching, the option of a manipulator from data base is worked out.

Keywords

Computer-Aided Design, Design Stages, Production Model, Programmatic Interface

1. Introduction

As known many technical objects (TO) (types of TO are showed in poster) are designed by means of standard stages (requirement specification, technical suggestion, draft design and detail design) with using CAD systems [1] [2]. Architecture of program interface is worked out in corresponding to the designing stages by the system's menus and special systems. Analysis worked out separate tool of the programmatic, informative and mathematical providing for the automated planning on the stages of planning that provided application and realization of technical, programmatic, informative, mathematical etc. A tool in the different areas of industry showed that they properly did not provide universality and complex planning of the difficult technical systems [3] [4].

In this connection, decision of problem on development of the intelligence programmatic system of computer-aided design, providing complex approach of realization of the automated procedures and operations on the basic stages of planning from the scientific point of view is an actual and priority problem and direction.

Being base on principles of universality, openness and planning flexibility, as a primary purpose of project is offered software for a complex computer-aided design, where up basis the structure of the stages of planning of the technical systems, programmatic procedures and executive directives, and also interface, takes with the programmatic modules of the database management system, by CAD of circulation of documents, processing of documents, subsystem of computer experiments, system of automation of mathematical calculations and in-plant, global computer system, computer graphics and multimedia.

2. Solution of the Problem

The primary purpose of the project is development of the single intellectually-operating system of designer for complex computer-aided of technical objects of the different setting design. Scheme of interface of the intellectually-operating system of designer (IOSD) of the complex designing is offered with the following blocks of menu:

1) Designing stages;
2) Project programs;
3) Programmatic-operating functions.

Work of IOSD is based on stage-by-stage approach since the entry of basic data and application of planning object domain to creation of detail design. IOSD allows controlling the universal programmatic interface on the basis of intellectually-operating functions of complex computer-aided of the technical systems of the different setting design.

In a depended on analyze of progressive status of CAD the following research problems have to solve:

1) Development of the common scheme of computing design with program interface;
2) Development of first panel of "Designing stages" with its menus and program operations;
3) Development of second panel of "Project programs" with its menus and program objects;
4) Development of third panel of "Programmatic-operating functions" with its menus and service operations;
5) Providing cooperation of the system with program functions of a corporate network;
6) Program control of design information of the IOSD on the level of technical provide.

The programmatic package of interface for a complex computer-aided of technical objects design is developed on the base of instrumental programmable environment of Delphi; procedures of logical search and choice of information on the stage of technical suggestion developed on the base of intellectual programmable environment of Prolog and database management systems, control system of base of the designed data; procedures of engineering design will be realized on the basis of the prepared designer and technological programmatic tool of AutoCAD [5].

The basic idea of project consists in the decision of different project tasks within the framework of one programmatic system with the use of comfortable programmatic interface.

2.1. Architecture of Programmatic Interface for Computer Designing Procedures

As it is generally known, great number of technical objects of planning, for development of that required application CAD on the stages of planning exist. In accordance with the stages of planning, the commands of system menu and types of the special programmatic systems are developed the architecture of programmatic interface with the automated procedures on design stages (**Figure 1**).

On the first stage project procedure of requirement specification will be realized where a standard template is filled depending on an application of technical object domain. The entered data are saved to the database of user-designer. For example, on the special system window design object area, a type of design element, characteristics of the design element and other parameters are chosen. After inputting all design data they are saved in the system base [6].

On the second stage project procedure of technical suggestion will be realized, where the problems of local (in the database of user-designer) and global search decide.

At a global search of the automated choosing a few variants of the projects (in accordance with data of requirement specification) comes true, their animation and video presentation. The final choice of the best variant

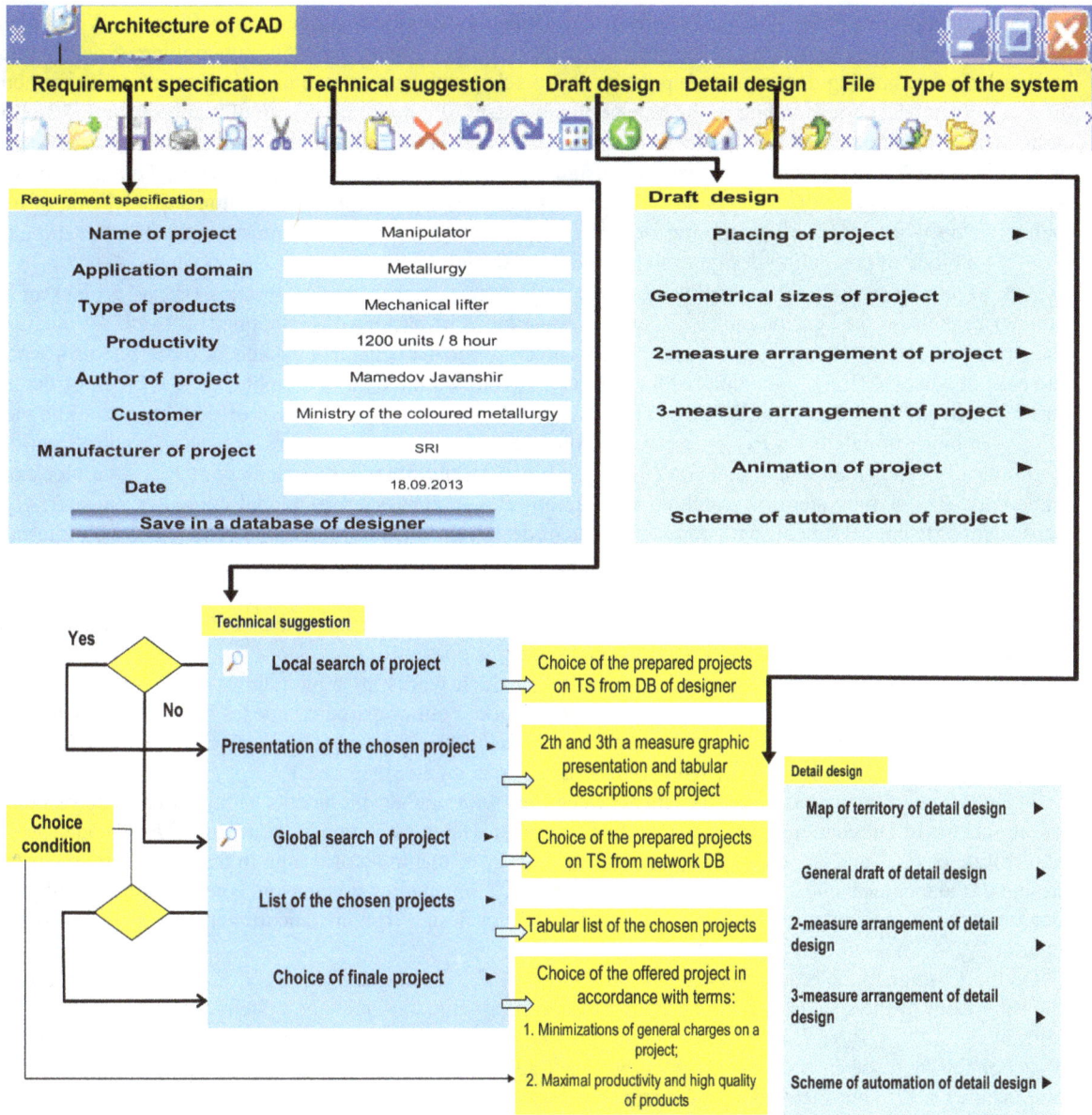

Figure 1. The architecture of programmatic interface with the programmatic procedures on design stages.

comes true in accordance with terms: applications of innovative technologies; to the high yield; producing quality of products; providing reliable automation of planning object; to minimization of material charges and other.

On the third stage procedure of the preliminary planning technical object will be realized. The basic project operations are drafts in two and three projections and 2 and 3-measure images (axonometric presentation of draft) and registration of these (specification and angular stamp of draft) these drafts, animation presentation, and also scheme of automation of planning object [7].

On the fourth stage of the working planning through a menu of final technical documentations get out accordingly for unsealing [8]. On the basis of documentations of detail design of technical objects are practically developed in accordance with productive terms of the design object.

2.2. Production Procedures of Program Interface of Designing

For simulation of all functions of the intelligence program interface designing procedures by means of production algorithm is needed to create. Every subsystem and block of architecture of program interface of computing design of manipulator of the flexible manufacture module (as shown in requirement specification) will designate as subsystems: *Pos*—a subsystem of the programmatic-operating system providing functioning of open complex subsystem of computing design of manipulator; *Pi*—a subsystem of interface including set of menu for activation of commands, corresponding to the blocks of subsystem of computing design of manipulator and auxiliary subsystems providing flexibility on the whole; *Pap_i* (where $i = \overline{1,5}$)—a subsystem of computing design of manipulator; (*Pap_1*—a block of computing design of composes structure of manipulator; *Pap_2*—a block of standard active elements of manipulator; *Pap_3*—a block of computing design of automation scheme of manipulator; *Pap_4*—a block of computing design of technical control and diagnostic system of manipulator; *Pap_5*—a block of computing design of non standard elements of manipulator); *Psp_i* (where $i = \overline{1,3}$)—a subsystem of special packets of the applied software and programming-language systems (P_{sp_1}—a block of constructor packets on the base of T-FLEX CAD; P_{sp_2}—a block of packets of the applied software for creation of data base and control system of data base; P_{sp_3}—a block of the mathematical packets of the applied software on the base of MathCAD; P_{sp_4}—a block of the applied software on the base of Delphi; P_{sp_5}—a block of the intelligence software on the base of PROLOG;); P_{bdz_i} (where $i = \overline{1,4}$)—a subsystem of data base and knowledge (P_{bdz_1}—a block of data base; P_{bdz_2}—a block of knowledge base); P_{arm_i} –(where $i = \overline{1,4}$)—a subsystem of algorithmically calculation and simulation (P_{arm_1}—a block of algorithmically calculation; P_{arm_2}—a block of simulation); P_{pv}—a subsystem of searching and option; P_{qs}—a subsystem of global computer network; P_{to}—a subsystem of technical support with periphery units of design document printing and connection with Internet.

The program operations of interface of computing design of manipulator and its option works are executed as follows:

1) After plugging (t_1) of the operating system P_{os} in the subsystem of hardware of *Pto*, through a main menu shell program of the same name of the complex subsystem is opened;

2) Designer using the subsystem of interface of *Pi*, which works after including (t_2^1), opens the library of subsystem of computer design of P_{ap_i} with the before worked out prepared composes structures, scheme of automation of every manipulator on the whole, and also data base of these standard active elements of P_{ap_2} and if necessary edits this file and saves the edited information in computer memory.

3) By means of command (t_2^2) a designer creates files with new developments of layout chart corresponding to the scheme of automation in the subsystem of computer design of composes structure of P_{ap_1} and scheme of automation of P_{ap_3} and technical control and diagnostically of manipulator design in the subsystem of computer design of the technical checking and diagnostically of P_{ap_4}, kinematics scheme and construction draft of gripper devices, special manipulators in the subsystem of computer design of non-standard active elements of manipulator P_{ap_5}.

On the 3rd stage in the process of new file, subsystem of P_{ap_1} creation, P_{ap_2}, P_{ap_3}, P_{ap_4}, P_{ap_5} co-operate with the subsystem of the special application packages and programmatic-language system P_{sp_1}, P_{sp_2}. Thus subsystems of P_{ap_1}, P_{ap_3}, P_{ap_4}, P_{ap_5} co-operate also with the subsystem of algorithmic calculations and design of P_{ar}, P_{am}. For this purpose in a control of open complex subsystem of computing designing of manipulator panel commands get (t_2^3) out for drawing of composes scheme or calculation of parameters of non-standard elements and design of scheme of automation of manipulator.

4) On this stage for the wearing-out (t_3) of subsystem of search and choice of P_{pv} of composes structures, active elements the commands of subsystem of database of P_{bdz_1} and knowledge of P_{bdz_2} get out on the basis of terms-queries. At the choice of database the software environment of the system Microsoft Access is opened and a query is set for a search and choice of the prepared types of active elements with their specifications. At the choice of files of base of knowledge the intellectual programmatic system PROLOG (subsystem of P_{sp_5}) is opened where a query is set for a search and choice of elements of control system, worked out non-standard active elements, composes structures and scheme of automation of manipulator on the basis of algorithms of calculation. In case of search of the prepared composes structures of manipulator with standard elements the command of the network including (P_{qs}-subsystem of global computer network) gets out from menu of search. Thus choosing the certificate menu (t_4) of subsystem of global computer network of P_{qs} it is possible to get additional information about open system of the computer designing of manipulator.

2.3. Procedures of Option of Manipulator on a Stage of the Technical Suggestion of the Program Interface of Designing

Option of the necessary manipulator project from data base (DB_i) of the designers within the framework of local workstation is worked out. A query for option of a project includes the variants of basic data of the requirement specification, which were before saved in DB_1. In accordance with the expressions from the basic data and technical parameters of a manipulator for more exact option, the following function is given: $f_1, f_2, \cdots, f_n$ are constituents of the function: force of clamping of manipulator F, its geometrical sizes.

Taking into account the values of mass of the manipulated object of m_{ij} and predominance of mass of the manipulator of m_{gij} from m_{ij} (coefficients taking into account mass of gripper of c_g and taking into account the type of drive of manipulator (M_i) of c_d) by means of the following terms, the manipulator gets out with the required carrying capacity (C_i):

$$\textit{If } \left([m_{i1}] \Rightarrow M_i \cup \left(C_i = m_{i1} c_g c_d \right) \right)$$

(where $m_{i1} \to c_g = 1.4 \wedge c_d = 1.1$ for hydraulic motor),

Then (on the basis of characteristic parameters from a database, the search of the required type of M_i is realized).

Thus, the following condition of limitation is taken into account:

$$C_i \leq C_{\max},$$

where $C_{\max}$—maximal value of carrying capacity of M_i, that gets out from a database. Accordingly depending on their carrying capacity, type and construction parameters of M_i is got out. Maximal tension in the cross runner of the manipulated object allows to choose a drive for manipulator.

Setting the following condition

$$\textit{If } \left(\sigma_{mo} \leq [s]_u \right),$$

Then (from the database of the standard drives of gripper of M_i, the most nearby value of forces of clamp of the manipulated object is got out ($F = \sigma_{mo}S$), S—area, which exposed to the clamp (mm^2)) & (($F - F_{st}$) $\to min$, where F_{cm}—standard value of clamping M_i).

Summarizing the conclusions of stage-by-stage realization of architecture of interface of the computer designing of manipulator and its option it should be noted that taking into account the basic requirements to creation of instrument of the computer design of manipulator, subsystems provide universality, adaptively and flexibility of the system at development of manipulator of different areas of metallurgical industry.

The subsystems of the algorithmic, mathematical providing, expert and information retrieval operations used as autonomous blocks provide authenticity, fast-acting and reliability of executable project operations and procedures.

As be obvious from a **Figure 1** in a control of computer-aided design panel for creation of new project, opening of existent projects, organization of work with the subsystems of architecture of instrument of the computer designing of manipulator is used procedures of programmatic block "*File*".

3. Conclusions

The basic results of the article are the following:

1) The worked out architecture of computer design and option of technical systems works by principle of the stage-by-stage designing on the basis of the programmatic interface. It provides the complex decision of planning design operations, since the entry of basic data and application of planning object domain to creation of detail design [9] [10].

2) It was given the logical procedures of production model for designing and option of manipulator which provided operative and flexible designing and searching-option operations on the level of one universal program interface.

3) The programmatic interface has practical application during a computer-aided and option of the technical systems for the different setting design.

The proffered intellectually-operating system of designer will allow in future:

1) To decide the problem of complex computer-aided design and option, since the entry of basic data and application of planning object domain to creation of detail design;

2) To manage effectively the universal programmatic interface on the basis of intellectually-operating functions of complex computer-aided of the technical systems of the different setting design;

3) From the presence of great number of logical queries for a search, choice and planning, to promote the lev-

el of intellectuality of this system;

    4) To adapt quickly with procedures of the project programs and the operation system;

    5) To use the date system in the mode of on-line dialogue with designers for acquisition necessary project information.

References

[1]   Bozdoc, M. (2003) The History of CAD.

[2]   Cunvu Li Basis CAD of CAD/CAM/CAE. "Peter", Moscow—Sankt Peterburg, 2004, 560 p.

[3]   Voronenco, V.P. (2001) Designing of the Manufacture Systems Is in an Engineer: Studies. Manual. Tiraspol: Rio Psu, 349 p.

[4]   Chelishev B.E. (2004) Computer-Aided of Design of Technology in Machine-Building. Electronic Magazine "Science and Education". Engineering Education of Association of Technical Universities. AL No. FC 77.

[5]   Polovincin, A.I. and Bobkov, N.K. (2004) Bush of G.Y. Automation of the Searching Constructing (Artificial Intelligence Is in the Machine Constructing). *Electronic Journal of "Science and Education"*. Engineering Formation of Association of Technical Universities. AL No. FC 77.

[6]   Aliyev, R.A., Mammadov, J.F. and Akhmadov, M.A. (2005) Development of Tool of the Automated Planning of Control System of FMS.-M.: Mechatronic, Automation, Management, No. 9, 27-35.

[7]   Mammadov J.F. (2002) Problems and Stages of Development of the Automated Choice and Planning of the Flexible Manufacture Systems. The Redaction of Technical Literature "House of Baku University" of Baku, 175 p.

[8]   Mamedov, J.F. and Huseynov, A.H. (2011) Algorithmic and Programmatic Providing of the Automated Layout of the Flexible Productive System. In: *Design Information Technologies in Planning and Production*, Scientific and Technical Magazine, Moscow, No. 1, 49-52.

[9]   Mamedov J.F. (2013) Development of Structure of Interface of Complex Automated Planning of the Technical Systems Software. Moscow. *Announcer of Computer and Informative Technologies*, No. 5, 18-21.

[10] Golovanov, N. (2014) Geometric Modeling: The Mathematics of Shapes. Create Space Independent Publishing Platform.

16

Robust Adaptive Control for a Class of Systems with Deadzone Nonlinearity

Nizar J. Ahmad*, Mahmud J. Alnaser, Ebraheem Sultan, Khuloud A. Alhendi

Faculty of Electronic Engineering Technology, College of Technological Studies, The Public Authority for Applied Education and Training (PAAET), Kuwait City, Kuwait
Email: *nj.ahmad@paaet.edu.kw

Abstract

This paper presents a robust adaptive control scheme for a class of continuous-time linear systems with unknown non-smooth asymmetrical deadzone nonlinearity at the input of the plant. The methodology is applied to handle input deadzone as well as unmeasurable disturbances simultaneously in strictly matched systems. The proposed controller robustly cancels any residual distortion caused by the inaccurate deadzone cancellation scheme. The scheme is shown to successfully cancel the deadzone's deleterious effect as well as eliminate other unmeasurable disturbances within the span of the input. The new controller ensures the global stability of all states and adaptations, and achieves asymptotic tracking. The asymptotic stability of the closed-loop system is proven by Lyapunov arguments, and simulation results confirm the efficacy of the control methodology.

Keywords

Adaptive Control, Non-Symmetric Deadzone, Hard Nonlinearity

1. Introduction

The significance of the deadzone problem lies in the fact that it affects many physical and practical systems. Examples of such systems are the ones containing hydraulic or pneumatic values, electronic circuits and devices, temperature regulation circuits, and in actuators such as servo valves and DC motors. In most cases, the parameters of the deadzone nonlinearity are unknown and continuously varying with time and temperature. Deadzones exist in a number of industrial applications specially the ones requiring high precision such as medical robots, semiconductor manufacturing, and precision machine tools. It has been shown that deadzone in actuators,

---

*Corresponding author.

such as hydraulic servo-valves, gives rise to limit cycling and instability. The advances reached in the area of adaptive compensation and control theory gave rise to increased interest in handling the deadzone problem. There have been many techniques that addressed the problem and have been shown to reduce if not eliminate the degradation of system performance resulting in an improved tracking accuracy and ensured stabilization of such systems.

One sensible approach to counter the effect of the deadzone, shown in **Figure 1**, was presented by [1] which involved designing an inverse deadzone function to cancel its effect. The approach of designing an inverse adaptive deadzone compensator was thoroughly investigated in [2] and [3] which was shown to improve performance. Lewis *et al.* in [4] proposed a fuzzy logic type inverse deadzone compensator, meanwhile, a neural network inverse compensator was designed in [5]. Both approaches show clear improvement in reducing the tracking error. In [6], a new adaptive controller of linear or nonlinear systems with deadzone is introduced without constructing a deadzone inverse. Global and asymptotic tracking was achieved and simulation results were presented.

An adaptive sliding mode control scheme used to offset a non-symmetrical deadzone nonlinearity in continuous time was presented in [7]. The problem of chattering inherent with sliding mode control is handled by allowing a small controlled tracking error.

In recent years, many researchers addressed the deadzone problem with encouraging results. In [8], a novel function was introduced to describe deadzone nonlinearity. To show the effectiveness of the proposed equivalent function the authors combined it with vibration of a cantilever beam.

In this paper we are motivated by the success of our earlier results deadzone compensation of DC motor presented in [9] and [10]. The extension involves combining the deadzone for a class of linear systems with uncertainties in the span of the input. The uncertainties are assumed to be bounded by a pth order polynomial in the state of the system. A robust adaptive controller will compensate for the unmeasurable disturbances as well as any mismatch error in estimating the deadzone parameters. The proposed method does not require any knowledge of the deadzone parameters or the specialized design of an inverse deadzone controller and only an upper bound of the deadzone spacing which is easily determined a priori.

2. The Problem Setup: Dynamics of a Non-Symmetrical Deadzone Nonlinearity

A common representation of a non-symmetrical deadzone nonlinearity, shown in **Figure 1**, described in [1] as follows

$$DZ(u(t)) = \begin{pmatrix} m(u - d_r), & \text{if} & u > d_r \\ 0, & \text{if} & -d_l < u < d_r \\ m(u + d_l), & \text{if} & u < -d_l \end{pmatrix} \tag{1}$$

where $DZ(u(t))$ denotes the output of deadzone function preceding a plant input, m is the slope of the lines, $(d_r - d_l)$ is the width of the deadzone distance, and $u(t)$ is the input of the deadzone block as shown in **Figure 1**. A more convenient representation of a non-symmetrical deadzone was presented in [6] as

$$DZ(u) = u - sat_d(u) \tag{2}$$

where $sat_d(u)$ represents a non-symmetrical saturation function given by

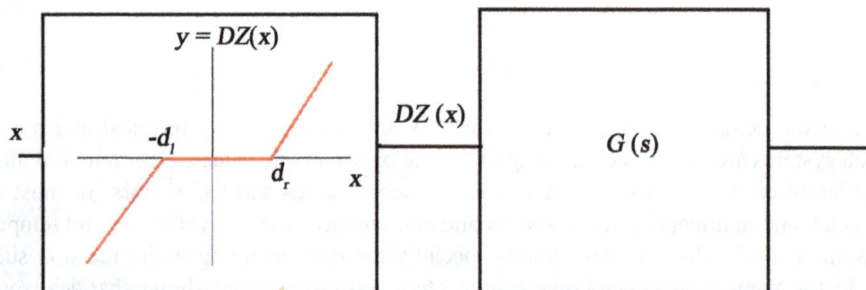

Figure 1. Non-symmetric deadzone nonlinearity at the input of a linear plant.

$$sat_d(u) = \begin{pmatrix} d_r & \text{if} & u > d_r \\ u, & \text{if} & -d_l < u < d_r \\ -d_l, & \text{if} & u < -d_l \end{pmatrix} \tag{3}$$

A novel representation for (2) was presented in [11] for an exact symmetrical deadzone by defining $(d = d_r - d_l)$ as the deadzone spacing function as

$$DZ(u) = m/2(2u + |u - d| - |u + d|) \tag{4}$$

Consequently, the saturation function can be written as

$$sat_d(u) = m/2(|u - d| - |u + d|) \tag{5}$$

In most applications the deadzone parameters are unknown or time and temperature varying. Instead, for developing an inverse deadzone function as in [9], we advocate a robust adaptive compensator that handles the variability of the deadzone parameters as part of an input disturbance. However, we outline some necessary and reasonable assumptions to be used in the proof of the efficacy of the proposed control methodology:

(A1) The deadzone parameters $d_r > 0$ and $d_l > 0$.

(A2) The deadzone parameters d_r, and d_l are bounded as follows:

$$d_l \in [d_l \min, d_l \max] \quad \text{and} \quad d_r \in [d_r \min, d_r \max].$$

(A3) Without any loss of generality, the slope of the deadzone m is positive and is set to 1.

(A4) The output of the deadzone block $DZ(u)$ is not available for measurement.

Remark 1. Assumptions (A1) and (A2) are the actual physical attributes of a real industrial deadzone and is adopted in the literature [6]. Therefore, the saturation function given by (4) can be shown to have an upper bound by closely analysing Equation (5). Using the general inequality rule $|a \pm b| < |a| + |b|$ we have

$$\|u - d\| \le \|u\| + \|d\| \tag{6}$$

$$\|u + d\| \le \|u\| + \|d\| \tag{7}$$

From Equations (6) from (7) we can state

$$(\|u - d\| - \|u + d\|) \le 0 \tag{8}$$

Multiplying by the slope $\dfrac{m}{2}$ gives

$$\frac{m}{2}(\|u - d\| - \|u + d\|) \le 0 \tag{9}$$

The left hand side of (9) is the saturation function given by (5). In case of a non-symmetrical deadzone function the upper-bound may be chosen by employing assumption (A2) as follows

$$sat_d(u) \le d_M \tag{10}$$

where

$$d_M = \frac{m}{2}(\|d_r \max\| - \|d_l \min\|) \ge 0 \tag{11}$$

The upper bounds will play a pivotal role to ensure the overall global stability of the close loop dynamics as will be demonstrated in the following section.

3. Robust Adaptive Controller Design

Considering the following nonlinear system with input deadzone nonlinearity described as

$$\dot{x} = Ax + B\{DZ(u) + \psi(x)\}$$
$$y = Cx, \tag{12}$$

where the matrices A and B are given by

$$A = \begin{pmatrix} 0 & 1 & \cdots & 0 \\ 0 & 0 & 1 & 0 \\ \vdots & \vdots & \ddots & \vdots \\ 0 & 0 & \cdots & 0 \end{pmatrix}, \quad B = \begin{pmatrix} 0 \\ \vdots \\ 0 \\ 1 \end{pmatrix}$$

And $\psi(x)$ represents the unmeasurable disturbance. The collective bounds can be expressed as

$$\|\psi(x)\| \leq \sum_{k=0}^{p} \gamma_k \|x\|^k \tag{13}$$

Let the reference model to be tracked given by

$$\dot{x}_m = Ax_m + B\{Kx_m + r\} \tag{14}$$

where $K \in R^{1 \times n}$ and r is a reference signal. The tracking error dynamics $\tilde{x} = x - x_m$ may be written as follows:

$$\dot{\tilde{x}} = A\tilde{x} + B\{DZ(u) + \psi(x) - Kx_m - r\} \tag{15}$$

Inserting the deadzone equation (2) into (15) yields

$$\dot{\tilde{x}} = A\tilde{x} + B\{u - sat(u) + \psi(\tilde{x}) - Kx_m - r\} \tag{16}$$

Therefore, for the class of systems described in (13) and deadzone given in (7), we use the result stated as Lemma RANDM in [12] and modify it to ensure asymptotic convergence. The modified controller is

$$u_d(t) = +Kx_m + r - \alpha B^T P\tilde{x} - \hat{\beta}B^T P\tilde{x} + \rho\tanh\left[(a + bt)B^T P\tilde{x}\right] \tag{17}$$

where α, a, $b > 0$, $\rho \geq d_M$, and P is the positive definite symmetric solution of the Algebraic Riccati equation (ARE). The adaptation $\hat{\beta}$ is used to ensure robustness of the controller. The combined output of the compensator and the deadzone nonlinearity may be written as

$$DZ(u_d) = u_d - sat(u_d) = sat(u_d) + Kx_m + r - \alpha B^T P\tilde{x} - \hat{\beta}B^T P\tilde{x} + \rho\tanh\left[(a + bt)B^T P\tilde{x}\right] \tag{18}$$

Inserting the proposed control laws (17) and the output of the deadzone block given by Equation (18) into the error dynamics (16) results in the closed loop dynamics

$$\dot{\tilde{x}} = A\tilde{x} + B\{-sat(u_d) - \alpha B^T P\tilde{x} - \hat{\beta}B^T P\tilde{x} + \rho\tanh\left[(a + bt)B^T P\tilde{x}\right] + \psi(\tilde{x})\} \tag{19}$$

The adaptation law for $\hat{\beta}$ given by

$$\dot{\hat{\beta}} = \Gamma B^T P\tilde{x}^2 \tag{20}$$

where $\Gamma > 0$ is a constant scalar.

Theorem. For the plant described by (12) with input deadzone (1), and the robust adaptive control law (17) along with the adaptive update law (20) will ensure the closed-loop stability and boundedness of tracking error, hence reducing the effects of deadzone.

Proof. Using the following positive definite control Lyapunov function

$$V = \tilde{x}^T P\tilde{x} + \frac{\Gamma^{-1}}{2}\tilde{\beta}^2 \tag{21}$$

where $\hat{\beta} = \tilde{\beta} + \beta^*$ differentiating along the trajectories of the system yields

$$\dot{V} = \dot{\tilde{x}}^T P\tilde{x} + \tilde{x}^T P\dot{\tilde{x}} + \Gamma^{-1}\tilde{\beta}\dot{\hat{\beta}} \tag{22}$$

$$= \left(A\tilde{x} + B\{DZ(u) + \psi(x)\}\right)^T P\tilde{x} + \tilde{x}^T P\left(A\tilde{x} + B\{DZ(u) + \psi(x)\}\right) + \Gamma^{-1}\tilde{\beta}\dot{\hat{\beta}} \tag{23}$$

Substituting for the closed loop dynamics given by (19) in (23) gives

$$\dot{V} = \left(A\tilde{x} + B\left\{ -sat(u_d) - \alpha B^T P\tilde{x} - \hat{\beta} B^T P\tilde{x} + \rho\tanh\left[(a+bt) B^T P\tilde{x} \right] + \psi(x) \right\} \right)^T P\tilde{x}$$
$$+ \tilde{x}^T P\left(A\tilde{x} + B\left\{ -sat(u_d) - \alpha B^T P\tilde{x} - \hat{\beta} B^T P\tilde{x} + \rho\tanh\left[(a+bt) B^T P\tilde{x} \right] + \psi(x) \right\} \right) + \Gamma^{-1}\tilde{\beta}\dot{\hat{\beta}}. \tag{24}$$

Collecting terms and simplifying

$$\dot{V} = \tilde{x}^T \left(A^T P + PA \right)\tilde{x} + 2\tilde{x}^T PB\left\{ -\alpha B^T P\tilde{x} - \hat{\beta} B^T P\tilde{x} + \psi(x) \right\}$$
$$+ 2\tilde{x}^T PB\left\{ -sat(u_d) + \rho\tanh\left[(a+bt) B^T P\tilde{x} \right] \right\} + \Gamma^{-1}\tilde{\beta}\dot{\hat{\beta}}. \tag{25}$$

Rearranging terms we get

$$\dot{V} = \tilde{x}^T \left(A^T P + PA - 2\alpha PBB^T P \right)\tilde{x} - \hat{\beta} B^T P\tilde{x}\tilde{x}^T PB + 2\tilde{x}^T PB\psi(x)$$
$$+ 2\tilde{x}^T PB\left\{ -sat(u_d) + \rho\tanh\left[(a+bt) B^T P\tilde{x} \right] \right\} + \Gamma^{-1}\tilde{\beta}\dot{\hat{\beta}}. \tag{26}$$

The first term can be simplified by solving the Algebraic Reccati Equation given by

$$A^T P + PA - 2\alpha PBB^T P = -Q \tag{27}$$

resulting in

$$\dot{V} = -\tilde{x}^T Q\tilde{x} - \hat{\beta}\tilde{x}^T PB^2 + 2\tilde{x}^T PB\ \psi(\tilde{x}) + \Gamma^{-1}\tilde{\beta}\dot{\hat{\beta}} - 2\tilde{x}^T PB\left(sat(u_d) - \rho\tanh\left[(a+bt) B^T P\tilde{x} \right] \right) \tag{28}$$

Replacing the adaptation law (20) and replacing $\hat{\beta} = \tilde{\beta} + \beta^*$ in (28) yields

$$\dot{V} = -\tilde{x}^T Q\tilde{x} - \left(\tilde{\beta} + \beta^* \right)\tilde{x}^T PB^2 + \tilde{\beta}\tilde{x}^T PB^2 + 2\tilde{x}^T PB\psi(x) - 2\tilde{x}^T PB\left(sat(u_d) - \rho\tanh\left[(a+bt) B^T P\tilde{x} \right] \right) \tag{29}$$

$$\dot{V} = -\tilde{x}^T Q\tilde{x} - \beta^*\tilde{x}^T PB^2 + 2\tilde{x}^T PB\left[(\psi(x)) \right] - 2\tilde{x}^T PB\left(sat(u_d) - \rho\tanh\left[(a+bt) B^T P\tilde{x} \right] \right) \tag{30}$$

So far the first two terms are negative. As for the third term we utilize the general inequality $2ab \le a^2 + b^2$ to establish proper bounds as follows

$$\left\| 2\tilde{x}^T PB \cdot \psi(x) \right\| \le \varsigma\left\| \tilde{x}^T PB \right\|^2 + \varsigma^{-1}\left\| \psi(x) \right\|^2 \tag{31}$$

Using the inequality (13) to modify (31) to become

$$\varsigma\left\| \tilde{x}^T PB \right\|^2 + \varsigma^{-1}\left\| \psi(x) \right\|^2 \le \varsigma\left\| \tilde{x}^T PB \right\|^2 + \varsigma^{-1}\gamma\left\| \tilde{x} \right\|^2 \tag{32}$$

Therefore, the inequality of (32) can be incorporated in (30) as

$$\dot{V} \le -\left(\lambda_{min}(Q) - \varsigma^{-1}\gamma \right)\left\| \tilde{x} \right\| - \beta^*\left\| \tilde{x}^T PB \right\|^2 + \varsigma\left\| \tilde{x}^T PB \right\|^2 - 2\tilde{x}^T PB\left(sat(u_d) - \rho\tanh\left[(a+bt) B^T P\tilde{x} \right] \right) \tag{33}$$

By choosing the degree of freedom ς satisfying the condition $\varsigma < \dfrac{\lambda_{min}(Q)}{\gamma}$ and choosing β^* to be greater than ς ensures that the first three terms of $\dot{V}$ negative. Meanwhile, the last term in (33) can be upper bounded as follow

$$-2\tilde{x}^T PB\left(sat(u_d) - \rho\tanh\left\{ (a+bt) B^T P\tilde{x} \right\} \right) \le -2\left\| \tilde{x}^T PB \right\|\left(\left\| sat(u_d) \right\| - \rho\left(\tanh\left\{ (a+bt)\left\| \tilde{x}^T PB \right\| \right\} \right) \right) \tag{34}$$

Utilizing the upper bounds on $\left\| sat(u_d) \right\| < d_M$ given by (10) and rewriting the right hand side of (34)

$$-2\left\| \tilde{x}^T PB \right\| \cdot \rho \cdot \left(\delta - \tanh\left[(a+bt)\left\| \tilde{x}^T PB \right\| \right] \right) \tag{35}$$

where $\delta = d_M/\rho < 1$. Therefore the last term in (35) insures that $\dot{V} \le 0$ as long as

$$\delta - \tanh\left[(a+bt)\left\| \tilde{x}^T PB \right\| \right] > 0 \tag{36}$$

or

$$\left\|\tilde{x}^{T} PB\right\| > \Pi(t) \triangleq \frac{1}{2(a+bt)} \ln\left(\frac{1+\delta}{1-\delta}\right) \tag{37}$$

To conclude, by choosing $\rho > d_M$ then $\dot{V}$ is rendered negative and $\left\|\tilde{x}^T PB\right\|$ converges to a closed and vanishing region as time increases. Therefore, since $\left\|\tilde{x}^T PB\right\| \to 0$ as $t \to 0$ and $\tilde{x}^T PB = c_1\tilde{x}_1 + c_2\tilde{x}_2 + \cdots + c_n\tilde{x}_n = 0$ implies that by choosing the c vector as the coefficient of a strictly Hurwitz polynomial will make the closed loop system error asymptotically stable. For a more through conclusion of the proof one may refer to [13].

4. Illustrative Example

In this section, we illustrate the proposed controller to compensate for a system with a deadzone nonlinearity presented in [14] as

$$\ddot{x} = a_1 \frac{(1-e^{-x})}{(1+e^{-x})} - a_2(\dot{x}^2 + 2x)\sin(\dot{x}) - 0.5a_3 x\sin(3t) + DZ(u) \tag{38}$$

The parameter used for the simulation is shown in **Table 1**. The plant (41) may be written in state space form by defining $x = x_1$ and $\dot{x} = x_2$ then

$$\dot{x}_1 = x_2 \tag{39}$$

$$\dot{x}_2 = a_1 \frac{(1-e^{-x_1})}{(1+e^{-x_1})} - a_2(x_2^2 + 2x_1)\sin(x_2) - 0.5a_3 x_1\sin(3t) + DZ(u)$$

Resulting in

$$A = \begin{pmatrix} 0 & 1 \\ 0 & 0 \end{pmatrix}, \quad B = \begin{pmatrix} 0 \\ 1 \end{pmatrix}$$

The solution of the ARE equation was chosen to be

$$P = \begin{pmatrix} 130.17 & 22.36 \\ 22.36 & 14.55 \end{pmatrix}, \quad Q = \begin{pmatrix} 200 & 0 \\ 0 & 40 \end{pmatrix}$$

Table 1. Parameters utilized in the example.

Item	Systems Physical Attributes		
	Parameter	Value	Unit
1	k_p	40.0	PD Gain
2	k_v	13.0	PD gain
3	d_r^*	20.0	Deadzone Right
4	d_l^*	15.0	Deadzone Left
6	Γ	10.0	Gains
7	b, a_1, a_2, a_3	1.0	Scalars
9	ρ	4.0	Gain
10	α	0.2	Gain
11	a, b	0.2	Scalars

The reference model to be tracked is

$$\dot{x}_m = \begin{pmatrix} 0 & 1 \\ -k_p & -k_v \end{pmatrix} \begin{pmatrix} x_{m1} \\ x_{m2} \end{pmatrix} + \begin{pmatrix} 0 \\ 1 \end{pmatrix} r(t)$$

for a sinusoidal reference trajectory given by

$$r(t) = 2 + \sin(\omega_d t) + \frac{1}{2}\sin(5\omega_d t)$$

Meanwhile, the unmeasurable disturbance $\psi(x)$ can be collectively bounded as

$$\|\psi(x)\| = a_1 \frac{(1 - e^{-x_1})}{(1 + e^{-x_1})} - a_2(x_2^2 + 2x_1)\sin(x_2) - 0.5a_3 x_1 \sin(3t)$$

$$\leq \sum_{k=0}^{p} \gamma_k \|x\|^k$$

(40)

Simulations of the system in (39) under the adaptive control law (17) and (20) have been performed. The upper bounds on actuator actual spacing $d_M = d_r - d_l = 35$ is assumed unknown. In order to demonstrate the performance improvement accomplished by our proposed method, the system under test given in (39) was used. The efficacy of the proposed method is proven by comparing its performance against the performance of a classic PD controller having equivalent gains. The complete parameters of the system under test and controller gains are listed in **Table 1**.

The simulation results presented in **Figure 2** and **Figure 3**, clearly show the tracking performance for x_1 and x_2 states along with their respective reference trajectory. **Figure 4** and **Figure 5** demonstrate the tracking error for the states $\tilde{x}_1$ and $\tilde{x}_2$ in blue in addition to the same tracking errors for the system under a PD controller in red. In both figures, the PD controller resulted in limit cycles where as the adaptive controller proved to be stable with no limit cycles and improved performance with a zero approaching tracking error. **Figure 6** shows the control effort $u_d(t)$ and $DZ(u_d)$. In **Figure 7** the evolution of the $\hat{\beta}$ which clearly demonstrates the boundedness of the adaptation. In **Figure 8**, the reference trajectory is changed to demonstrates a superior a step response performance when compared with PD controller. While the step error is approaching zero for the system under the proposed adaptive controller, a steady state error is persistent with the PD controller.

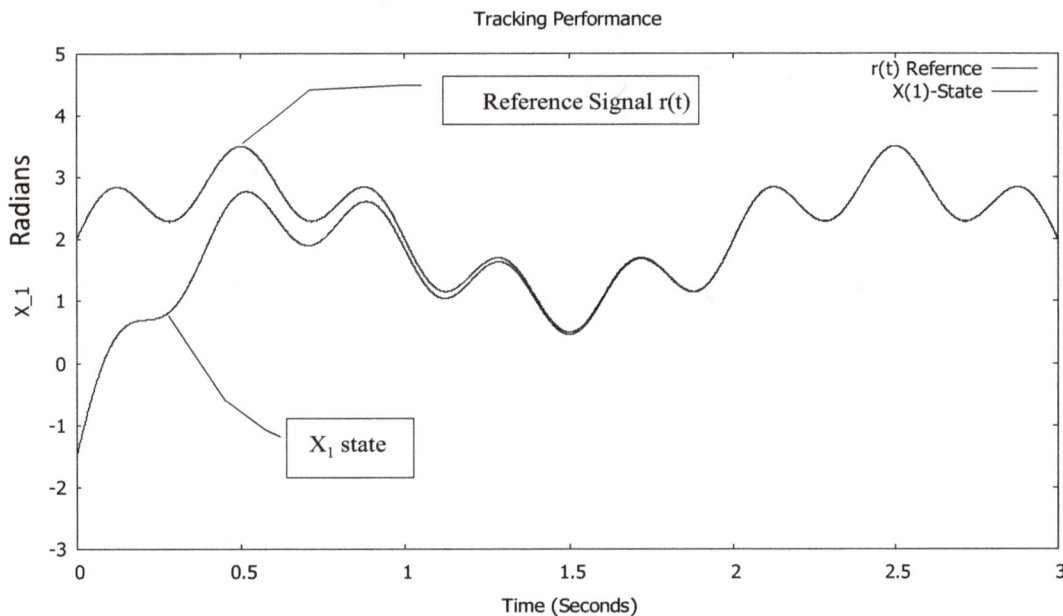

Figure 2. The tracking performance of x_1 state of the system under the proposed robust controller.

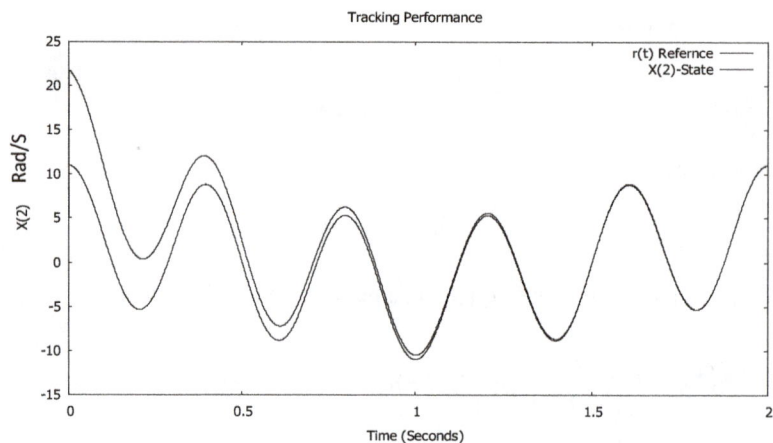

Figure 3. The tracking performance of x_2 state of the system.

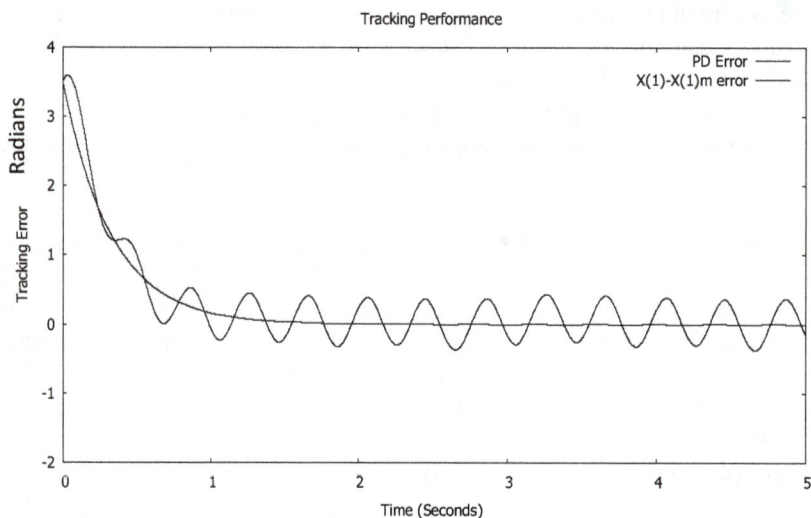

Figure 4. The adaptively compensated tracking error $\tilde{x}_1$ (blue) vs. the same tracking error of the system under a PD controller (red).

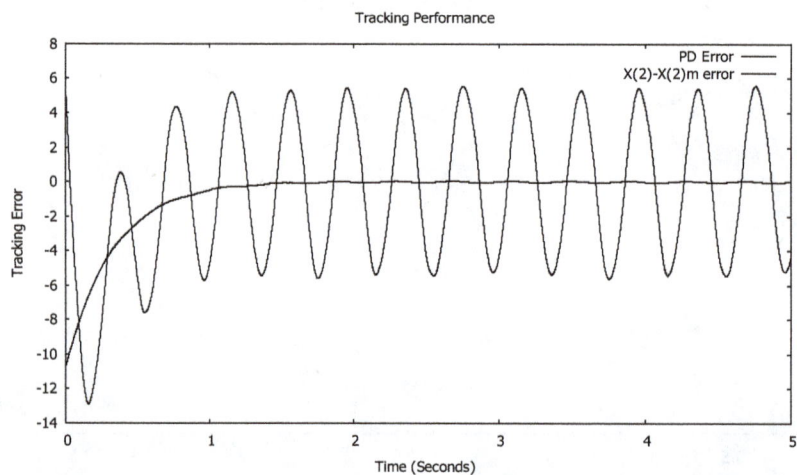

Figure 5. The adaptively compensated tracking error $\tilde{x}_2$ (blue) vs. the same tracking error of the system under a PD controller (red).

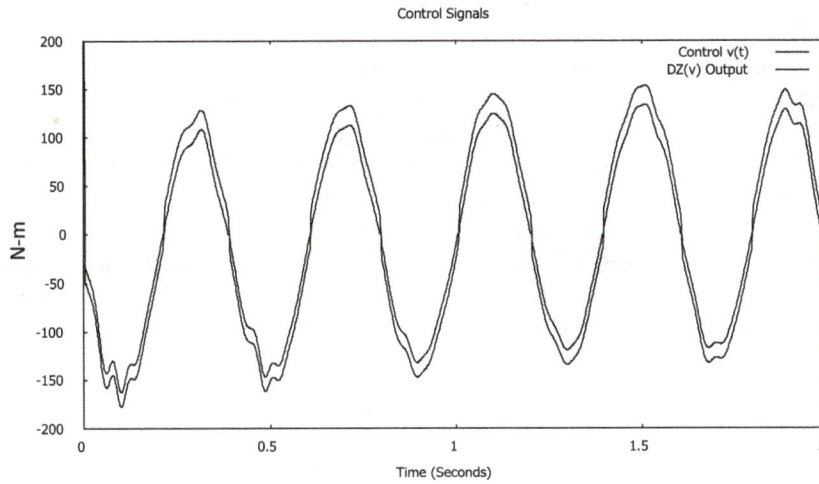

Figure 6. The control effort $DZ(u_d)$ in red vs. $v = u_d$ for the deadzone compensated system.

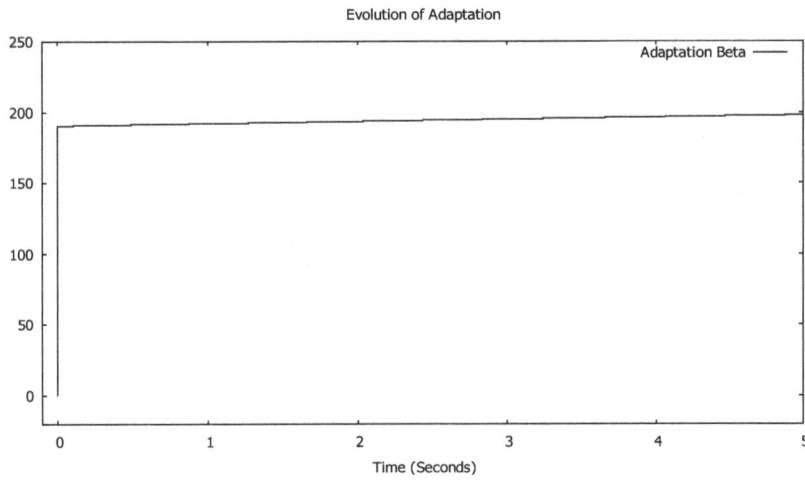

Figure 7. Evolution of the adaptation $\hat{\beta}$.

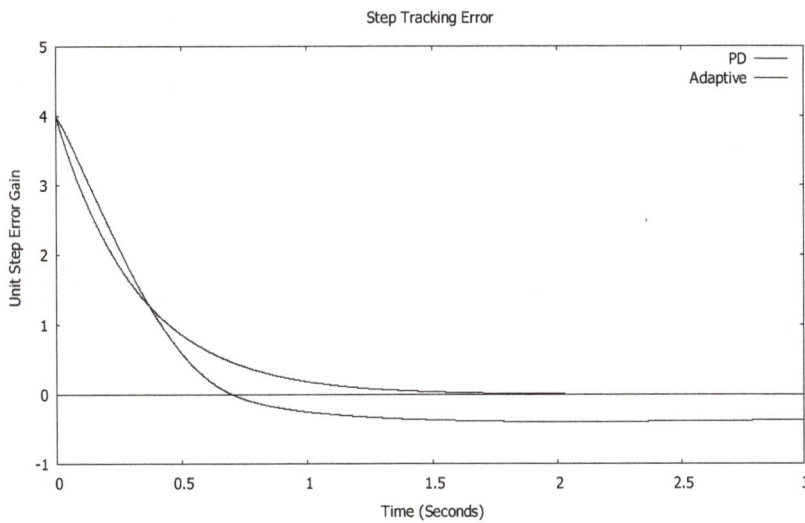

Figure 8. Step tracking performance error for $\tilde{x}_1$ for input $r(t) = 1$.

5. Conclusion

A robust adaptive controller is used to control a class of nonlinear systems with input deadzone nonlinearity at the input. The robust controller was shown to be superior in performance when compared to a more conventional control method such as a PD controller. The system under the proposed scheme has been shown to not only effectively stabilize a second order complex nonlinear system with disturbance, but also achieve asymptotic tracking. The advantage of the proposed method lies in the fact that no knowledge of the deadzone parameters needed and only an upper bound for the deadzone spacing is required. The adaptive deadzone inverse controller is smoothly differentiable and can easily be combined with any of the advanced control methodologies. The asymptotic stability of the closed-loop system has been proven by using Lyapunov arguments and simulation results confirmed the efficacy of the control methodology.

Funding

This work is supported and funded by the Public Authority of Applied Education and Training, Research Project No. (TS-14-03) t, Research Title (Adaptive Control of Systems with Output Deadzone).

References

[1]   Recker, D.A. and Kokotovic, P.V. (1993) Indirect Adaptive Nonlinear Control of Discrete-Time Systems Containing a Deadzone. *Proceedings of the 32nd Conference on Decision and Control*, 15-17 December 1993, San Antonio, 2647, 2653.

[2]   Toa, G. and Kokotovic, P. (1994) Adaptive Control of Plants with Unknown Dead-Zones. *IEEE Transactions on Automatic Control*, **39**, 59-68. http://dx.doi.org/10.1109/9.273339

[3]   Tao, G. and Kokotovic, P. (1995) Discrete-Time Adaptive Control of Systems with Unknown Dead-Zones. *International Journal of Control*, **61**, 1-17. http://dx.doi.org/10.1080/00207179508921889

[4]   Lewis, F.L., Tim, W.K., Wang, L.-Z. and Li, Z.X. (1999) Deadzone Compensation in Motion Control System Using Adaptive Fuzzy Logic Control. *IEEE Transactions on Control Systems Technology*, **7**, 731-742. http://dx.doi.org/10.1109/87.799674

[5]   Selmic, R.R. and Lewis, F.L. (2000) Deadzone Compensation in Motion Control Systems Using Neural Networks. *IEEE Transaction of Automatic Control*, **45**, 602-613.

[6]   Wang, X.-S., Su, C.-Y. and Hong, H. (2004) Robust Adaptive Control of a Class of Nonlinear Systems with Unknown Dead-Zone. *Automatica*, **40**, 407-413.

[7]   Zhonghua, W., Bo, Y., Lin, C. and Shusheng, Z. (2006) Robust Adaptive Deadzone Compensation of DC Servo System. *IEE Proc.-Control Theory Application*, **153**, 709-713.

[8]   Sedighi, H.M., Shirazi, K.H. and Zare, J. (2012) Novel Equivalent Function for Deadzone Nonlinearity: Applied to Analytical Solution of Beam Vibration Using He's Parameter Expanding Method. *Latin American Journal of Solids and Structures*, **9**, 1-10.

[9]   Ahmad, N.J., Alnaser, M.J. and Alsharhan, W.E. (2013) Asymptotic Tracking of Systems with Non-Symmetrical Input Deadzone Nonlinearity. *International Journal of Automation and Power Engineering*, **2**, 287-292.

[10] Ahmad, N.J., Ebraheem, H., Alnaser, M. and Alostath, J. (2011) Adaptive Control of a DC Motor with Uncertain Deadzone Nonlinearity at the Input. *Control and Decision Conference (CCDC)*, 4295-4299.

[11] Sedighi, H.M., Shirazi, K.H. and Zare, J. (2012) Novel Equivalent Function for Deadzone Nonlinearit: Appled to Analytical Solution of Beam Vibration Using He's Parameter Expanding Method. *Latin American Journal of Solids and Structures*, **9**, 443-451. http://dx.doi.org/10.1590/S1679-78252012000400002

[12] Jain, S. and Khorrami, F. (1995) Robust Adaptive Control of a Class of Nonlinear Systems: State and Output Feedback. *Proceeding of the* 1995 *American Control Conference*, Seatle, 1580-1584.

[13] Sankaranarayanan, S., Melkote, H. and Khorrami, F. (1999) Adaptive Variable Structure Control of a Class of Nonlinear Systems with Nonvanishing Perturbations via Backstepping. *Proceedings of the American Control Conference*, San Diego, 4491-4495.

[14] Zhang, T.P. and Feng, C.B. (1997) Adaptive Fuzzy Sliding Mode Control for a Class of Nonlinear Systems. *ACTA Automatica Sinica*, **23**, 361-369.

17

No-Wait Flowshops to Minimize Total Tardiness with Setup Times

Tariq Aldowaisan, Ali Allahverdi

Department of Industrial and Management Systems Engineering, Kuwait University, Kuwait City, Kuwait
Email: tariq.aldowaisan@ku.edu.kw, ali.allahverdi@ku.edu.kw

Abstract

The m-machine no-wait flowshop scheduling problem is addressed where setup times are treated as separate from processing times. The objective is to minimize total tardiness. Different dispatching rules have been investigated and three were found to be superior. Two heuristics, a simulated annealing (SA) and a genetic algorithm (GA), have been proposed by using the best performing dispatching rule as the initial solution for SA, and the three superior dispatching rules as part of the initial population for GA. Moreover, improved versions of SA and GA are proposed using an insertion algorithm. Extensive computational experiments reveal that the improved versions of SA and GA perform about 95% better than SA and GA. The improved version of GA outperforms the improved version of SA by about 3.5%.

Keywords

No-Wait Flowshop, Scheduling, Setup Times, Total Tardiness, Simulated Annealing, Genetic Algorithm

1. Introduction

In a no-wait flowshop environment, jobs are processed continuously from start to end without interruptions either on or between machines. Therefore, the start of a job on a given machine may be delayed to ensure that its operation completion coincides with the start of its subsequent operation on the next machine. Applications of such an environment are found in several industries; e.g., metal, plastic, chemical, and semiconductor. Ref. [1] surveyed the applications and research on the no-wait flowshop environment.

The m-machine no-wait flowshop problem has been addressed with respect to several performance measures including makespan and total completion time. For the makespan, examples of research are Refs. [2]-[9]. For separate setup times and makespan performance measure, see Ref. [10]. As for the total completion time, some of the works include Refs. [5] [11]-[13]. Ref. [14] proposed a constructive heuristic for minimizing total flow-time where setup times were considered as separate.

While makespan and total completion time are directly related to job completion times, other measures such as number of tardy jobs, maximum tardiness, and total tardiness are more focused on job due dates. Recent research on the latter type of measures includes Refs. [15]-[17]. Ref. [17] proposed six different heuristics for minimizing total tardiness. Ref. [16] considered both makespan and maximum tardiness where the objective is to find approximations of Pareto front by using simulating annealing. Ref. [15] proposed several heuristics for the m-machine no-wait flowshop problem with respect to the number of tardy jobs. It is important to note that the minimization of the number of tardy jobs directly affects costs associated with late delivery as well as the percentage of on-time shipments, which is often used to appraise managers' performance. Unlike the number of tardy jobs, the total tardiness measure accounts for the duration of delay which has financial as well as reputation and good will impacts. In many practical situations, the cost increases with the delay duration. This increase may take the form of financial penalties corresponding to different time period delays. Therefore, total tardiness is a more appropriate measure for such situations.

Ref. [18] addressed the m-machine no-wait scheduling problem with respect to the total tardiness performance measure, where they proposed several efficient heuristics. However, they did not consider setup time which was relevant in some industries. Ref. [19] pointed out the significance of considering setup times and Refs. [20] and [21] provided a survey of the research on scheduling with setup times.

2. Notation

For the m-machine no-wait flowshop, we assume that the set of n jobs are ready for processing at time zero. Each job $j=1,\cdots,n$ has to be processed continuously in the same order on a set of $i=1,\cdots,m$ machines without interruptions either on or between machines. The following notation will be used.

E: tardiness factor.

R: due date range factor.

d_j: due date of job j; which is generated using the E and R factors.

C_j: completion time of job j on the last machine.

T_j: tardiness of job j; where $T_j = \max\{C_j - d_j, 0\}$.

P_{ij}: processing time of job j on machine i.

S_{ij}: setup time of job j on machine i.

The objective is to sequence all jobs to minimize the total tardiness.

3. Dispatching Rules

Dispatching rules are simple heuristics for building a schedule. Their popularity is due to their ability to rapidly provide good solution in practical production settings. They are also used as initial sequences for the proposed heuristics in the next section. A number of well-known dispatching rules have been initially investigated with regard to the considered criterion. Three of the considered dispatching rules outperformed the others. The following is an explanation of the three.

Let s denote the sequence of jobs which are scheduled so far and t denote the time at which jobs need to be selected. Moreover, let $C_j(s)$ denote the completion time of job j considered for scheduling (not in s yet) if it is scheduled as the last job in the sequence s.

1. Earliest due date with processing and setup times (EDDP); where jobs are sequenced in non-decreasing order of $d_j / \sum_{i=1}^{m}(S_{ij} + P_{ij})$.

2. Modified due date (MDD); where jobs are sequenced in non-decreasing order of $\max\{d_j, C_j(s)\}$.

A third dispatching rule is proposed as follows:

Sequence the jobs with the minimum total tardiness of jobs based on averaging the setup and processing time for each job across all machines (TOTA), reducing it to a single machine problem. The average of the setup and processing time of each job across all machines is first determined, then the sequence is obtained by finding the optimal sequence based on the average setup and processing time for each job. The due date for TOTA is calculated by dividing the due dates associated with the m-machine by the number of machines.

When comparing the three rules of MDD, EDDP, and TOTA, it was found that (see **Figure 1**) MDD performed best followed by TOTA. For all the considered experimental cases, which will be detailed later, MDD performed about 60% better than TOTA; while TOTA outperformed EDDP by about 15%.

In the next section, we use the MDD as an initial sequence to develop an improved simulated annealing heuristic; and use the three dispatching rules MDD, TOTA, and EDDP as part of the initial generation population to develop an improved genetic algorithm heuristic.

4. Proposed Heuristics FISA2 and FIGA2

[18] proposed a simulated annealing algorithm, called ISA, and a genetic algorithm, called IGA. The same algorithms ISA and IGA of [18] are used here after modifying the various input parameters, including initial sequences, to account for the setup time.

The parameters of ISA and IGA have also been investigated for the separate setup time environment. Experimentations revealed that the parameter values of ISA and IGA which were found to do well in Ref. [18] were also found to perform well in the separate setup environment. **Table 1** and **Table 2** summarize the tested and selected parameters' values for the two heuristics.

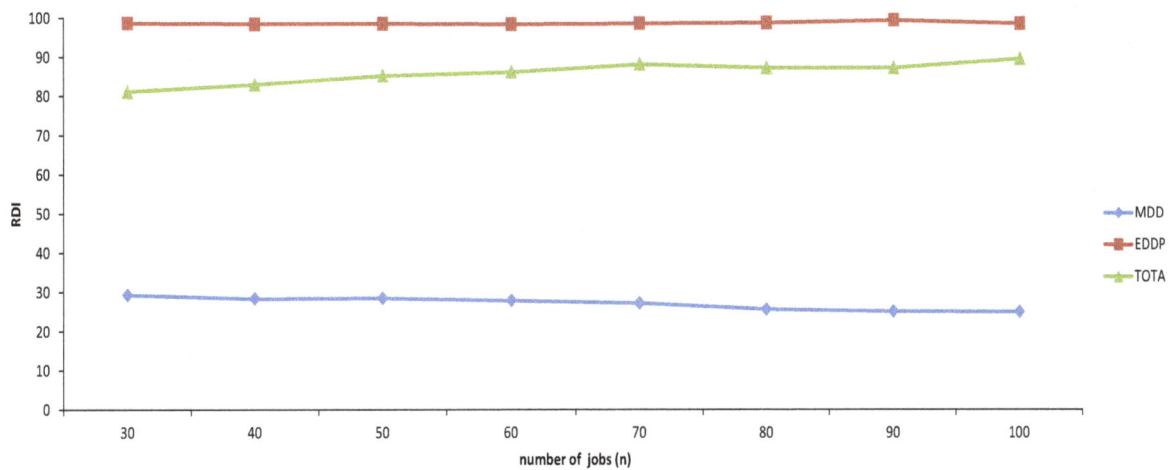

Figure 1. Comparison of dispatching rules.

Table 1. Selected parameters' values for ISA.

Parameter	Description	Tested Values	Selected Values
MAM	Max Accepted Moves	$n/2, n, 2n, 3n, 4n$	$2n$
MTM	Max Total Moves	$n/2, n, 2n, 3n, 4n, 5n$	$5n$
MFC	Max Freezing Counter	2, 5, 10, 15, 20	10
FT	Freezing Temperature	5, 10, 20, 30, 150	20
TRF	Temperature Reduction Factor	0.60, 0.70, 0.80, 0.80, 0.90, 0.95	0.90
AMP	Accepted Move Percentage	0.05, 0.10, 0.15, 0.20, 0.25, 0.30, 0.35	0.10

Table 2. Selected parameters' values for IGA.

Parameter	Description	Tested Values	Selected Values
POP	Population Size	$n/2, n, 2n, 3n$	n
GEN	Number of Generations	$n/2, n, 2n, 3n$	$2n$
PC	Probability of Crossover	0.70, 0.75, 0.80, 0.85, 0.90	0.75
PM	Probability of Mutation	0.00, 0.01 0.02, 0.03, 0.04, 0.05, 0.10	0.05

Ref. [18] applied an insertion and exchange procedures to ISA and IGA and obtained better heuristics called FISA and FIGA with about 90% improvement over ISA and IGA.

In this paper, in addition to the one-block procedure used by [18], we propose a two-block insertion procedure. The following are the steps of the two-block insertion procedure.

Step 1: Use ISA as initial sequence s.

Step 2: Set $k = 1$.

Step 3: Pick the first two jobs from s and select the better with regard to total tardiness of the two permutations as the current solution.

Step 4: Set $k = k + 1$.

Step 5: Generate candidate sequences by selecting the next two jobs from s and inserting the two permutations of these two jobs into each position of the current solution. Among the generated candidates, the sequence with the least total tardiness value becomes the current solution.

Step 6: Go to Step 4 until all jobs in s have been considered.

For a problem with odd number of jobs, the last job to be selected will be inserted as a single job in all possible positions in the current solution.

We have investigated the one-block and two-block insertion procedures for the considered problem, the results indicated that the one-block outperforms the two-block one. The resulting two heuristics of applying the one-block insertion and exchange procedures are denoted as FISA2 and FIGA2. The same insertion is used for the FIGA2 by replacing ISA with IGA in Step 1.

5. Computational Analysis

In this section we conduct experiments to first study the effect of the various parameters such as E and R, and evaluate the performances of the proposed heuristics against each other. The experiments were performed on a PC running Windows 7 32-bits, with an Intel Dual Core CPU 2.26 GHz and 3 GB RAM. The data generation and testing application were developed using the C# programming language which runs on the top of Microsoft's NET Framework 3.5. **Table 3** describes the experiment parameters and their values.

The considered values E, R, p_{ij}, and s_{ij} are commonly used in the scheduling literature; e.g. [22]. The due dates of jobs are generated using the parameters E and R with a uniform distribution between $LB(1 - E - R/2)$ and $LB(1 - E + R/2)$, where LB is a tight lower bound on the makespan.

$$LB = \max_{1 \leq j \leq m} \left\{ \sum_{i=1}^{n} \left(S_{ij} + P_{ij} \right) + \min_{1 \leq i \leq n} \sum_{k=1}^{j-1} \left(S_{ij} + P_{ij} \right) + \min_{1 \leq i \leq n} \sum_{k=j+1}^{m} \left(S_{ij} + P_{ij} \right) \right\}$$

where, $\sum_{k=1}^{0} \left(S_{ij} + P_{ij} \right) = 0$ and $\sum_{k=m+1}^{m} \left(S_{ij} + P_{ij} \right) = 0$.

The total number of all possible combinations is 1296 based on the considered values of n, m, E, R, and k; where each combination is replicated 30 times.

The performance measure used for evaluating the heuristics is the relative deviation index (RDI); calculated as follows:

Table 3. Experiment parameters.

Parameter	Values
n	30, 40, …, 100
m	5, 8, …, 20
E	0.2, 0.4, 0.6
R	0.2, 0.6, 1.0
k	0.5, 1.0, 1.5
p_{ij}	Uniform (1, 100)
s_{ij}	Uniform (1, 100 k)

$$\text{RDI} = (\text{Heuristic Solution} - \text{Best Solution}) / (\text{Worst Solution} - \text{Best Solution}) \times 100$$

where the Heuristic Solution is the solution obtained by a given heuristic, Best Solution and Worst Solution are the best and worst solutions obtained from among the compared heuristics. The RDI produces a result between 0 and 100; where the best solution will have RDI values closest to 0.

Figure 2 shows the results of the RDI comparison for different number of jobs. The ISA heuristic performed better than IGA by about 6%; however both of them are significantly outperformed by FIGA2 and FISA2 by over 95%. The difference between FIGA2 and FISA2 is not significant; though FIGA2 slightly outperforms FISA2 by less than 3.5%. **Figure 3**, which shows the performance for different number of machines, yielded similar results. In both **Figure 3** and **Figure 4**, the increase in number of jobs and machines did not show any significant impact on the RDI performance.

Figure 4 shows that using various E/R combinations has generally low impact on RDI; except for a slight improvement in ISA performance for high E values. With respect to varying the k value to control the setup time duration, it didn't generally have a significant impact.

As shown from **Figure 5**, the computational time for IGA and FIGA2 are higher than the ISA and FISA2 with FIGA2 being the highest and ISA the lowest. For the largest number of jobs $n = 100$, the computational time for all heuristics is less than 2 seconds. Therefore, computational time is considered negligible.

6. Conclusion

The m-machine no-wait flowshop scheduling problem, where setup times are considered as separate from processing times, with the objective of minimizing total tardiness has been considered. Different dispatching rules

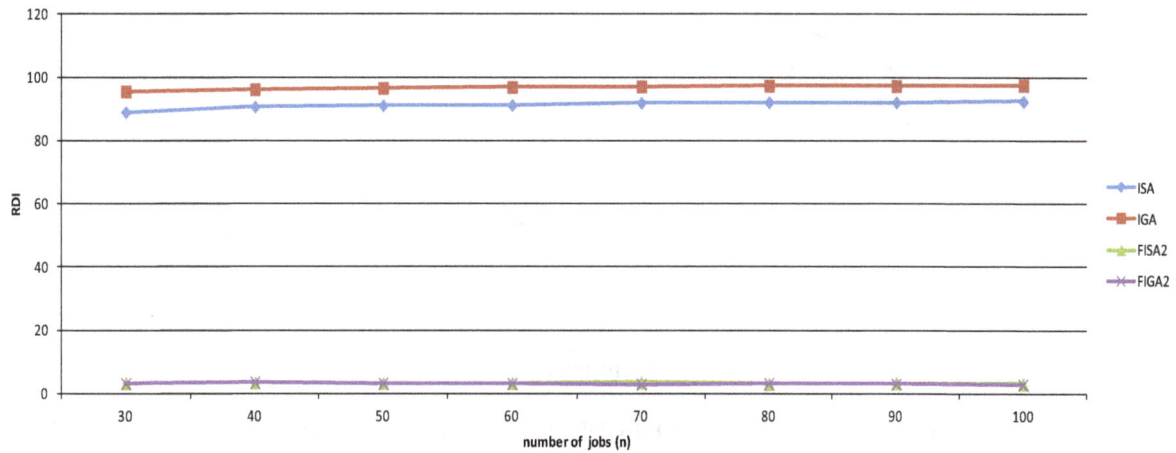

Figure 2. Comparison of heuristics for RDI criterion vs. n; for $\overline{E}$, $\overline{R}$, and $\overline{m}$.

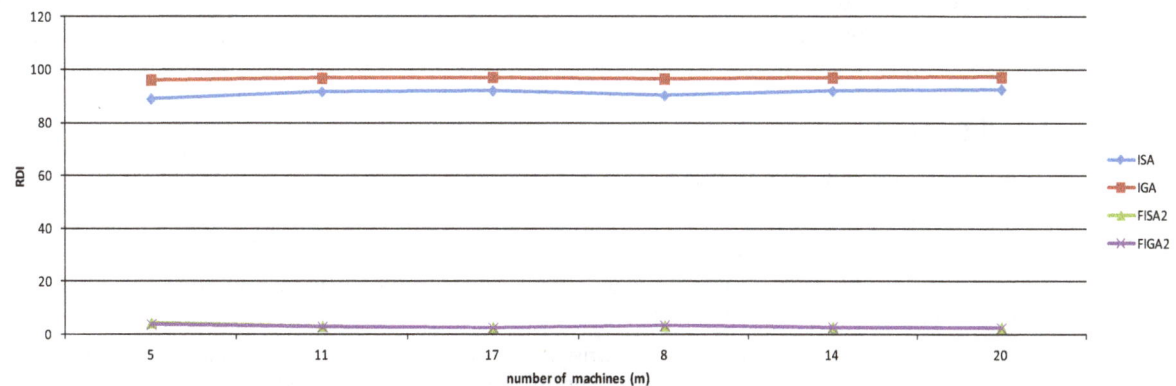

Figure 3. Comparison of heuristics for RDI criterion vs. m; for $\overline{E}$, $\overline{R}$, and $\overline{n}$.

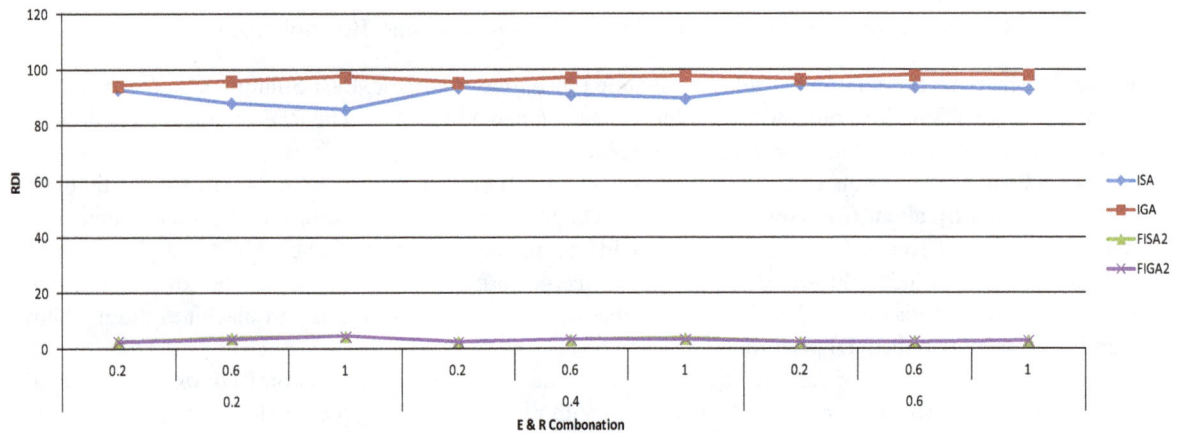

Figure 4. Comparison of heuristics for RDI criterion for different *E/R* combinations.

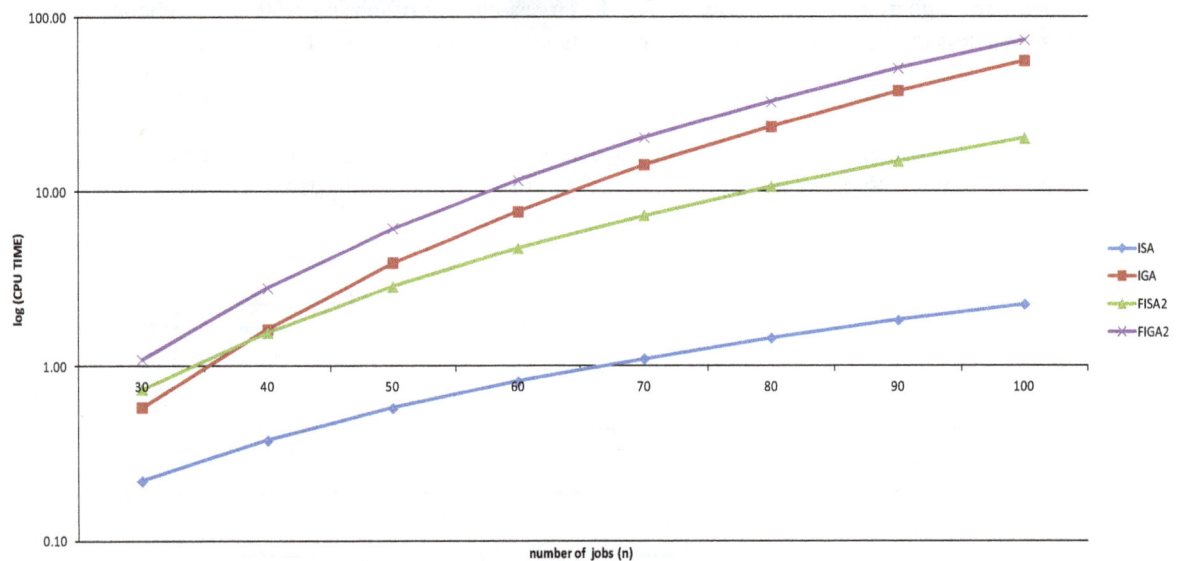

Figure 5. The computational time of heuristics for different number of jobs.

have been proposed and three of which, MDD, EDDP, TOTA, have been found performing well. The dispatching rule MDD outperforms TOTA about 60%, which outperforms EDDP about 15%. Furthermore, a simulated annealing algorithm (ISA) and a genetic algorithm (IGA) have been proposed. It has been found that IGA outperforms ISA about 6%. Moreover, improved versions of ISA and IGA have been proposed and it has been found that improved versions outperform ISA and IGA about 95%. Of the improved versions, FIGA2 outperforms FISA2 about 3.5%.

Acknowledgements

This research was supported by Kuwait University Research Administration project number Grant No. EI01/12.

References

[1]   Hall, N.G. and Sriskandarajah, C. (1996) A Survey of Machine Scheduling Problems with Blocking and No-Wait in Process. *Operations Research*, **44**, 510-525. http://dx.doi.org/10.1287/opre.44.3.510

[2]   Aldowaisan, T. and Allahverdi, A. (2003) New Heuristics for No-Wait Flowshops to Minimize Makespan. *Computers & Operations Research*, **30**, 1219-1231. http://dx.doi.org/10.1016/S0305-0548(02)00068-0

[3]   Allahverdi, A. and Aldowaisan, T. (2004) No-Wait Flowshops with Bicriteria of Makespan and Maximum Lateness.

European Journal of Operational Research, **152**, 132-147. http://dx.doi.org/10.1016/S0377-2217(02)00646-X

[4]   Framinan, J.M. and Nagano, M.S. (2008) Evaluating the Performance for Makespan Minimisation in No-Wait Flow-shop Sequencing. *Journal of Materials Processing Technology*, **197**, 1-9. http://dx.doi.org/10.1016/j.jmatprotec.2007.07.039

[5]   Pan, Q.K., Fatih Tasgetiren, M. and Liang, Y.C. (2008) A Discrete Particle Swarm Optimization Algorithm for the No-Wait Flowshop Scheduling Problem. *Computers & Operations Research*, **35**, 2807-2839. http://dx.doi.org/10.1016/j.cor.2006.12.030

[6]   Kalczynski, P.J. and Kamburowski, J. (2007) On No-Wait and No-Idle Flow Shops with Makespan Criterion. *European Journal of Operational Research*, **178**, 677-685. http://dx.doi.org/10.1016/j.ejor.2006.01.036

[7]   Davendra, D., Zelinka, I., Bialic-Davendra, M., Senkerik, R. and Jasek, R. (2013) Discrete Self-Organising Migrating Algorithm for Flow-Shop Scheduling with No-Wait Makespan. *Mathematical and Computer Modelling*, **57**, 100-110. http://dx.doi.org/10.1016/j.mcm.2011.05.029

[8]   Zhu, J., Li, X. and Wang, Q. (2009) Complete Local Search with Limited Memory Algorithm for No-Wait Job Shops to Minimize Makespan. *European Journal of Operational Research*, **198**, 378-386. http://dx.doi.org/10.1016/j.ejor.2008.09.015

[9]   Tseng, L.Y. and Lin, Y.T. (2010) A Hybrid Genetic Algorithm for No-Wait Flowshop Scheduling Problem. *International Journal of Production Economics*, **128**, 144-152. http://dx.doi.org/10.1016/j.ijpe.2010.06.006

[10] Nagano, M.S., Da Silva, A.A. and Nogueira Lorena, L.A. (2014) An Evolutionary Clustering Search for the No-Wait Flow Shop Problem with Sequence Dependent Setup Times. *Expert Systems with Applications*, **41**, 3628-3633. http://dx.doi.org/10.1016/j.eswa.2013.12.013

[11] Chen, C.L., Neppalli, R.V. and Aljaber, N. (1996) Genetic Algorithms Applied to the Continuous Flow Shop Problem. *Computers & Industrial Engineering*, **30**, 919-929. http://dx.doi.org/10.1016/0360-8352(96)00042-3

[12] Aldowaisan, T. and Allahverdi, A. (2004) New Heuristics for m-Machine No-Wait Flowshop to Minimize Total Completion Time. *Omega*, **32**, 345-352. http://dx.doi.org/10.1016/j.omega.2004.01.004

[13] Chang, J.L., Gong, D.W. and Ma, X.P. (2007) A Heuristic Genetic Algorithm for No-Wait Flowshop Scheduling Problem. *Journal of China University of Mining and Technology*, **17**, 582-586. http://dx.doi.org/10.1016/S1006-1266(07)60150-3

[14] Nagano, M.S., Miyata, H.H. and Araújo, D.C. (2014) A Constructive Heuristic for Total Flowtime Minimization in a No-Wait Flowshop with Sequence-Dependent Setup Times. *Journal of Manufacturing Systems*. http://dx.doi.org/10.1016/j.jmsy.2014.06.007

[15] Aldowaisan, T.A. and Allahverdi, A. (2012) No-Wait Flowshop Scheduling Problem to Minimize the Number of Tardy Jobs. *The International Journal of Advanced Manufacturing Technology*, **61**, 311-323. http://dx.doi.org/10.1007/s00170-011-3659-x

[16] Jolai, F., Asefi, H., Rabiee, M. and Ramezani, P. (2013) Bi-Objective Simulated Annealing Approaches for No-Wait Two-Stage Flexible Flow Shop Scheduling Problem. *Scientia Iranica*, **20**, 861-872.

[17] Liu, G., Song, S. and Wu, C. (2013) Some Heuristics for No-Wait Flowshops with Total Tardiness Criterion. *Computers & Operations Research*, **40**, 521-525. http://dx.doi.org/10.1016/j.cor.2012.07.019

[18] Aldowaisan, T. and Allahverdi, A. (2012) Minimizing Total Tardiness in No-Wait Flowshops. *Foundations of Computing and Decision Sciences*, **37**, 149-162. http://dx.doi.org/10.2478/v10209-011-0009-6

[19] Allahverdi, A. and Soroush, H.M. (2008) The Significance of Reducing Setup Times/Setup Costs. *European Journal of Operational Research*, **187**, 978-984. http://dx.doi.org/10.1016/j.ejor.2006.09.010

[20] Allahverdi, A., Gupta, J.N. and Aldowaisan, T. (1999) A Review of Scheduling Research Involving Setup Considerations. *Omega*, **27**, 219-239. http://dx.doi.org/10.1016/S0305-0483(98)00042-5

[21] Allahverdi, A., Ng, C.T., Cheng, T.E. and Kovalyov, M.Y. (2008) A Survey of Scheduling Problems with Setup Times or Costs. *European Journal of Operational Research*, **187**, 985-1032. http://dx.doi.org/10.1016/j.ejor.2006.06.060

[22] Vallada, E., Ruiz, R. and Minella, G. (2008) Minimising Total Tardiness in the m-Machine Flowshop Problem: A Review and Evaluation of Heuristics and Metaheuristics. *Computers & Operations Research*, **35**, 1350-1373. http://dx.doi.org/10.1016/j.cor.2006.08.016

18

On the Analysis of PLC Programs: Software Quality and Code Dynamics

Mohammed Bani Younis

Faculty of Engineering, Philadelphia University, Amman, Jordan
Email: mbaniyounis@philadelphia.edu.jo

Abstract

As a result of sudden failure in the Programmable Logic Control (PLC) controlled process, the need of diagnosis arises. Diagnosis problem plays an important role to monitor failures in PLC used to control the whole process. Nowadays, due to the lack of the needed tools available to perform this action automatically, it is accomplished manually. Usually, the time consuming method is used by back-tracking the failure on an actuator due to the corresponding sensors. This paper analyzes the software quality metrics and their application on the PLC programs. Aiming to implement metrics that gives predictive information about diagnosability of an Instruction List (IL) PLC programs, this could minimize the needed effort to check the program in case of mistakes. Furthermore, to get a better prediction about diagnosability, new metrics are introduced which are able to give more information about the semantics of a program. But they are not yet fully developed and have to be analyzed.

Keywords

PLC Programs, Diagnosability, Dependency, Program Slicing Component

1. Introduction

Programmable Logic Controllers (PLCs) are special type of computer with a cyclic behavior used for controlling machines and process. PLC is a good example of discrete event system or logic control with extensions to be able to implement and control analog controllers and deal with continuous signals like PID control algorithms.

The cycle of the PLC starts by reading the sensor values from the controlled process or machine then processing the control algorithm and finally controlling the actuators according to the given specifications.

To maintain the efficiency of the PLC in manufacturing and production, it is necessary in case of modernizations and re-structuring of machines to modernize the PLC to meet these requirements. Another important task related to PLC is the possibility to diagnose the PLC in case of errors or mistaken behavior. Diagnosis is aimed at finding the source of a failure in a system. Many research ideas tackling this problem with advanced methods for failure diagnosis for discrete event systems are available in ([1]-[4]). Although these methods are rich of research ideas have not yet found wide acceptance in industry.

The standard approach in industry is still manual diagnosis. This is usually performed in case unexpected situation in the manufacturing process. In this case the diagnosis obviously starts by checking the actuator responsible for the missing action. If this actuator is found to function properly, the next step is to back track the reasons through the PLC. The reasons behind this unusual behavior can be because of the non proper or broken sensors. It is generally justified due to the absence of errors in the algorithms implemented on the PLC. Hence, the task is to track back the dependency of the output signal corresponding to the failing actuator—possibly over several internal variables of the PLC—to the corresponding input signals. After these are found, the respective sensors have to be checked for failures. This issue can in turn be very costly and time consuming in case of complex and huge PLC applications.

Fault diagnosis and fault tolerance are important aspects of maintenance which improve the system reliability. These important issues can be implemented in different ways. The main PLCs have self-diagnostic capabilities in order to detect internal failure. Power supply failure is generally shown with a led.

A fault on input/output (units, connecting cable, terminal,...) can be diagnosed with led indicators that show the signal logic value or run specific diagnosis routines.

A run fault (system program termination due to some anomalous operations) is also normally shown with a led.

A complete diagnostic tool within one whole package with the PLC machines or systems is preferable by many testing engineers. However, there're several tools available, such as:
• Profibus diagnostic tools;
• BT200 diagnostic tools from Siemens;
• xEPI for Ethernet, Profibus, Interface system.

A fault on the application program is more difficult to check, and different solutions have been proposed but not mature. The aim of the presented work is to investigate how to use known software quality measures to determine the connections inside a PLC program. This allows a better understanding of the program and hence the effort for manual diagnosis. The PLC programs under consideration are assumed to be given in Instruction List language (IL). This PLC language is used for the application of the approach because it is the most commonly used PLC language in Europe.

Since IL is a special form of Assembly language, for several metrics to be applied, it is necessary to transform the program to a higher level description. To this end, the presented method utilizes a reverse engineering approach for IL programs presented in [4]. During the calculation of the metrics graphical representations of dependency relations are derived. These can be used as an aid for the engineer in the actual diagnosis process.

The paper is structured as follows. Section 2 discusses software quality to measure the structure of the PLC program. New methods related to the measures of the dynamic and semantic of a given PLC program are presented in Section 3. Section 4 concludes this paper and gives and outlook on future work.

2. Software Quality and Diagnosis (Former Works)

In the Software Engineering community, the field of measures or metrics to measure software characteristics is maturing, see e.g. [5] and [6] for overviews. Some of the best-known software measures were already date back to the late seventies ([7] [8]). However, in the area of PLC programs software quality is rarely studied. Frey [9] introduced the concept of Transparency to measure the understandability of PLC programs described by a special form of Petri Net. Dandachi *et al.* [10] provided similar metrics to be applied to PLC programs given in Sequential Function Chart language. A different application of software measures is presented by Lucas and Tilbury in [11]. There, the complexity of solutions to the same problem described with different PLC languages is measured and compared. A systemic method dedicated through the modeling of PLC system architecture and PLC features as components was proposed in [12] and [13]. This model is used for the construction of verification model.

There are many measures which are normally used nowadays to determine the quality of software were applied to investigate the software quality of the PLC programs as Presented in [13]. Most measures presented concentrate more on the program's structure than on the contents. The size measure belongs to the most often used measures, namely the LOC (Lines of Code) and the Halstead measure with which the software will be measured in terms of the operators and operands. Another classical complexity measure shows the "Cyclomatic Complexity" from McCabe which refers to the flow chart of the code. A measure which determines the complexity of a program by the investigation of its graph is the "Tree Impurity" [6].

The following table briefly summarizes the evaluation of the measures presented above and shows their applicability to IL PLC programs (cf. **Table 1**) [13].

3. Code Dynamic and Semantic Measures

3.1. Algorithm to Establish Formulas from IL

The code is perused from above where the first line of code is processed and the individual conditions according to their operations are recorded. The brackets in the specified conditions are also registered in case available. Each condition has a set, reset or an assignment "*i.e.*:=".

IL of the form

```
Condition 1
    set X
condition 2
    reset X
UX                      ; load the x
UA                      ; And A (a  Boolean variable)
    :=Y                 ; assign the result to Y
```

The resulting formulas:

$$X \leftarrow \left[(\text{Condition } 1) + X \right] \cdot \left[\overline{(\text{Condition } 2) + \overline{X}} \right] \tag{1}$$

$$Y \leftarrow \left[X \cdot A \right] \tag{2}$$

$$Y \leftarrow \left\{ \left[(\text{Condition } 1) + X \right] \cdot \left[\overline{(\text{Condition } 2) + \overline{X}} \right] \right\} \cdot A \tag{3}$$

The resulting formulas in case of set and reset, for X to be set to 1, the condition to set has to be fulfilled or X was set to 1 in the previous cycle $(\text{Condition } 1) + X$. To keep X in this state the condition for resetting is not allowed to be fulfilled or X was 0 in the previous cycle $\overline{(\text{Condition } 2) + \overline{X}}$. These formulas are used to set X to 1 in case the condition is true otherwise it will be reset to 0.

Set and reset the variable used in the formulas is necessary because by setting a variable it is stored as value 1 or True until the conditions for resetting are met. This can, however, be an advantage to check the programming. When in the formula (*i.e.* only an X is appeared) which means that for X, the conditions for set or reset is missing. By the assignment operation the processing is different, because no storage for the variable takes place. And the previous state of the variable is not known, this justifies the non-appearance of Y in the formulas.

Once the individual formulas for each variable are constructed, these formulas are substituted into each other and an overall expression for the variables is formulated. If this total expression met, the output variable is set to 1, otherwise it is zero. The above described formulas are expressed with the help of finite state machines (cf. **Figure 1** for the first formula, **Figure 2** for the second formula, and **Figure 3** after the merging of both formulas).

Table 1. Evaluation of the discussed measures (+ = good; 0 = fair, − = bad).

Measure	Software feasibility	Applicability to IL	Significance for diagnosability
Size	++	+	− (Serves for the coarse appraisal)
Halstead	++	+	0 (Overview about operators and operands)
McCabe	+	− (Graph is necessary)	− (Information about conditional jumps)
Tree Impurity	0	0 (Graph is necessary)	+ (Blocks invoke by other blocks)

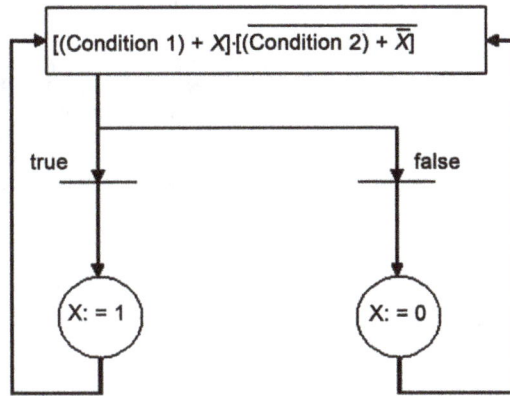

Figure 1. First formula.

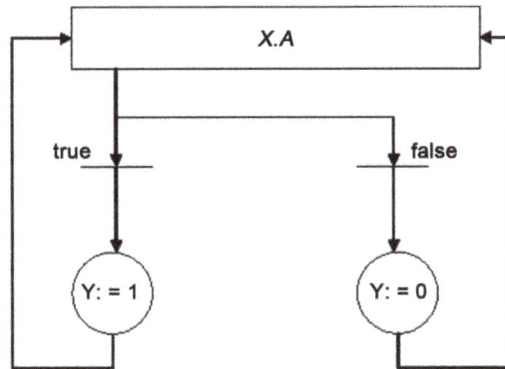

Figure 2. Second formula.

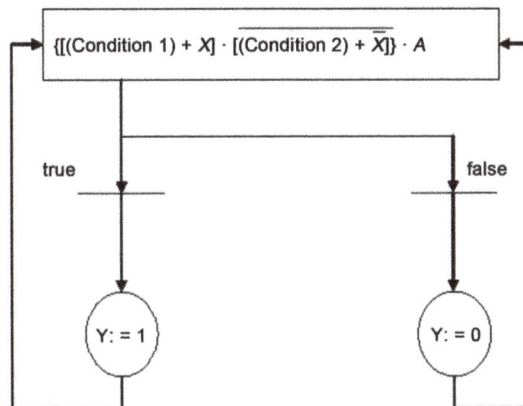

Figure 3. Formula resulting from the merge of both formulas.

One recognizes here the correctness of the formula, because no matter how many times the loop is executed, the correct output appears accordingly as long as the conditions do not change.

This formula can now be used to determine the software quality of the program. Once you have set all the variables that are represented in the formula for the output calculation, all the variables of which this output depends are obtained. In the formula not only the input variables are yielded, but also it shows which variables are used, such as Merker (used as Flags in Step 5 from Siemens) or other outputs are used to deliver this output. In case outputs or merker from other networks or other function blocks are used, these should be noted using brackets in the used formula. This allows expressing the dependency between the distinct modules. For the software quality, these formulas are used to count the number of outputs and merker. This gives a measure of how far dependency on the individual outputs from each other (cf. Example shown in **Figure 4**).

```
NETZWERK   1                        NETZWERK   2
0000       :U    A   2.4            0014       :U(
0001       :U(                      0015       :U    A   2.5
0002       :UN   A   2.5            0016       :UN   M   101.6
0003       :L    KT  003.2          0017       :O
0004       :SE   T   43             0018       :U    M   1.0
0005       :NOP  0                  0019       :U    M   101.6
0006       :NOP  0                  001A       :O    E   0.2
0007       :NOP  0                  001B       :)
0008       :U    T   43             001C       :U    T   5
0009       :)                       001D       :=    A   2.4
000A       :U    E   6.1            001E       :S    A 2.6
000B       :U    A   4.5            001F       :BE
000C       :S    A   2.5
000D       :ON   T   5
000E       :O    E   6.0
000F       :O    M   101.6
0010       :ON   A   4.5
0011       :R    A   2.5
0012       :NOP  0
0013       :***
```

Figure 4. Program example.

Network 1:

$$A2.5 = 1 \leftarrow \left[\left(A2.4 + \left(\overline{A2.5} \cdot T43 \right) \cdot E6.1 \cdot A4.5 \right) + A2.5 \right] \cdot \overline{\left[\left(\overline{T5} + E6.0 + M101.6 + \overline{A4.5} \right) + \overline{A2.5} \right]}$$

Network 2:

$$A2.4 = 1 \leftarrow \left[\left(A2.5 \cdot \overline{M101.6} \right) + \left(M1.0 \cdot M101.6 \right) + E0.2 \right] \cdot T5$$

$$A2.6 = 1 \leftarrow \left[\left(\left(A2.5 \cdot \overline{M101.6} \right) + \left(M1.0 \cdot M101.6 \right) + E0.2 \right) \cdot T5 \right] + A2.6$$

After the replacement of $A2.5$ in $A2.4$ and $A2.6$ the Formula for $A2.4$ become as follows. From the Formula of $A2.4$ it shows that the output depends on the input variables $E6.1$; $E6.0$; $E0.2$. It is clear that network 2 depends on network 1. The output $A2.6$ is expressed only one time in the formula which means that the reset condition is not available for this output.

$$A2.6 \leftarrow \left[\left(\left\{ \left[\left(A2.4 + \left(\overline{A2.5} \cdot T43 \right) \cdot E6.1 \cdot A4.5 \right) + A2.5 \right] \cdot \overline{\left[\left(\overline{T5} + E6.0 + M101.6 + \overline{A4.5} \right) + \overline{A2.5} \right]} \right\}_{Netzwerk1} \right. \right.$$
$$\left. \left. \cdot \overline{M101.6} \right) + \left(M1.0 \cdot M101.6 \right) + E0.2 \right] \cdot T5$$

3.2. Dependency Analysis through Program Slicing

In order to prove properties in a program, it is useful to look at only the part of the program *i.e.* slice, which causes these properties. This slice is then converted into a model in a type of dependency graph [14]. This graph corresponds to the data flow graph, which additionally edges for paths of all conditions to directly controlled instructions whose execution depends directly on the evaluation of the considered condition inserted [13]-[15]. This provides information on cause-effect relationships in an observed section of the program.

There are two types of a slicing along a dependency graph. The backward slicing, which describes the instructions influence a value under consideration. In a dependency graph all paths that refer to a variable and enter the node are considered. Through this method, irrelevant parts of the program can be hidden during debugging. The forward slicing, describes which instructions are influenced by an observed value. In a dependency graph, all the paths that define a variable in out of a node are considered. Through this slice method the consequence modifications of the program can be estimated [16] and [17].

The following summarizes the information available when applying the slicing method:

• Instructions, which may affect the value of a variable at a certain location under consideration;

- Involved variables and data flows;
- Control influences [6].

This method shows the dependence of variables has been already transferred to the ladder diagram of PLC in [18]. In this work an algorithm is used to translate the program step by step into a time-machine model by which each conditional statement are assigned to two transitions. The first time automata represent the fulfillment of the condition, which then leads to the next instruction. Other transition is required, if the condition is not satisfied, a negative transition. This then leads directly to the final state of the automaton.

As an application of slicing, data dependency and control can be analyzed on the PLC program. Data dependence and control dependence are known in the literature in terms of the Control Flow Graph (CFG) of a program. CFG is a representation using graph notation which contains nodes representing a basic block without jumps and control predicate in the program, an edge connecting two nodes together which represents the possible flow from former node to another [19]. Special nodes are used in the CFG to represent the starting and the end of the program labeled START and STOP respectively. The sets of the DEF (i) and REF(i) are used to represent the sets of defined and referenced variables. CFG is used to express several types of dependencies, *i.e.* flow dependence, control dependence, output dependence and etc. Flow dependence takes place when an instruction depends on the previous instruction, while control dependence occurs when an instruction depends on a preceding instruction to be executed or not.

Program Dependence Graph (PDG) [20] [21] as a slicing methods is used for the reachability. This method can be also applied on the PLC program. PDG is a directed graph with vertices corresponding to statements and control predicates, and edges corresponding to flow and control dependences.

To be able to use these methods of dependency on the PLC program, it should be written on a high level language *i.e.* Structured Text (ST). Otherwise, if the program is written in IL or any other low level language the Program has to be abstracted first. The abstraction can be applied using the methods and algorithms introduced in [22] and [23]. The depicted code in **Figure 5** is used to apply the slicing dependency. The program is abstracted using the methods introduced in [22] and [23] (cf. **Figure 6**). **Figure 7** expresses the CFG, where the edges from the IF conditions to the successive nodes are control dependent, whereas the remaining edges in the figure are flow dependent. **Figure 8** elucidates the PDG where thick edges represent control dependence and thin edges represent flow dependencies.

The following table briefly summarizes the evaluation of the measures presented above and shows their applicability to IL PLC programs (cf. **Table 2**).

```
NETZWERK 1
0000    :U     E 38.1      //AND Operation
0002    :U     E 38.2
0004    :O                 //OR Operation
0006    :U     E 38.1
0008    :U     E 38.3
000A    :O
000C    :U     E 38.2
000E    :U     E 38.3
0010    :=     M 100.0     at least two Fans running
0012    :UN    E 38.1      // ANDN Operation
0014    :UN    E 38.2
0016    :UN    E 38.3
0018    :=     M 100.1     no running Fan
001A    :U (
001C    :O     M 100.0     Continuous Light
001E    :O
0020    :U     M 100.1
0022    :U     M 99.1      Flashing with 2 Hz
0024    :O
0026    :UN    M 100.0
0028    :UN    M 100.1
002A    :U     M 99.2      Flashing with 0,5 Hz
002C    :)
002E    :U     A 42.4      „Active"
0030    :=     A 51.7      LCD lamp
0032    BE
```

Figure 5. PB program as IL in Step 5.

```
IF          (E 38.1 AND E 38.2) OR (E 38.1 AND E 38.3)
            OR    (E 38.2 AND E 38.3)
   THEN     M 100.0=1
   ELSE     M 100.0=0
IF          NOT E 38.1 ANDN E 38.2 ANDN E 38.3
   THEN     M 100.1=1
   ELSE     M 100.1=0
IF          (M 100.0) OR (M 100.1 AND M 99.1) OR
            (NOT M 100.0 ANDN M 100.1 AND M 99.2) AND A 42.4
   THEN     A 51.7 =1
   ELSE     A 51.7=0
```

Figure 6. Abstraction of the PB program.

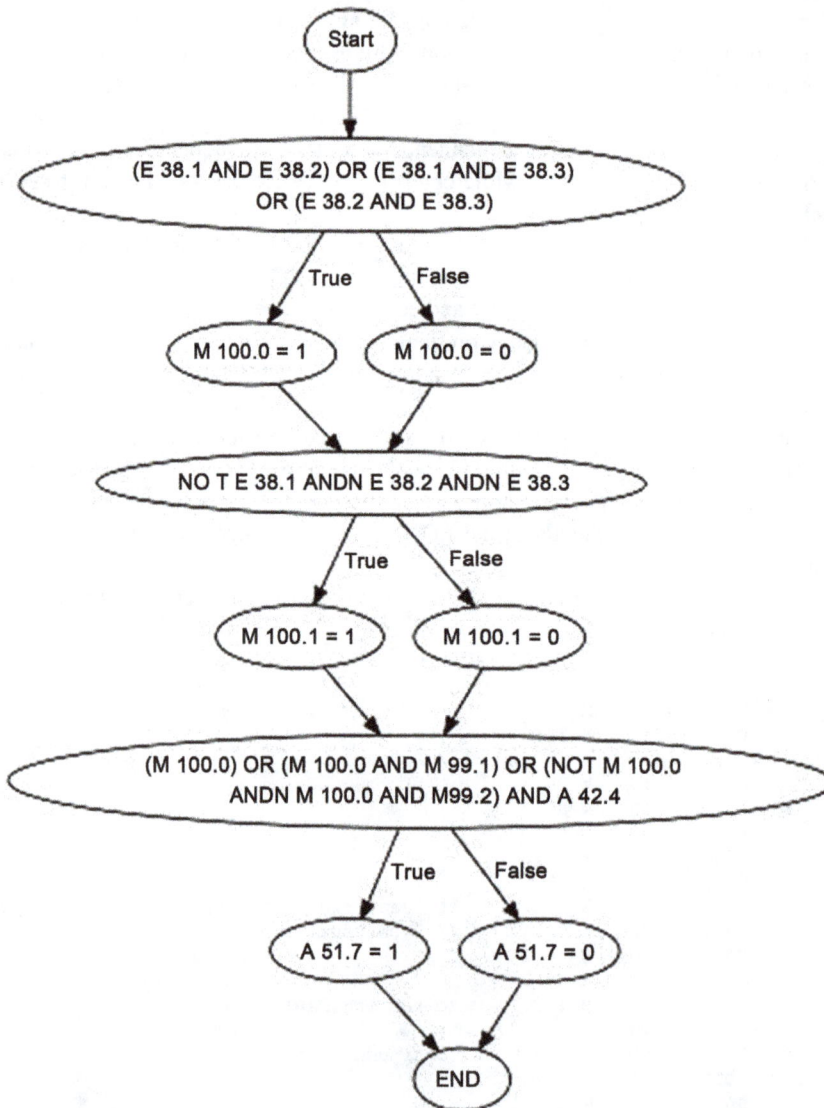

Figure 7. Control flow diagram.

Table 2. Evaluation of the dynamic and semantic measures (+ = good; 0 = fair, − = bad).

Measure	Software feasibility	Applicability to IL	Significance for diagnosability
Formulas	0	+	++ (Depending on the variables)
Slicing	-	+	+ (Depending on the state of the variables)

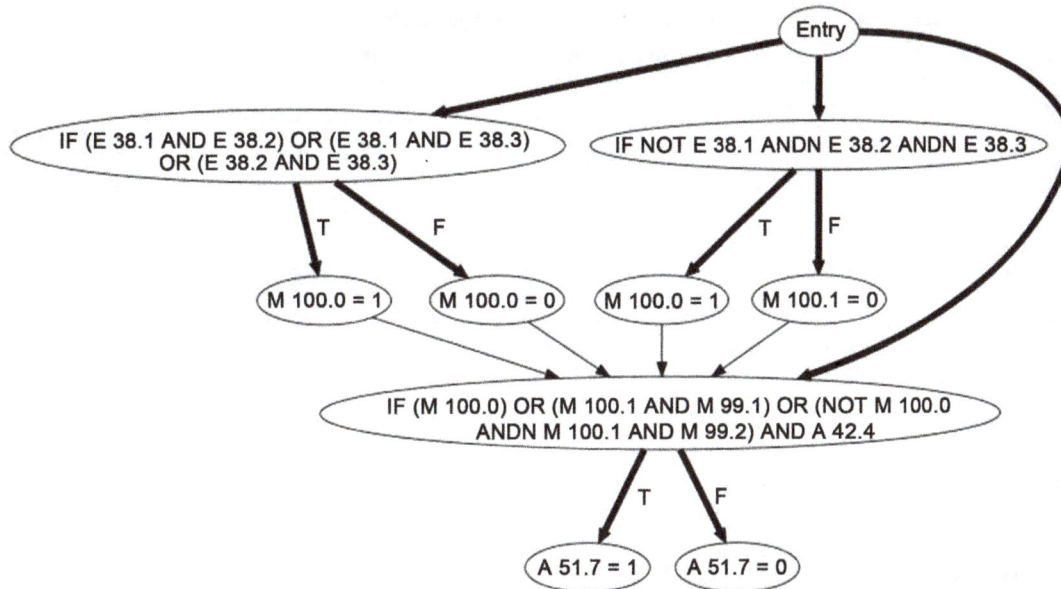

Figure 8. Program flow diagram.

4. Conclusions and Outlook

To allow easy diagnosis in the case of failures in a production system automated by PLCs, the quality of the PLC software considered is a crucial factor. As a by-product of the measures calculation several dependence relations can by visualized. It is expected that this will help an engineer in performing the diagnosis task.

The metrics most commonly used nowadays only examine the structure of a program, where it eases the application to the respective programs, as well as on the PLC IL program. They can be implemented directly and provide a rough guideline for the quality of the program. However, the significance of the diagnostic capability with respect to a program could be considered a drawback. It is very difficult, for example, from the length of the code to infer the degree of difficulty of the code.

It is quite different with metrics that focus on the code dynamics. These try to present the relations between individual modules or variables, and thereby find out the complexity of a program. Applying for example the formulas on the PLC, the dependence of the variable, or better expressed: the dependence of the output are determined by the inputs. This gives a better indication than standard metrics, but this is much more difficult to accomplish, because of the effort needed to express the formulas.

The semantic metrics dedicated through the building of the formulas which represent the dependency between the outputs and the inputs in the PLC are investigated. After the building and structuring of all formulas these can be merged together to allow finding the final expressions related to the PLC outputs.

The dependency analysis as a new metric was performed and presented. This analysis explained the best way the variables are changed in a program. In this metric, the dependencies of the entire program were formed. However, these were devoid of information content.

In the next steps, especially the dependency analysis should be further investigated and implemented since through them the best statements about the diagnosability of PLC IL program could be made. Here the focus should be mainly on the slicing, which represents a simplification of the dependence graph by considering only relevant variables for the problem. Once there is an appropriate implementation of the algorithms, it is important to investigate how these models can be used by the inclusion of data models of real plant data for online diagnosis.

References

[1]   Lunze, J. and Schröder, J. (2001) State Observation and Diagnosis of Discrete-Event Systems Described by Automata. *Discrete Event Dynamic Systems—Theory and Applications*, **11**, 319-396.

[2]   Papadopoulus, Y. and McDermid, J. (2001) Automated Safety Monitoring: A Review and Classification of Methods.

International Journal of Condition Monitoring and Diagnostic Engineering Management, **4**, 14-32.

[3] Sampath, M., Sengutpa, R., Lafortune, S., Sinnamohideen, K. and Tenekeztis, D. (1996) Failure Diagnosis Using Discrete Event Models. *IEEE Transactions on Control Systems Technology*, **4**, 105-124. http://dx.doi.org/10.1109/87.486338

[4] Bani Younis, M. (2006) Re-Engineering Approach for PLC Programs Based on Formal Methods. Dissertation, University of Kaiserslautern, Kaiserslautern.

[5] Höcker, H., Itzfeld, W.D., Schmidt, M. and Timm, M. (1994) Comparative Descriptions of Software Quality Metrics. *GMD-Studien Nr. 81, GMD*, Bonn.

[6] Kann, S.H. (2003) Metrics and Models in Software Quality Engineering. 2nd Edition, Addison Wesley Professional.

[7] Halstead, M.H. (1977) Elements of Software Science. Elsevier, New York.

[8] McCabe, T. (1976) A Complexity Measure. *IEEE Transactions on Software Engineering*, **SE-2**, 308-320. http://dx.doi.org/10.1109/TSE.1976.233837

[9] Frey, G. (2002) Software Quality in Logic Controller Design. *Proceedings of the IEEE SMC* 2002, Tunisia, 6-9 October 2002, 515-520.

[10] Dandachi, A., Lohmann, S. and Engell, S. (2007) Complexity of Logic Controllers. *Preprints of 1st IFAC Workshop on Dependable Control of Discrete Systems*, Cachan, 13-15 June 2007, 279-284.

[11] Lucas, M. and Tilbury, D. (2002) Quantitative and Qualitative Comparisons of Plc Programs for a Small Testbed. *Proceeding of the American Control Conference*, Alaska, 8-10 May 2002, 4165-4171.

[12] Wang, R., *et al.* (2013) Component-Based Formal Modeling of PLC Systems. *Journal of Applied Mathematics*, **2013**, Article ID: 721624.

[13] Bani Younis, M. and Frey, G. (2007) Software Quality Measures to Determine the Diagnosability of PLC Applications. *Proceedings of the of the* 12*th IEEE International Conference on Emerging Technologies and Factory Automation*, Patras, 25-28 September, 368-375.

[14] Horwitz, S., Reps, T. and Binkley, D. (1990) Interprocedural Slicing Using Dependence Graphs. *ACM Transactions on Programming Languages and Systems*, **12**, 26-60. http://dx.doi.org/10.1145/77606.77608

[15] Weiser, M. (1979) Program Slices: Formal, Psychological, and Practical Investigations of an Automatic Program Abstraction Method. Ph.D. Thesis, University of Michigan, Ann Arbor.

[16] Weiser, M. (1982) Programmers Use Slices When Debugging. *Communications of the ACM*, **25**, 446-452. http://dx.doi.org/10.1145/358557.358577

[17] Weiser, M. (1984) Program Slicing. *IEEE Transactions on Software Engineering*, **10**, 352-357. http://dx.doi.org/10.1109/TSE.1984.5010248

[18] Zoubek, B., Roussel, J.-M. and Kwiatkowska, M. (2003) Towards Automatic Verification of Ladder Logic Programs. *Proceedings of IMACS-IEEE CESA*'03 *Computational, Engineering in Systems Applications*, Lille, 9-11 July 2003, 6 p.

[19] Tip, F. (1994) A Survey of Program Slicing Techniques. *Journal of Programming Languages—JPL*, **3**.

[20] Kuck, D.J., Kuhn, R.H., Padua, D.A., Leasure, B. and Wolfe, M. (1981) Dependence Graphs and Compiler Optimizations. *Conference Record of the 8th ACM Symposium on Principles of Programming Languages*, New York, 207-218.

[21] Ferrante, J., Ottenstein, K.J. and Warren, J.D. (1987) The Program Dependence Graph and Its Use in Optimization. *ACM Transactions on Programming Languages and Systems*, **9**, 319-349. http://dx.doi.org/10.1145/24039.24041

[22] Bani Younis, M. and Frey, G. (2005) Formalization and Visualization of Non-Binary PLC Programs. *Proceedings of the* 44*th IEEE Conference on Decision and Control (CDC* 2005) *and European Control Conference (ECC* 2005), Seville, 12-15 December 2005, 8367-8372.

[23] Frey, G. and Bani Younis, M. (2004) A Re-Engineering Approach for PLC Programs using Finite Automata and UML. *Proceedings of* 2004 *IEEE International Conference on Information Reuse and Integration*, *IRI*-2004, Las Vegas, 8-10 November, 24-29.

19

Sigma-Point Filters in Robotic Applications

Mohammad Al-Shabi

Department of Mechatronics Engineering, Philadelphia University, Amman, Jordan
Email: maqas2002@yahoo.com

Abstract

Sigma-Point Kalman Filters (SPKFs) are popular estimation techniques for high nonlinear system applications. The benefits of using SPKFs include (but not limited to) the following: the easiness of linearizing the nonlinear matrices statistically without the need to use the Jacobian matrices, the ability to handle more uncertainties than the Extended Kalman Filter (EKF), the ability to handle different types of noise, having less computational time than the Particle Filter (PF) and most of the adaptive techniques which makes it suitable for online applications, and having acceptable performance compared to other nonlinear estimation techniques. Therefore, SPKFs are a strong candidate for nonlinear industrial applications, *i.e.* robotic arm. Controlling a robotic arm is hard and challenging due to the system nature, which includes sinusoidal functions, and the dependency on the sensors' number, quality, accuracy and functionality. SPKFs provide with a mechanism that reduces the latter issue in terms of numbers of required sensors and their sensitivity. Moreover, they could handle the nonlinearity for a certain degree. This could be used to improve the controller quality while reducing the cost. In this paper, some SPKF algorithms are applied to 4-DOF robotic arm that consists of one prismatic joint and three revolute joints (PRRR). Those include the Unscented Kalman Filter (UKF), the Cubature Kalman Filter (CKF), and the Central Differences Kalman Filter (CDKF). This study gives a study of those filters and their responses, stability, robustness, computational time, complexity and convergences in order to obtain the suitable filter for an experimental setup.

Keywords

Sigma Point, Unscented Kalman Filter, Cubature Kalman Filter, Centeral Difference Kalman Filter, Filtering, Estimation, Robotic Arm, PRRR

1. Introduction

Robotic applications, especially robotic arm, become widely used in industries due to their simplicity and the ability to do multi-task/multi-function with few numbers of settings and/or arrangements. The problem with such applications is the necessary to apply nonlinear control signals to achieve the desired trajectories. The latter is not easy to be implemented and has several limitations [1]-[3]. For example, Sliding Mode Control (SMC) [1] is one of the robust control approaches. However, it suffers from chattering. Although several researches have proposed to eliminate the chattering, the problem is still not fully solved. The limitation of such controllers increases as uncertainties present, *i.e.* modeling uncertainties and noise. This becomes worse when the number of measurement is less than the number of states.

Filters, especially model based filters [2]-[7], have been used to remove some of those constrains. It is a cheap method that could be used to obtain the unmeasured-hidden-states, and/or it could be used to reduce the noise effect. The optimal solution for such applications in their linear case is the Kalman Filter (KF) [7]-[12]. When the system is nonlinear, the KF is modified to be applicable for such applications. Several researches have been developed to overcome this limitation. Those include linearizing the system by Taylor Series Approximation (TSA) up to the first order such as the Perturbation Kalman Filter [9] [13] [14], the Extended Kalman filter (EKF) [8] [15]-[17], and the Iterated Extended Kalman filter (IEKF) [7] [15] [18]-[20], or up to higher order such as the Higher Order Extended Kalman Filter (HOEKF) [15] [21]-[23]. The later shows that in order to increase the accuracy of high nonlinear application, TSA is not a suitable approach as it takes long computation time with complicated structure [24]. Therefore, different approaches were developed including the combination of KF with intelligent techniques such as [25]-[29], or finding different approaches to approximate the nonlinearity such as the Sigma-Point Kalman Filter (SPKF) [2] [4] [5] and the Particle Filter (PF) [30]. The rest of the paper will be divided as the following: Section two includes an introduction to the SPKF including the algorithms used in this paper, UKF, CKF and CDKF. The mathematical model of the PRRR robotic arm application is showed in Section three. Results, discussion and conclusion are listed and discussed in Sections four and five.

2. The Sigma-Point Kalman Filter

The SPKFs linearize the nonlinear models statistically using weighted linear regression method. This is done by obtaining a certain number of points, referred to as sigma points, from the state neighborhood using the probability distribution function as shown in **Figure 1**. Those points are projected through the system model, and then combined together using appropriate weights as shown in **Figure 2**. This provides with a mechanism that covers

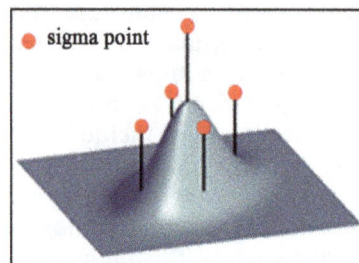

Figure 1. Sigma-Points for n = 2 [31].

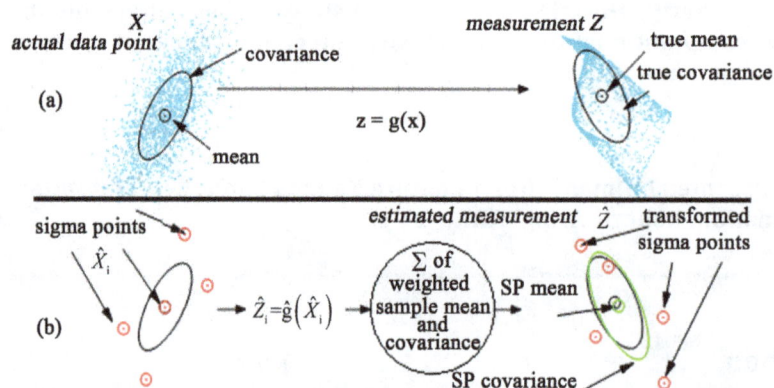

Figure 2. (a) The actual system states and their nonlinear measurement; (b) The Sigma-Points KF's estimates [31].

the actual mean and covariance without the need to linearize the model by TSA and calculate the Jacobian matrices. Moreover, it accommodates noise disturbances that are not Gaussian [4] [5] [15] [31]-[34].

Several algorithms have been created using the above principle. Although, different approaches were used to derive those algorithms, the general outline remain the same as will be proven in the next subsections. The major differences between those methods could be summarized to the number of the sigma points, how to choose them, and what are the appropriate weights for the combining step. Moreover, they may differ on calculating the covariance matrices [35]. Some SPKFs algorithm will be described on the next subsections.

2.1. The Unscented Kalman Filter

The Unscented Kalman Filter is a SPKF that has been developed using the unscented transformations. The latter has several form including *general unscented* [15], *simplex unscented* [35] [36], and *spherical unscented* [36] [37], transformations. The structures of the resulting filters are similar and could be summarized by the pseudo code of **Table 1**, where ϱ_1 and ϱ_2 are parameters used to select the sigma points for the a priori and a posteriori estimates, respectively. Those differ from a filter to another and it result on obtaining different sigma points. Consequently, different number of sigma points and different associated weights are obtained. Those are illustrated by **Table 2**.

Table 1. Thepseudocode of the unscented kalman filter [2] [3] [15] [24].

$k = 0 \rightarrow$ Initialize $\hat{x}_{0|0}$ and $P_{0|0}$

Start $\quad k = k+1$

$\qquad$ for $i = 0,1,\cdots,q$

$$\hat{X}_{i_{k-1|k-1}} = \hat{x}_{k-1|k-1} + (\varrho_1)_i$$

Calculate W_i $\qquad\qquad\qquad\qquad\qquad$ ////Comments

$$\hat{X}_{i_{k|k-1}} = \hat{f}\left(\hat{X}_{i_{k-1|k-1}}, u_{k-1}\right)$$ $\qquad$ //// q is the number of the sigma point

End $\qquad\qquad\qquad\qquad\qquad\qquad\qquad$ //// draw the sigma points and their weights using **Table 2**

$$\hat{x}_{k|k-1} = \sum_{i=0}^{q} W_i \hat{X}_{i_{k|k-1}}$$ $\qquad$ //// propagate the points through the filter

//// combining the sigma points to obtain the a priori estimate

$$P_{k|k-1} = \sum_{i=0}^{q} W_i \left(\hat{X}_{i_{k|k-1}} - \hat{x}_{k|k-1}\right)\left(\hat{X}_{i_{k|k-1}} - \hat{x}_{k|k-1}\right)^{\mathrm{T}} + Q_{k-1}$$ $\qquad$ //// calculating the a priori covariance matrix

$\qquad$ for $i = 0,1,\cdots,q$

$$\hat{X}_{i_{k|k-1}} = \hat{x}_{k|k-1} + (\varrho_2)_i$$ $\qquad$ //// Redefine the sigma point and their weight from **Table 2** to obtain their a priori measurements

Calculate W_i

$$\hat{Z}_{i_{k|k-1}} = \hat{g}\left(\hat{X}_{i_{k|k-1}}\right)$$ $\qquad$ //// combining the sigma points' measurements to obtain the a priori measurement

End $\qquad\qquad\qquad\qquad\qquad\qquad\qquad$ //// Calculating the output's error covariance matrix

$$\hat{z}_{k|k-1} = \sum_{i=0}^{q} W_i \hat{Z}_{i_{k|k-1}}$$

$$P_{zz} = \sum_{i=0}^{q} W_i \left(\hat{Z}_{i_{k|k-1}} - \hat{z}_{k|k-1}\right)\left(\hat{Z}_{i_{k|k-1}} - \hat{z}_{k|k-1}\right)^{\mathrm{T}} + R_k$$ $\qquad$ //// The correction gain
//// Updating the estimate and its covariance matrix

$$P_{xz} = \sum_{i=0}^{q} W_i \left(\hat{X}_{i_{k|k-1}} - \hat{x}_{k|k-1}\right)\left(\hat{Z}_{i_{k|k-1}} - \hat{z}_{k|k-1}\right)^{\mathrm{T}}$$ $\qquad$ //// Repeat Stages

$$K_k = P_{xz} P_{zz}^{-1}$$

$$\hat{x}_{k|k} = \hat{x}_{k|k-1} + K_k \left(z_k - \hat{z}_{k|k-1}\right)$$

$$P_{k|k} = \left(P_{k|k-1} - K_k P_{zz} K_k^{\mathrm{T}}\right)$$

Go back to Start

Table 2. The differences between the UKF methods [15].

Method	$(\varrho_j)_i, j = 1,2; i = 1,2,\cdots,q$	$W_i, i = 1,2,\cdots,q$	q
UKF	$(\varrho_j)_i = \begin{cases} 0 & i = 0 \\ \left(\sqrt{nP_j}\right)_i^T & 1 \le i \le n \\ -\left(\sqrt{nP_j}\right)_i^T & n+1 \le i \le 2n \end{cases}$ $P_1 = P_{k-1|k-1}, \quad P_2 = P_{k|k-1}$	$W_i = \begin{cases} 0 & i = 0 \\ \dfrac{1}{2n} & i \ne 0 \end{cases}$	$2n+1$

$(\varrho_j)_i = \sqrt{P_j}\rho_i^n, \quad P_1 = P_{k-1|k-1}, \quad P_2 = P_{k|k-1}$

As ρ_i^n is obtained recursively as follows:

$\rho_0^1 = 0$ and $\rho_1^1 = \rho_2^1 = \dfrac{-1}{\sqrt{2W_1}}$,

(the superscript is the recursive index)
for $l = 2,\cdots,n$ (number of the states)

Method		W	q
Simplex UKF	$\rho_i^l = \begin{cases} \begin{bmatrix} \rho_0^{l-1} \\ 0 \end{bmatrix} & i = 0 \\ \begin{bmatrix} \rho_i^{l-1} \\ \dfrac{-1}{\sqrt{2W_{l+1}}} \end{bmatrix} & 1 \le i \le l \\ \begin{bmatrix} \mathbf{0}_{l-1\times 1} \\ \dfrac{l}{\sqrt{2W_{l+1}}} \end{bmatrix} & i = l+1 \end{cases}$ End	W_0 is chosen as $W_0 \in [0,1)$ $W_i = \begin{cases} 2^{-n}(1-W_0) & 1 \le i \le 2 \\ 2^{l-n}(W_1) & i > 2 \end{cases}$	$n+2$

Similar to the simplex UKF except that

Method		W	q
Spherical UKF	$\rho_i^l = \begin{cases} \begin{bmatrix} \rho_0^{l-1} \\ 0 \end{bmatrix} & i = 0 \\ \begin{bmatrix} \rho_i^{l-1} \\ \dfrac{-1}{\sqrt{l(l+1)W_1}} \end{bmatrix} & 1 \le i \le l \\ \begin{bmatrix} \mathbf{0}_{l-1\times 1} \\ \dfrac{l}{\sqrt{l(l+1)W_1}} \end{bmatrix} & i = l+1 \end{cases}$	W_0 is chosen as $W_0 \in [0,1)$ $W_i = \dfrac{1-W_0}{n+1}$	$n+2$

The statistical regression used in unscented filters provides with better approximation that the Jacobian matrices. It has been proven that UKFs approximates up to a third order TSA for Gaussian distributions [15], and second order TSA for non-Gaussian distributions [31]. Both, the simplex and the spherical unscented KFs are used to reduce the computational time; as they use less sigma points. However, their stability is limited for few order of TSA [15] [37]. The general UKF provide with better estimation compared to the previous two. However, it has a larger computational time.

2.2. The Cubature Kalman Filter

The Cubature Kalman filter (CKF) is derived by using the third-degree cubature rule to numerically approximate the Gaussian-weighted integrals defined as [38] [39]:

$$\int_R F(x)W(x)dx \tag{2.1}$$

where W is the weight function and it is Gaussian with the form $\mathcal{N}\left(x;\bar{x};\sigma\right)$, $\bar{x}$ and σ are the Gaussian's mean and standard deviation. Assuming that the states are Gaussian as well, a scheme similar to the UKF could be obtained. However, due to the Gaussian Nature, the covariance matrices will differ from those obtained from UKF. Those are illustrated by **Table 3**.

2.3. The Central Difference Kalman Filter

The Central Difference Kalman Filter (CDKF), described in [40]-[42], was derived in two major stages. The first stage was to linearize the system model using TSA. In the second stage, the derivatives were replaced with their numerical Stirling's polynomial interpolation forms (NSPI) [43], that is defined as the follow [44]:

Table 3. Thepseudocode of the cubature kalman filter [38] [39].

$$k = 0 \rightarrow \text{Initialize } \hat{x}_{00} \text{ and } P_{00}$$

Start $k = k+1$

$$\text{for } i = 0,1,\cdots,q$$

$$\hat{X}_{i_{k-1|k-1}} = \hat{x}_{k-1|k-1} + \begin{cases} 0 & i=0 \\ \left(\sqrt{nP_{k-1|k-1}}\right)_i^{\mathrm{T}} & 1 \le i \le n \\ \left(\sqrt{nP_{k-1|k-1}}\right)_i^{\mathrm{T}} & n+1 \le i \le 2n \end{cases}$$ //// Comments

//// q is the number of the sigma point

//// draw the sigma points

$$\hat{X}_{i_{k|k-1}} = \hat{f}\left(\hat{X}_{i_{k-1|k-1}}, u_{k-1}\right)$$

end   //// propagate the points through the filter

$$\hat{x}_{k|k-1} = \frac{1}{2n}\sum_{i=1}^{q}\hat{X}_{i_{k|k-1}}$$

//// combining the sigma points to obtain the a priori estimate

$$P_{k|k-1} = \frac{1}{2n}\sum_{i=1}^{q}\left(\hat{X}_{i_{k|k-1}}\hat{X}_{i_{k|k-1}}^{\mathrm{T}} - \hat{x}_{k|k-1}\hat{x}_{k|k-1}^{\mathrm{T}}\right) + Q_{k-1}$$

//// calculating the a priori covariance matrix

$$\text{for } i = 0,1,\cdots,q$$

$$\hat{X}_{i_{k|k-1}} = \hat{x}_{k|k-1} + \begin{cases} 0 & i=0 \\ \left(\sqrt{nP_{k|k-1}}\right)_i^{\mathrm{T}} & 1 \le i \le n \\ \left(\sqrt{nP_{k|k-1}}\right)_i^{\mathrm{T}} & n+1 \le i \le 2n \end{cases}$$ //// Redefine the sigma point to obtain their a priori measurements

$$\hat{Z}_{i_{k|k-1}} = \hat{g}\left(\hat{X}_{i_{k|k-1}}\right)$$

end   //// combining the sigma points' measurements to obtain the a priori measurement

//// Calculating the output's error covariance matrix

$$\hat{z}_{k|k-1} = \frac{1}{2n}\sum_{i=1}^{q}\hat{Z}_{i_{k|k-1}}$$

$$P_{zz} = \frac{1}{2n}\sum_{i=1}^{q}\left(\hat{Z}_{i_{k|k-1}}\hat{Z}_{i_{k|k-1}}^{\mathrm{T}} - \hat{z}_{k|k-1}\hat{z}_{k|k-1}^{\mathrm{T}}\right) + R_k$$

//// The correction gain

$$P_{xz} = \frac{1}{2n}\sum_{i=1}^{q}\left(\hat{X}_{i_{k|k-1}}\hat{Z}_{i_{k|k-1}}^{\mathrm{T}} - \hat{x}_{k|k-1}\hat{z}_{k|k-1}^{\mathrm{T}}\right)$$

//// Updating the estimate and its covariance matrix

$$K_k = P_{xz}P_{zz}^{-1}$$

//// Repeat Stages

$$\hat{x}_{k|k} = \hat{x}_{k|k-1} + K_k\left(z_k - \hat{z}_{k|k-1}\right)$$

$$P_{k|k} = \left(P_{k|k-1} - K_k P_{zz} K_k^{\mathrm{T}}\right)$$

Go back to Start

$$\partial f^{(n)}(x) = \frac{1}{2}\left(f^{(n-1)}\left(x + \frac{T_s}{2}\right) - f^{(n-1)}\left(x - \frac{T_s}{2}\right)\right)$$ (2.2)

The previous stages result on a scheme that is similar to the weighted regression of the UKF as shown in **Table 4**. However, it differs from the UKF on how to obtain the sigma points, how to calculate the weights, and how to calculate the covariance matrices. The CDKF has been found to have a superior performance among the other SPKFs [15] [30] [45]. Moreover, the CDKF uses one control parameter, T_{cd}, which derived in [45] to have a value of $\sqrt{3}$ for Gaussian distributions.

3. PRRR-Mathematical Model

The algorithms in section two are applied to a four DOF robotic arm that consists of one prismatic joint and three revolute joints (PRRR) that is presented by **Figure 3** and **Figure 4**. The model has been derived in [1] and [2], and is summarized as follow.

$$\tau = M(\theta)\ddot{\theta} + V(\theta, \dot{\theta}) + G(\theta)$$ (3.1)

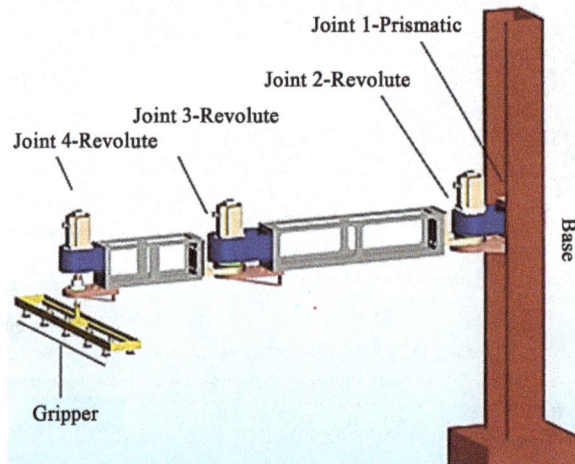

Figure 3. Four-DOFPRRR Robotic Arm [1] [2].

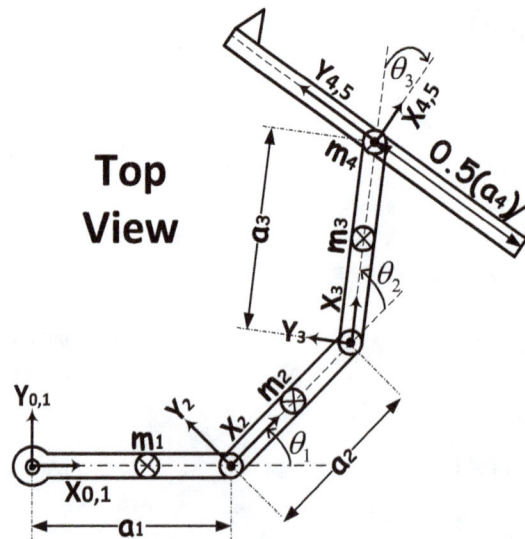

Figure 4. Top view of the PRRR Robotic Arm [1] [2].

Table 4. The pseudocode of sigma-point central difference kalman filter [45].

$$k = 0 \rightarrow \text{Initialize } \hat{x}_{0|0} \text{ and } P_{0|0}$$ //// Comments

$$\text{Start} \quad k = k+1$$

$$\text{for } i = 0,1,\cdots,(q = 2n)$$ //// draw the sigma points

$$\hat{X}_{i_{k-1|k-1}} = \hat{x}_{k-1|k-1} + \begin{cases} 0 & i = 0 \\ T_{cd}\left(\sqrt{P_{k-1|k-1}}ki\right)_i^{\mathrm{T}} & 1 \le i \le n \\ -T_{cd}\left(\sqrt{P_{k-1|k-1}}\right)_i^{\mathrm{T}} & n+1 \le i \le 2n \end{cases} \quad \hat{X}_{i_{k|k-1}}$$

$$= \hat{f}\left(\hat{X}_{i_{k-1|k-1}}, u_{k-1}\right)$$ //// propagate the points through the filter

end //// combining the sigma points to obtain the a priori estimate

$$\hat{x}_{k|k-1} = \sum_{i=0}^{q}\left[\hat{X}_{i_{k|k-1}} \times \begin{cases} \dfrac{T_{cd}^2 - n}{T_{cd}^2} & i = 0 \\ \dfrac{1}{2T_{cd}^2} & i \ne 0 \end{cases}\right]$$ //// calculating the a priori covariance matrix

$$P_{k|k-1} = \sum_{i=1}^{n}\frac{1}{4T_{cd}^2}\left(\hat{X}_{i_{k|k-1}} - \hat{X}_{i+n_{k|k-1}}\right)\left(\hat{X}_{i_{k|k-1}} - \hat{X}_{i+n_{k|k-1}}\right)^{\mathrm{T}}$$

$$+ \sum_{i=1}^{n}\frac{T_{cd}^2 - 1}{4T_{cd}^4}\left(\begin{array}{c}\hat{X}_{i_{k|k-1}} + \hat{X}_{i+n_{k|k-1}} \\ -2\hat{X}_{0_{k|k-1}}\end{array}\right)\left(\begin{array}{c}\hat{X}_{i_{k|k-1}} + \hat{X}_{i+n_{k|k-1}} \\ -2\hat{X}_{0_{k|k-1}}\end{array}\right)^{\mathrm{T}} + Q_{k-1}$$ //// Redefine the sigma point to obtain their a priori measurements

$$\text{for } i = 0,1,\cdots,q$$

$$\hat{X}_{i_{k|k-1}} = \hat{x}_{k|k-1} + \begin{cases} 0 & i = 0 \\ T_{cd}\left(\sqrt{P_{k|k-1}}\right)_i^{\mathrm{T}} & 1 \le i \le n \\ -T_{cd}\left(\sqrt{P_{k|k-1}}\right)_i^{\mathrm{T}} & n+1 \le i \le 2n \end{cases}$$

$$\hat{Z}_{i_{k|k-1}} = \hat{g}\left(\hat{X}_{i_{k|k-1}}\right)$$ //// combining the sigma points' measurements to obtain the a priori measurement

end //// Calculating the output's error covariance matrix

$$\hat{z}_{k|k-1} = \sum_{i=0}^{q}\left[\hat{Z}_{i_{k|k-1}} \times \begin{cases} \dfrac{T_{cd}^2 - n}{T_{cd}^2} & i = 0 \\ \dfrac{1}{2T_{cd}^2} & i \ne 0 \end{cases}\right]$$

$$P_{zz} = \sum_{i=1}^{n}\frac{1}{4T_{cd}^2}\left(\hat{Z}_{i_{k|k-1}} - \hat{Z}_{i+n_{k|k-1}}\right)\left(\hat{Z}_{i_{k|k-1}} - \hat{Z}_{i+n_{k|k-1}}\right)^{\mathrm{T}}$$

$$+ \sum_{i=1}^{n}\frac{T_{cd}^2 - 1}{4T_{cd}^4}\left(\begin{array}{c}\hat{Z}_{i_{k|k-1}} + \hat{Z}_{i+n_{k|k-1}} \\ -2\hat{Z}_{0_{k|k-1}}\end{array}\right)\left(\begin{array}{c}\hat{Z}_{i_{k|k-1}} + \hat{Z}_{i+n_{k|k-1}} \\ -2\hat{Z}_{0_{k|k-1}}\end{array}\right)^{\mathrm{T}} + R_k$$ //// The correction gain
//// Updating the estimate and its covariance matrix
//// Repeat Stages

$$P_{xz} = \frac{1}{2T_{cd}}\sqrt{P_{k|k-1}}\left(\begin{bmatrix}\hat{Z}_{1_{k|k-1}}^{\mathrm{T}} \\ \vdots \\ \hat{Z}_{n_{k|k-1}}^{\mathrm{T}}\end{bmatrix} - \begin{bmatrix}\hat{Z}_{1+n_{k|k-1}}^{\mathrm{T}} \\ \vdots \\ \hat{Z}_{2n_{k|k-1}}^{\mathrm{T}}\end{bmatrix}\right)$$

$$K_k = P_{xz}P_{zz}^{-1}$$

$$\hat{x}_{k|k} = \hat{x}_{k|k-1} + K_k\left(z_k - \hat{z}_{k|k-1}\right)$$

$$P_{k|k} = \left(P_{k|k-1} - K_k P_{zz} K_k^{\mathrm{T}}\right)$$

Go back to Start

$$\begin{bmatrix} F_z \\ \tau_1 \\ \tau_2 \\ \tau_3 \end{bmatrix} = \begin{bmatrix} m_T & 0 & 0 & 0 \\ 0 & A_1 & A_4 & A_5 \\ 0 & A_4 & A_2 & A_6 \\ 0 & A_5 & A_6 & A_3 \end{bmatrix} \begin{bmatrix} \ddot{d}_1 \\ \ddot{\theta}_1 \\ \ddot{\theta}_2 \\ \ddot{\theta}_3 \end{bmatrix} + \begin{bmatrix} 0 \\ A_7 \\ A_8 \\ 0 \end{bmatrix} + \begin{bmatrix} -gm_T \\ 0 \\ 0 \\ 0 \end{bmatrix} \tag{3.2}$$

where;

$$A_1 = \left[\frac{1}{4} m_2 a_2^2 + m_3 \left(a_2^2 + \frac{a_3^2}{4} + a_2 a_3 c_2 \right) + (m_4 + m_5)\left(a_2^2 + a_3^2 + 2a_2 a_3 c_2 \right) + \left(I_{z2} + I_{z3} + I_{z4} + I_{z5} \right) \right] \tag{3.3}$$

$$A_2 = \left[\frac{1}{4} m_3 a_3^2 + (m_4 + m_5) a_3^2 + \left(I_{z3} + I_{z4} + I_{z5} \right) \right] \tag{3.4}$$

$$A_3 = A_5 = A_6 = \left[I_{z4} + I_{z5} \right] \tag{3.5}$$

$$A_4 = \left[m_3 \left(\frac{a_3^2}{2} + a_2 a_3 c_2 \right) + 2(m_4 + m_5)\left(a_3^2 + a_2 a_3 c_2 \right) + \left(I_{z3} + I_{z4} + I_{z5} \right) \right] \tag{3.6}$$

$$A_7 = -\left[(m_3 + 2m_4 + 2m_5)\dot{\theta}_1 \dot{\theta}_2 + (m_3 + m_4 + 2m_5)\dot{\theta}_2^2 \right] a_2 a_3 s_2 \tag{3.7}$$

$$A_8 = -\left[2(m_3 + m_4 + m_5)\dot{\theta}_1 \dot{\theta}_2 + \frac{1}{2}(m_3 + 2m_4 + 2m_5)\dot{\theta}_1^2 \right] a_2 a_3 s_2 \tag{3.8}$$

$$m_T = m_1 + m_2 + m_3 + m_4 + m_5 \tag{3.9}$$

The system is discretized using the following definition

$$\dot{x}_k = (x_{k+1} - x_k)/T_s \tag{3.10}$$

where T_s is the sampling time and it is equal to 0.001 sec. If the states defined as the following.

$$\begin{aligned} X_k &= \begin{bmatrix} d_k & \dot{d}_k & \theta_{1k} & \dot{\theta}_{1k} & \theta_{2k} & \dot{\theta}_{2k} & \theta_{3k} & \dot{\theta}_{3k} \end{bmatrix}^T \\ &= \begin{bmatrix} X_{1k} & \dot{X}_{1k} & X_{2k} & \dot{X}_{2k} & X_{3k} & \dot{X}_{3k} & X_{4k} & \dot{X}_{4k} \end{bmatrix}^T \\ &= \begin{bmatrix} X_{1k} & X_{2k} & X_{3k} & X_{4k} & X_{5k} & X_{6k} & X_{7k} & X_{8k} \end{bmatrix}^T \end{aligned} \tag{3.11}$$

And knowing that

$$\begin{bmatrix} \ddot{d}_1 \\ \ddot{\theta}_1 \\ \ddot{\theta}_2 \\ \ddot{\theta}_3 \end{bmatrix}_k = \begin{bmatrix} m_T & 0 & 0 & 0 \\ 0 & A_1 & A_4 & A_5 \\ 0 & A_4 & A_2 & A_6 \\ 0 & A_5 & A_6 & A_3 \end{bmatrix}_k^{-1} \begin{bmatrix} F_z \\ \tau_1 \\ \tau_2 \\ \tau_3 \end{bmatrix}_k - \begin{bmatrix} 0 \\ A_7 \\ A_8 \\ 0 \end{bmatrix}_k - \begin{bmatrix} -gm_T \\ 0 \\ 0 \\ 0 \end{bmatrix}_k = \begin{bmatrix} f_1^* \\ f_2^* \\ f_3^* \\ f_4^* \end{bmatrix}_k \tag{3.12}$$

$$\begin{bmatrix} \dot{d}_1 \\ \dot{\theta}_1 \\ \dot{\theta}_2 \\ \dot{\theta}_3 \end{bmatrix}_{k+1} = \begin{bmatrix} \dot{d}_1 \\ \dot{\theta}_1 \\ \dot{\theta}_2 \\ \dot{\theta}_3 \end{bmatrix}_k + T_s \left(\begin{bmatrix} m_T & 0 & 0 & 0 \\ 0 & A_1 & A_4 & A_5 \\ 0 & A_4 & A_2 & A_6 \\ 0 & A_5 & A_6 & A_3 \end{bmatrix}_k^{-1} \begin{bmatrix} F_z \\ \tau_1 \\ \tau_2 \\ \tau_3 \end{bmatrix}_k - \begin{bmatrix} 0 \\ A_7 \\ A_8 \\ 0 \end{bmatrix}_k - \begin{bmatrix} -gm_T \\ 0 \\ 0 \\ 0 \end{bmatrix}_k \right) \rightarrow \begin{bmatrix} X_2 \\ X_4 \\ X_6 \\ X_8 \end{bmatrix}_{k+1} = \begin{bmatrix} f_2 \\ f_4 \\ f_6 \\ f_8 \end{bmatrix}_k \tag{3.13}$$

$$\begin{bmatrix} d_1 & \theta_1 & \theta_2 & \theta_3 \end{bmatrix}_{k+1}^T = \begin{bmatrix} X_1 & X_3 & X_5 & X_7 \end{bmatrix}_{k+1}^T = \begin{bmatrix} X_1 & X_3 & X_5 & X_7 \end{bmatrix}_k^T + T_s \begin{bmatrix} X_2 & X_4 & X_6 & X_8 \end{bmatrix}_k^T = \begin{bmatrix} f_1 & f_3 & f_5 & f_7 \end{bmatrix}_k^T \tag{3.14}$$

Then the overall state space could be defined as

$$\dot{X}_{k+1} = \dot{X}_k + T_s \begin{bmatrix} f_1 & f_2 & f_3 & f_4 & f_5 & f_6 & f_7 & f_8 \end{bmatrix}_k^T \tag{3.15}$$

Equations (3.3)-(3.9) have several parameters. Those are summarized by **Table 5**.

4. Results

The system in section 3 was simulated several time -for each filter including UKF, CKF and CDKF-. Four cases were obtained as follows:

1. Assuming all the states were measured.
2. Assuming that the position and angles were measured while their derivatives were not measured.
3. Similar to the first case. However, modeling uncertainties were injected; e.g. the masses were multiplied by 1.5.
4. Similar to the second case. However, modeling uncertainties were injected; e.g. the masses were multiplied by 1.5.

4.1. Results for System without Uncertainties; Cases 1 and 2

The results of cases 1 and 2 were summarized by **Table 6** and **Table 7**. The results showed that the filters gave similar performance for all the states when no modeling presented, refer to **Figure 5** and **Figure 6**. The performance of the filters for measured states were better than those obtained for non-measured states.

4.2. Results with Uncertainties

When modeling errors presented, the RMSE increased as shown in **Table 8** and **Table 9**. However, their effect became large, and maybe unstable, for the states that were not measured as shown in **Figure 7**. In such cases, the CDKF showed the superior performance; the filter remained stable. However, the UKF and CKF had a poor performance. The errors were bounded. However, they were high, refer to **Figures 8-10**.

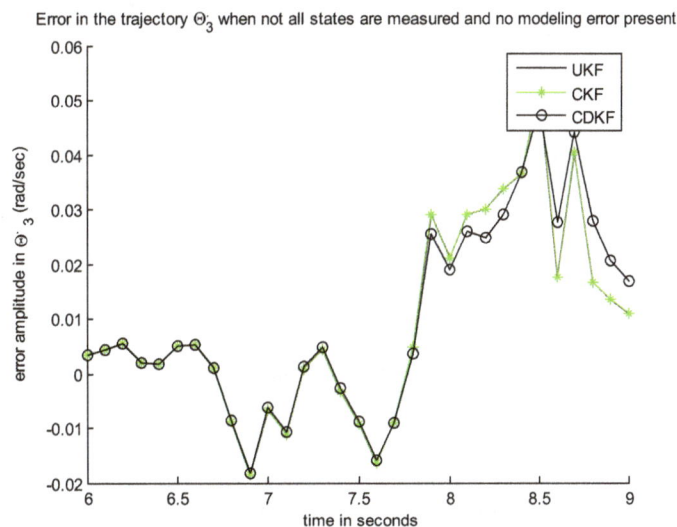

Figure 5. The performance of the filters for the third angler velocity, cases 1 and 2.

Table 5. Parameters' Value for the robotic arm.

Parameter	Value	Parameter	Value	Parameter	Value
m_1	21.5 kg	I_1	1.042 kg·m^2	a_1	0.25 m
m_2	16 kg	I_2	13 kg·m^2	a_2	1.2 m
m_3	8.5 kg	I_3	3.12 kg·m^2	a_3	0.8m
m_4	7.9 kg	I_4	1 kg·m^2	a_4	1.2
m_5	6.3 kg	I_5	0.84 kg·m^2	g	-9.81 m/s^2

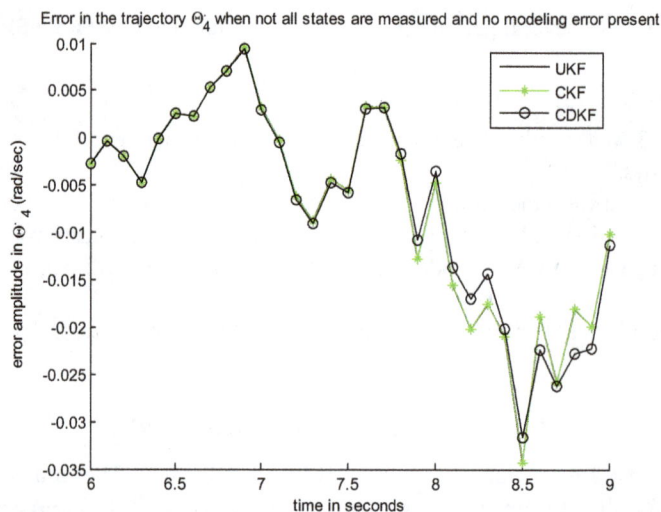

Figure 6. The performance of the filters for the fourth angler velocity, cases 1 and 2.

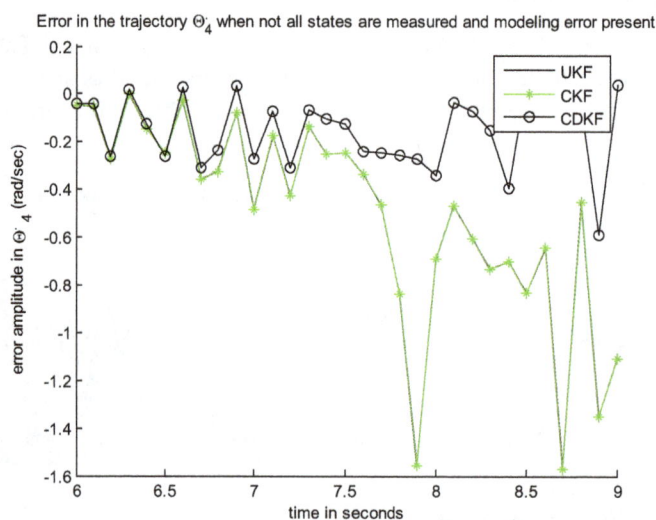

Figure 7. The performance of the filters for the fourth angler velocity, case 4.

Table 6. The root mean square error for the filters UKF, CKF and CDKF for case 1.

RMS in	UKF ×10⁻⁶	CKF ×10⁻⁶	CDKF ×10⁻⁶
d	32.7	32.7	32.7
$\dot{d}$	29.2	29.2	29.2
θ_1	31	31	31
$\dot{\theta}_1$	35.8	35.8	35.8
θ_2	22.2	22.2	22
$\dot{\theta}_2$	46.1	46.1	46.1
θ_3	25.6	25.6	25.6
$\dot{\theta}_3$	44.1	44.1	44.1

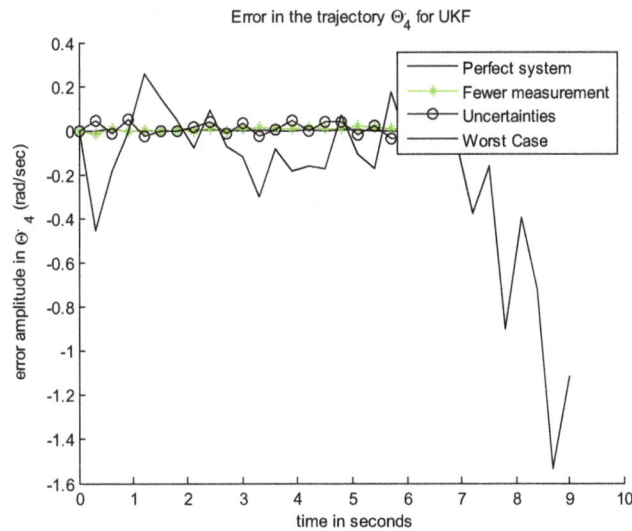

Figure 8. The error in estimating the fourth angular velocity using UKF for all cases.

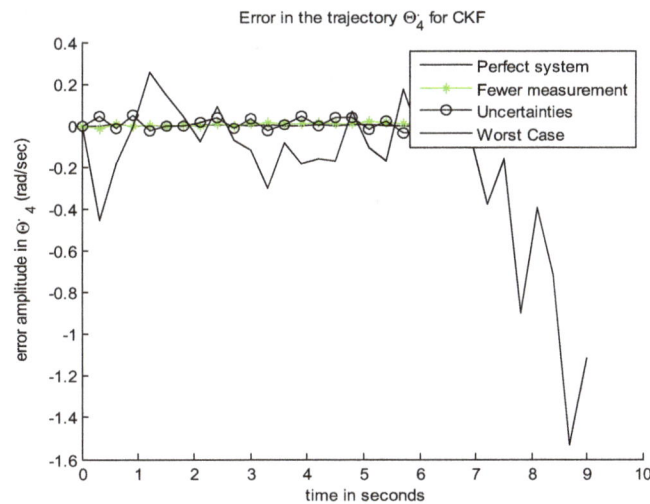

Figure 9. The error in estimating the fourth angular velocity using CKF for all cases.

Table 7. The root mean square error for the filters UKF, CKF and CDKF for case 2.

RMS in	UKF $\times 10^{-6}$	CKF $\times 10^{-6}$	CDKF $\times 10^{-6}$
d	35.9	35.9	35.9
$\dot{d}$	203.4	203.4	203.4
θ_1	41.4	41.4	41.3
$\dot{\theta}_1$	162.8	162.8	161.6
θ_2	30.6	30.6	31.9
$\dot{\theta}_2$	171.4	171.4	171.1
θ_3	49.2	49.2	49.1
$\dot{\theta}_3$	291.7	291.7	291.4

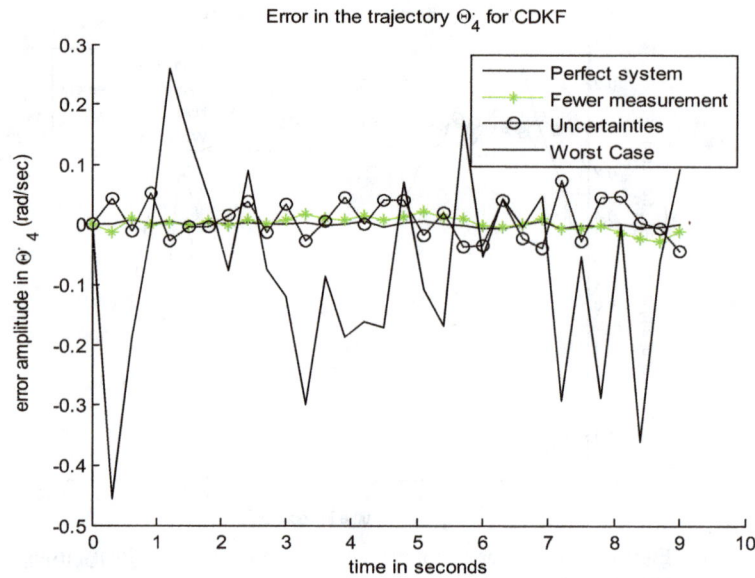

Figure 10. The error in estimating the fourth angular velocity using CDKF for all cases.

Table 8. The root mean square error for the filters UKF, CKF and CDKF for case 3.

RMS in	UKF $\times 10^{-6}$	CKF $\times 10^{-6}$	CDKF $\times 10^{-6}$
d	44.6	44.6	44.6
$\dot{d}$	375.8	375.8	375.8
θ_1	44.5	44.5	44.5
$\dot{\theta}_1$	367.4	367.4	367.4
θ_2	34.9	34.9	34.9
$\dot{\theta}_2$	365.1	365.1	365.1
θ_3	38.9	38.9	38.9
$\dot{\theta}_3$	366.5	366.5	366.5

Table 9. The root mean square error for the filters UKF, CKF and CDKF for case 4.

RMS in	UKF $\times 10^{-6}$	CKF $\times 10^{-6}$	CDKF $\times 10^{-6}$
d	146.9	146.9	146.8
$\dot{d}$	4588.5	4588.5	4588.9
θ_1	111.3	111.3	103.8
$\dot{\theta}_1$	2710.8	2710.8	1693.2
θ_2	120.3	120.3	112.2
$\dot{\theta}_2$	5590.7	5590.7	2447.4
θ_3	133.3	133.3	109.3
$\dot{\theta}_3$	4079.6	4079.6	1982.2

5. Conclusion

This work discussed the benefits of using Sigma-Point Kalman Filters in nonlinear application, *i.e.* PRRR robotic arm. Three types of SPKFs were used, namely Unscented, Cubature, and Central difference Kalman Filters. Four cases were used: the first and the second cases involved with system with no modeling errors; the third and the fourth cases involved with system injected with uncertainties. The first and the third cases assumed all the states were measured which was not the case in the other cases. The results showed that the filters gave good performance when all the states were measured. Reducing the number of measurements affected the results a little bit. The errors became larger than 10 times of those obtained in case 1 when modeling errors were presented and not all the states were measured. However, the CDKF showed stable performance in all cases. The latter gave an indication to use the CDKF in such applications.

References

[1] Hatamleh, K., Al-Shabi, M., Khasawneh, Q.A. and Al-Asal, M.A. (2014) Application of SMC and NLFC into a PRRR Robotic. *ASME* 2014 *International Mechanical Engineering Congress and Exposition*, Montreal, 14-20 November 2014, Paper No. IMECE2014-39136.

[2] Al-Shabi, M. and Hatamleh, K. (2014) The Unscented Smooth Variable Structure Filter Application into a Robotic Arm. *ASME* 2014 *International Mechanical Engineering Congress and Exposition*, Montreal, 14-20 November 2014, Paper No. IMECE2014-40118. http://dx.doi.org/10.1115/imece2014-40118

[3] Al-Shabi, M., Hatamleh, K. and Asad, A. (2013) UAV Dynamics Model Parameters Estimation Techniques: A Comparison Study. 2013 *IEEE Jordan Conference on Applied Electrical Engineering and Computing Technologies*, Amman, 3-5 December 2013.

[4] Ali, Z., Deriche, M. and Landolsi, M. (2009) Sigma Point Kalman Filters For Multipath Xhannel Estimation In CDMA Networks. *Proceedings of the* 2009 *6th International Symposium on Wireless Communication Systems*, Tuscany, 7-10 September 2009, 423-427. http://dx.doi.org/10.1109/ISWCS.2009.5285296

[5] Ambadan J. and Tang, Y.M. (2009) Sigma-Point Kalman Filter Data Assimilation Methods for Strongly Non-Linear Systems. *Journal of the Atmospheric Sciences*, **66**, 261-285. http://dx.doi.org/10.1175/2008JAS2681.1

[6] Anderson B. and Moore, J. (1979) Optimal Filtering. Prentice-Hall.

[7] Bar-Shalom, T. Li X.and Kirubarajan, T. (2001) Estimation with Applications to Tracking and Navigation—Theory, Algorithm and Software. John Wiley & Sons, Inc. http://dx.doi.org/10.1002/0471221279

[8] Grewal M. and Andrews, A. (2001) Kalman Filtering—Theory and Practice Using MATLAB. John Wiley & Sons, Inc.

[9] Barker, A., Brown, D. and Martin, W. (1995) Bayesian estimation and the Kalman Filter. *Computers & Mathematics with Applications*, **30**, 55-77.

[10] Welch, G. and Bishop, G. (2006) An Introduction to the Kalman Filter. Department of Computer Science, University of North Carolina, Chapel Hill, TR 95-041.

[11] Maybeck, P. (1979) Stochastic Models, Estimation, and Contro. Mathematics in Science and Engineering, Volume 141, Part 1, Academic Press, Waltham, iii-xix, 1-423.

[12] Kalman, R. (1960) A New Approach to Linear Filtering and Prediction Problems. *ASME Journal of Basic Engineering*, **82**, 35-45. http://dx.doi.org/10.1115/1.3662552

[13] Ormsby, C., Raquet, J. and Maybeck, P. (2006) A Generalized Residual Multiple Model Adaptive Estimator of Parameters and States. *Mathematical and Computer Modelling*, **43**, 1092-1113. http://dx.doi.org/10.1016/j.mcm.2005.12.003

[14] Negenborn, R. (2003) Robot Localization and Kalman Filters—On Finding Your Position in a Noisy World. MS Thesis, Utrecht University, Utrecht.

[15] Simon, D. (2006) Optimal State Estimation: Kalman, H [Infinity] and Nonlinear Approaches. Wiley-Interscience.

[16] Lary, D. and Mussa, H. (2004) Using an Extended Kalman Filter Learning Algorithm for Feed-Forward Neural Networks to Describe Tracer Correlations. *Atmospheric Chemistry and Physics Discussion*, **4**, 3653-3667. http://dx.doi.org/10.5194/acpd-4-3653-2004

[17] Leu, G. and Baratti, R. (2000) An Extended Kalman Filtering Approach with a Criterion to Set Its Tuning Parameters—Application to a Catalytic Reactor. *Computers & Chemical Engineering*, **23**, 1839-1849. http://dx.doi.org/10.1016/S0098-1354(00)00298-2

[18] Shojaie, K., Ahmadi, K. and Shahri, A. (2007) Effects of Iteration in Kalman Filter Family for Improvement of Estimation Accuracy in Simultaneous Localization and Mapping. *IEEE/ASME International Conference on Advanced In-*

telligent Mechatronics, Zurich, 4-7 September 2007, 1-6. http://dx.doi.org/10.1109/AIM.2007.4412453

[19] Zhang, Y., Zhou, D. and Duan, G. (2006) An Adaptive Iterated Kalman Filter. *IMACS Multiconference on Computational Engineering in Systems Applications*, Beijing, 4-6 October 2006, 1727-1730. http://dx.doi.org/10.1109/CESA.2006.4281916

[20] Hyland, J. (2002) An Iterated-Extended Kalman Filter Algorithm for Tracking Surface and Sub-Surface Targets. *OCEANS'02 MTS/IEEE*, **3**, 1283-1290. http://dx.doi.org/10.1109/oceans.2002.1191824

[21] Dungate, D., Theobald, R. and Nurse, F. (1999) Higher-Order Kalman Filter to Support Fast Target Tracking in a Multi-Function Radar System. *IEE ColloquiumTarget Tracking*: *Algorithms and Applications*, London, 11-12 November 1999, 14/1-14/3.

[22] Bayard, D. and Kang, B. (2003) A High-Order Kalman Filter for Focal Plane Calibration of NASA's Space Infrared Telescope Facility (SIRTF). *AIAA Guidance, Navigation and Control Conference and Exhibit*, Austin, 11-14 August 2003. http://dx.doi.org/10.2514/6.2003-5824

[23] Athans, M., Wishner, R. and Bertolini, A. (1968) Suboptimal State Estimation for Continuous-Time Nonlinear Systems from Discrete Noisy Measurements. *IEEE Transactions on Automatic Control*, **13**, 504-514. http://dx.doi.org/10.1109/TAC.1968.1098986

[24] Al-Shabi, M. (2012) The General Toeplitz/Observability Smooth Variable Structure Filter: Fault Detection and Parameter Estimation. LAP Lambert Academic Publishing, Saarbrücken.

[25] Nguyen, H. and Walker, E. (1996) A First Course in Fuzzy Logic. CRC Press, Boca Raton.

[26] Yager, R. and Zadeh, L. (1992) An Introduction to Fuzzy Logic Applications in Intelligent Systems. Kluwer Academic, location.

[27] Carrasco, R., Cipriano, A. and Carelli, R. (2005) Nonlinear State Estimation in Mobile Robots Using a Fuzzy Observer. *Processing of the 16th IFAC World Congress*, Vol. 16, Part 1, Czech Republic.

[28] Simon, D. (2003) Kalman Filtering for Fuzzy Discrete Time Dynamic Systems. *Applied Soft Computing Journal*, **3**, 191-207. http://dx.doi.org/10.1016/S1568-4946(03)00034-6

[29] Matia, F., Jimenez, A., Rodriguez-Losada, D. and Al-Hadithi, B.M. (2004) A Novel Fuzzy Kalman Filter for Mobile Robots Localization. *IPMU 2004, 10th International Conference on Information Processing and Management of Uncertainty in Knowledge Based Systems,* Volume II, Perugia.

[30] Chen, Z. (2003) Bayesian Filtering: From Kalman Filters to Particles and Beyond. McMaster University.

[31] Van Der Merwe, R. and Wan, E. (2004) Sigma-Point Kalman Filters for Integrated Navigation. *Proceedings of the Annual Meeting*, Institute of Navigation, 641-654.

[32] Wang, L., Wang, L., Liao, C. and Liu, J. (2009) Sigma-Point Kalman Filter Application on Estimating Battery SOC. *5th IEEE Vehicle Power and Propulsion Conference*, Dearborn, 7-10 September 2009, 1592-1595.

[33] Sadhu, S., Mondal, S., Srinivasan, M. and Ghoshal, T. (2006) Sigma Point Kalman Filter for Bearing Only Tracking. *Signal Processing*, **86**, 3769-3777. http://dx.doi.org/10.1016/j.sigpro.2006.03.006

[34] Schenkendorf, R., Kremling, A. and Mangold, M. (2009) Optimal Experimental Design with the Sigma Point Method. *IET Systems Biology*, **3**, 10-23. http://dx.doi.org/10.1049/iet-syb:20080094

[35] Tang, X., Zhao, X. and Zhang, X. (2008) The Square-Root Spherical Simplex Unscented Kalman Filter for State and Parameter Estimation. *9th International Conference on Signal Processing*, Beijing, 26-29 October 2008, 260-263.

[36] Kim, J. and Shin, D. (2005) Joint Estimation of Time Delay and Channel Amplitude by Simplex Unscented Filter without Assisted Pilot in CDMA Systems. *The 7th International Conference on Advanced Communication Technology*, **1**, 233-238.

[37] Julier, S. (2003) The Spherical Simplex Unscented Transformation. *Proceedings of the American Control Conference*, **3**, 2430-2434. http://dx.doi.org/10.1109/acc.2003.1243439

[38] Gadsden, S., Al-Shabi, M., Arasaratnam, I. and Habibi, S. (2010) Estimation of an Electrohydrostatic Actuator Using a Combined Cubature Kalman and Smooth Variable Structure Filter. *International Mechanical Engineering Congress and Exposition* (*IMECE*), American Society of Mechanical Engineers, Vancouver, British Columbia.

[39] Gadsden, A., Al-Shabi, M., Arasaratnam, I. and Habibi, S. (2014) Combined Cubature Kalman and Smooth Variable Structure Filtering: A Robust Nonlinear Estimation Strategy. *Signal Processing*, **96**, 290-299. http://dx.doi.org/10.1016/j.sigpro.2013.08.015

[40] Nrgaard, M., Poulsen, N. and Ravn, O. (2000) New Developments in State Estimation for Nonlinear Systems. *Automatica*, **36**, 1627-1638. http://dx.doi.org/10.1016/S0005-1098(00)00089-3

[41] Zhang, U., Gao, F. and Tian, L. (2008) INS/GPS Integrated Navigation for Wheeled Agricultural Robot Based on Sigma-Point Kalman Filter. 2008 *Asia Simulation Conference-7th International Conference on System Simulation and*

Scientific Computing, Beijing, 10-12 October 2008, 1425-1431. http://dx.doi.org/10.1109/ASC-ICSC.2008.4675598

[42] Zhu, J.H., Zheng, N.N., Yuan, Z.J. and Zhang, Q. (2009) A SLAM Algorithm Based on the Central Difference Kalman Filter. 2009 *IEEE Intelligent Vehicles Symposium*, Xi'an, 3-5 June 2009, 123-128. http://dx.doi.org/10.1109/IVS.2009.5164264

[43] Sadati, N. and Ghaffarkhah, A. (2007) POLYFILTER: A New State Estimation Filter for Nonlinear Systems. *International Conference on Control, Automation and Systems*, Seoul, 17-20 October 2007, 2643-2647.

[44] Henrici, P. (1964) Elements of Numerical Analysis. John Wiley and Sons, New York.

[45] Van Der Merwe, R. (2004) Sigma Point Kalman Filters for Probabilistic Inference in Dynamic State-Space Models. PhD Thesis, OGI School of Science & Engineering, Oregon Health & Science University, USA.

Nomenclature

$^{-1}$, ^{T}	Inverse, and transpose, respectively.
$(a)_i$	The i row of a.
a_{i-1}, α_{i-1}, d_i and θ_i	Link-i's length (m), twist (rad), and offset (m), and joint-i angle (rad), respectively.
c_i and s_i	$\cos(\theta_i)$ and $\sin(\theta_i)$, respectively.
c_{ij} and s_{ij}	$\cos(\theta_i + \theta_j)$ and $\sin(\theta_i + \theta_j)$, respectively.
e_m	The estimation error vectors in m.
$f(.)$	The system's model function.
F_z and τ_i	Prismatic joint-1 motor force (N) and Revolute joint-i motor torque (N. M), respectively.
g	Gravity acceleration (m/s^2).
$g(.)$	The sensor's model function.
i, j	Subscripts used to identify elements.
$I_{n \times n}$	The identity matrix with dimensions of $n \times n$.
k	Time step value.
$k \mid k-1$	The a priori value at time k.
$k \mid k$	The a posteriori value at time k.
K_X	The correction gain of the filter X.
$M(\Theta)$	Inertia matrix.
$m_1, m_2, \cdots, m_5$	Masses of links 1, 2, 3 and 4 respectively (kg).
m, n	Number of measurements and states, respectively.
P_{xx}	The state's error covariance matrix.
P_{zz}	The output's error covariance matrix.
P	The error covariance matrix.
q	The number of the sigma points.
Q	The process noise covariance matrix.
R	The measurements noise covariance matrix.
$\sum$	The summation operator.
T_s	Sampling time, and is equal to 0.001 sec.
τ	Joints force and torques vector.
$V(\Theta, \dot{\Theta})$	Viscous friction vector.
v, w	The measurement and system noise, respectively.
W_i	The assigned weight.
x	The state vector.
z	The output vector.
X_i and Z_i :	The estimate and its measurement for the ith sigma point, respectively.

20

Dual Dynamic PTZ Tracking Using Cooperating Cameras

Mohammed A. Eslami, John R. Rzasa, Stuart D. Milner, Christopher C. Davis*

Department of Electrical and Computer Engineering, University of Maryland, College Park, USA
Email: meslami@umd.edu, rzasaman@umd.edu, *davis@umd.edu

Abstract

This paper presents a real-time, dynamic system that uses high resolution gimbals and motorized lenses with position encoders on their zoom and focus elements to "recalibrate" the system as needed to track a target. Systems that initially calibrate for a mapping between pixels of a wide field of view (FOV) master camera and the pan-tilt (PT) settings of a steerable narrow FOV slave camera assume that the target is travelling on a plane. As the target travels through the FOV of the master camera, the slave cameras PT settings are then adjusted to keep the target centered within its FOV. In this paper, we describe a system we have developed that allows both cameras to move and extract the 3D coordinates of the target. This is done with only a single initial calibration between pairs of cameras and high-resolution pan-tilt-zoom (PTZ) platforms. Using the information from the PT settings of the PTZ platform as well as the precalibrated settings from a preset zoom lens, the 3D coordinates of the target are extracted and compared to those of a laser range finder and static-dynamic camera pair accuracies.

Keywords

Surveillance, PTZ Cameras, Cooperating Cameras

1. Introduction

Investigating the use of cooperating camera systems for real-time, high definition video surveillance to detect and track anomalies over time and adjustable fields of view is moving us towards the development of an automated, smart surveillance system. The master-slave architecture for surveillance, in which a wide field-of-view camera scans a large area for an anomaly and controls a narrow field of view camera to focus in on a particular target is commonly used in surveillance setups to track an object [1]-[3]. The static camera solution [4]-[6], or

---
*Corresponding author.

the master-slave system architecture with static master camera [6] [7] are well-researched problems, but is limited by the field of view of the master camera.

In particular, due to the computational complexity arising from object identification, having such systems operate in real-time is a hurdle within itself [1] [2] [8]. These setups often use background subtraction to detect a target within the FOV of the static camera and use a homography mapping between the pixels of the static camera to the pan/tilt (PT) settings of the slave camera to focus on the target. Look-up tables [3] and interpolation functions [9]-[11] are common tools used to navigate through the different settings to find the optimum setting for target tracking [6]. Essentially, a constraint is placed on the target such as the percentage of the image it must cover, or the centering of the target within the image at all times, or a combination of the two, and the intrinsic/extrinsic parameters are varied to find the optimum setting that best satisfies these constraints.

This paper presents a dual-dynamic camera system that uses in-house designed, high-resolution gimbals [12], and commercial-off-the-shelf (COTS) motorized lenses with position encoders on their zoom and focus elements to "recalibrate" the system as needed to track a target. The encoders on the lenses and gimbals of the master camera control the slave camera to zoom in and follow a target as well as extract its 3D coordinate relative to the position of the master camera. This system interpolates the homography matrix between pixels of the master camera and angles on the slave camera for different pan/tilts of the master camera. The master camera will keep a target in a specific region within the image and adjust its angle based on the trajectory of the target to force the target to stay within that region.

The homography mapping between the master and salve camera is updated anytime the master camera moves, so as to keep the control between the master-slave cameras continuous. The master camera turns off background subtraction every time it detects that it needs to move and reinitializes it after it has completed its movement. This system operates in real-time, and since the encoder settings are in absolute coordinates it can potentially be used to provide a 3D reconstruction of the trajectory of the target.

2. System Architecture

2.1. General Overview

The goal of this system is to use high definition, uncompressed video to resolve a target of length 5 m, such as a car, at ranges of 100 s of meters. This involves choosing appropriate hardware to be able to meet these requirements and the necessary control algorithms to allow the system to operate in real-time.

Mathematically, this situation can be modeled for a camera with w-pixel resolution in the horizontal direction imaging an object at a distance l from the camera with a FOV of θ_{FOV}:

$$w_{obj} = \frac{w}{l\theta_{FOV}} \tag{1}$$

which provides an accuracy of 2 cm at a target range of 100 m and object size of 10 m (with the lens at maximum zoom, corresponding to a FFOV of 2) which should suffice in traffic surveillance applications.

2.2. Offline/One-Time Calibrations

To minimize the amount of image processing needed and thereby reduce the computational complexity of the problem, the processing needed for detecting the features to identify the target should be done only in one camera. These calibrations can be divided into two parts: 1) Computer vision algorithms and toolboxes to extract optical parameters and initialize the control algorithm, and 2) Generation and interpolation of the optical parameters extracted at various zoom settings for target localization. These two calibration parts are shown in **Figure 1**.

The Matlab toolbox [13] was used to extract the calibration parameters of the cameras and generate the look-up table (LUT) of the lens at the various zoom/focus settings, which were then interpolated in the same manner as [14]. The initial homography between the master camera's pixels and the slave camera's pan/tilt settings was found by corresponding nine pixel points in the master camera to nine pan/tilt settings of the slave camera. The pair of (x, y) coordinates retrieved from the master camera and (p, t) coordinates from the slave camera form a calibration point which obeys Equation (2), where H is a linear mapping (3×3 homography matrix) and s is a constant scale factor.

```
┌─────────────────────────────────────┐
│   OFFLINE/ONE-TIME COMPUTATIONS      │
└─────────────────────────────────────┘
```

Computer Vision	Optical

Calibrate cameras and save initial settings	Generate LUT for optical parameters to zoom setting for each camera

Generate initial homography from Master to Slave	Interpolate LUT to get function for all setting in between

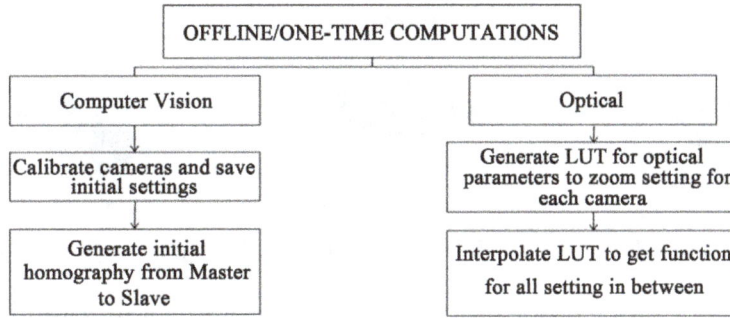

Figure 1. Calibration phase of the surveillance system.

$$\begin{bmatrix} p_{slave} \\ t_{slave} \\ 1 \end{bmatrix} = \frac{1}{s} H \begin{bmatrix} x_{master} \\ y_{master} \\ 1 \end{bmatrix} \tag{2}$$

At selected zoom settings of each camera, singular value decomposition (SVD) is used to find the homography between the pixels of the master camera and the pan/tilt settings of the slave camera to bring the (x, y) calibration point to its center. Errors will arise from the fact that all lenses exhibit some shift in their optic axis [3] as they zoom in on a target.

This procedure is repeated for the various regions that are defined for the master camera. That is, the master camera is held to a particular pan/tilt setting that defines a region and one homography mapping is attributed to it. Then, the pan/tilt settings of the master camera are changed to define the next region and a new homography mapping is applied to the new region. Once all regions are defined, the elements of the various homography mappings are interpolated linearly to be able to control the slave camera with the appropriate homography mapping of the target in a specified region. **Figure 2** shows the elements of the homography matrix for various pan/tilt settings of a master/slave camera setup with a baseline of 1.5 m, focal lengths of 33 mm and 100 mm, respectively, for an area that is 70 × 70 m at a range of 150 m. The surface plot is a linear interpolation through the angles that were chosen for calibration. The reason for choosing a linear interpolation for the data will be explained in the description of the tracking phase of the system.

2.3. Real-Time Tracking

A large region of interest about the center defined in the master camera on every frame checks to ensure the target stays within its boundaries. The pan/tilt settings of the master camera are adjusted as the target moves above/below or to the left/right of this region of interest. The increment of adjustment used is the same as that of the linear calibration to ensure accuracy in the homography being used. Ideally, the master camera should not be moving too much since it has a wide field of view and thus using the linear interpolation between these schemes is satisfactory. **Figure 3** shows the real-time tracking phase of the surveillance system.

Although the cameras are set in a master-slave relationship, the gimbal encoders from each camera are independent of one another. This amounts to having two independent, different viewpoints of the same scene, which provides stereovision. The range from such a setup can be approximated by a homogeneous linear method of triangulation, which often provides acceptable results. Its advantage over other methods is that it can be easily extended when additional cameras are added, a requirement of this system [15].

3. Simulations and Experimental Results

3.1. Simulations

Simulations model a perfect world with no noise in choosing the pairs of pixel points and pan/tilt settings to calibrate the homography matrices between the master and slave. Essentially, points in the world are mapped to the image of the master camera via a projection matrix and rotation matrices are chosen for the slave camera to have that point fall in the center of its image. To achieve this, an initial guess can be derived for the slave camera to point at the world point by finding the vector C from Equation (3) and is shown in **Figure 4**.

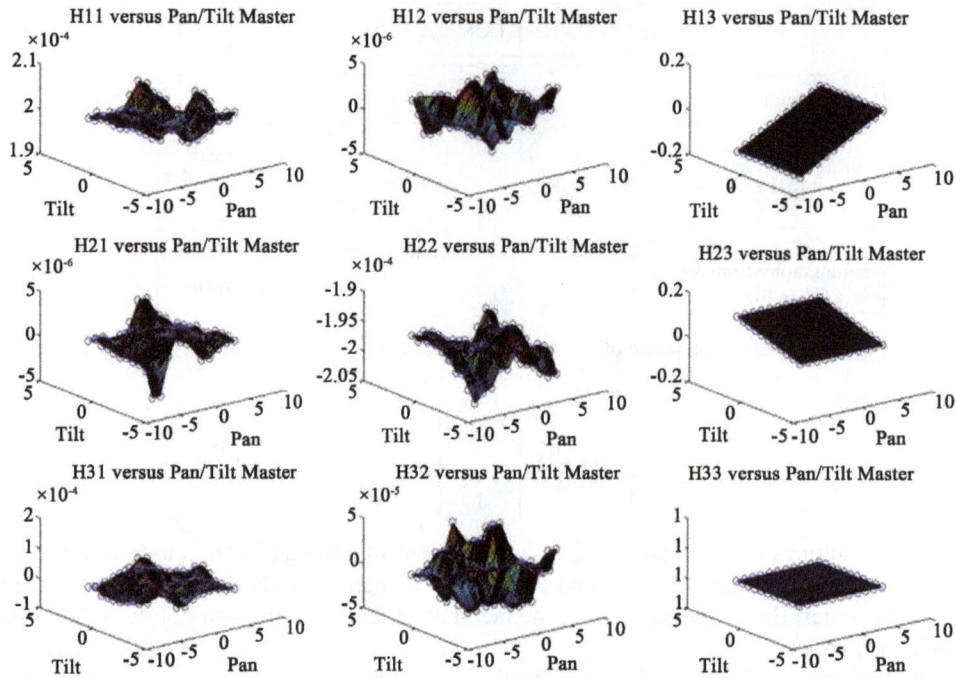

Figure 2. Homography matrix elements for various pan/tilt settings of the master camera.

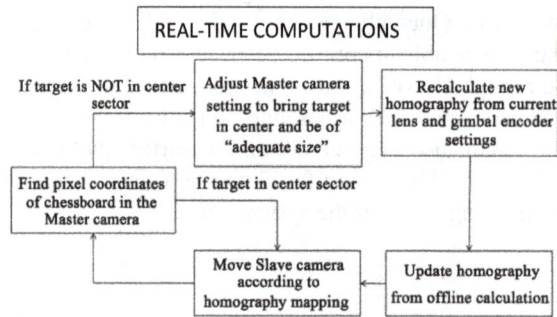

Figure 3. Real-time tracking of surveillance system.

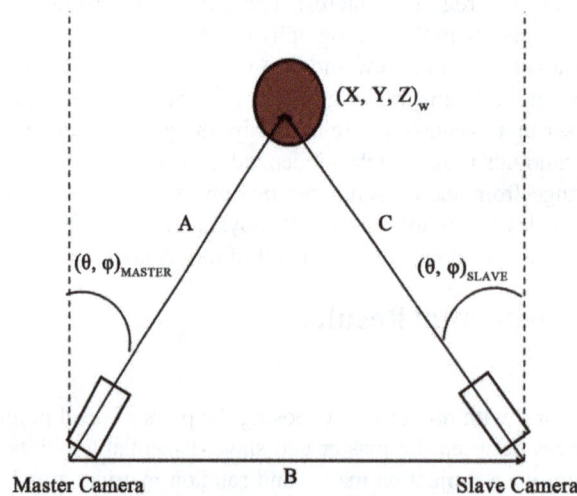

Figure 4. Initial guess to find the slave camera angles from master camera and baseline.

$$C = A - B \qquad (3)$$

The vector A is the fixed orientation of the master camera, and the vector B is the baseline vector between the master and the slave cameras. The projection matrix of the slave camera is then optimized to bring that world point into a region of within 10 pixels of the center of the slave image. A maximum of five bounces is allowed if the camera begins to hover around the world point as it tries to bring it within the center of its image. The pair of (x, y) coordinates retrieved from the master camera and (p, t) coordinates from the slave camera form a calibration point and this is repeated nine times.

Once the calibration stage is complete and all of the homographies are found, a target world coordinate is imaged into the master camera and the control algorithm is simulated to control the slave camera and localize the target. The simulated results are compared to pre-coded target coordinates and this is shown in **Figure 5**.

The y-coordinate shows the worse localization error at 10% of a surveyed area that is 50 × 50 × 150 m (Z coordinate × x coordinate × Y coordinate) large. This localization was extracted assuming the target is exactly centered within the slave camera. Another simulation recalculated the average relative error if the there was some noise added to the system that would cause the control algorithm to fail to bring the target exactly to the center of the slave camera FOV. This simulation is shown in **Figure 6**. It can be seen that increasing the baseline between the master and slave cameras reduces the localization error.

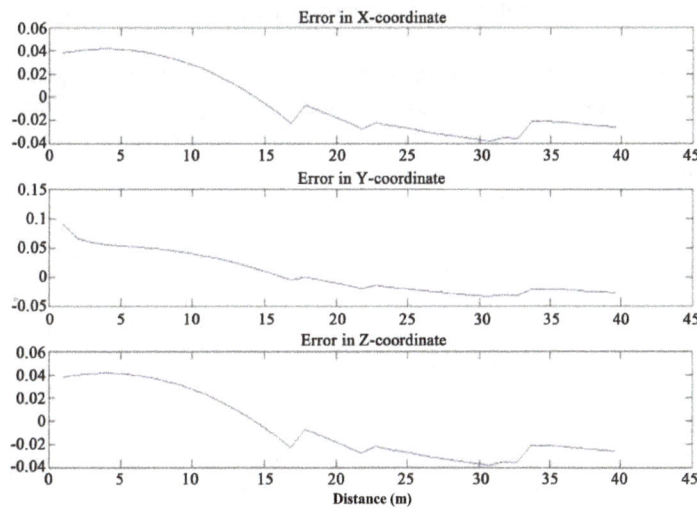

Figure 5. Relative error in the coordinates for a random walk in the calibrated environment. The x-axis is the iteration number while the y-axis is the relative error in localizing the target using stereovision versus the known simulated coordinate.

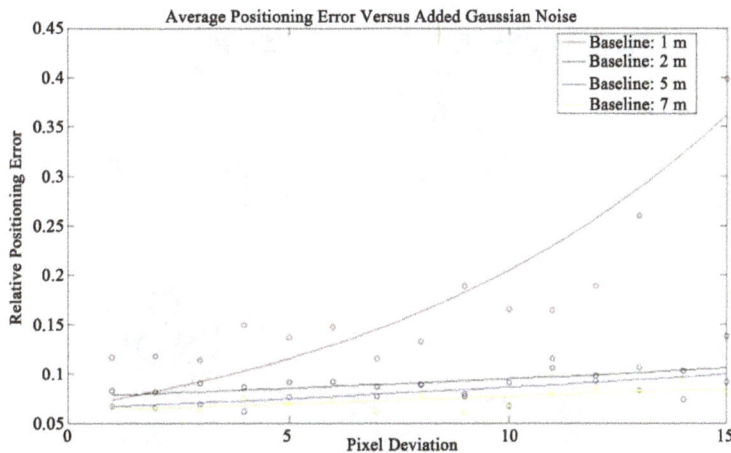

Figure 6. Positioning error with additive noise at different baseline measurements. Larger baselines compensate for the error produced by the noise.

An advantage of this system is that it does not need to corresponding features between cameras since the homographies will all be precalibrated manually. So long as the target is found in a single camera, the second camera will follow the target, without the need for image segmentation and identification algorithms.

3.2. Experimental Results

The experimental setup consisted of two Fujinon C22X23R2D-ZP1 motorized zoom lenses with digital preset to ensure that the precise position of the zoom and focus elements were known. The lenses were equipped with 16-bit encoders to accurately calibrate for the focal length by using the MATLAB camera calibration toolbox [13] at a number of zoom settings fitting the model to the commonly used exponential model between zoom/focus settings and focal length. The plots retrieved are similar to those shown by Wilson [14] and other surveillance papers that have motorized zoom capabilities [1] [2]. The cameras used in the stereo setup are Allied Vision Technologies GC1600CH, 2-Megapixel, 25 fps, Gigabit Ethernet machine vision cameras streaming uncompressed video data. The gimbals used are designed in-house [12] with a common yolk-style platform giving 360° continuous pan range and ±40° tilt range. The gimbals are driven with two direct-drive brushless AC servo motors with 20bit absolute encoders giving 0.000343° readout resolution. They are equipped to hold 50 lbs and have a 0.002° positioning repeatability with the optical system used in this work. The full system is shown in **Figure 7**. There were two experiments that were conducted: 1) Target localization and 2) Real-time surveillance tracking.

Figure 8 and **Figure 9** show experimental results obtained from the ranging experiment with the hardware setup. The Biomolecular Services Building across from the Kim Engineering Building on the Universtiy of Maryland campus was used as the plane for calibration, and points were selected in the tracking phase to center the slave camera. Google Earth was used to find the distance of the building relative to our laboratory and these were compared to the results given from the camera system. Google Earth's numbers were also confirmed with a GLR225 Bosch laser range finder by giving a measured distance between the buildings on the order of 170 m. The (X, Y) positions are roughly estimated based on the size of the windows on the building, which are 1.2 m wide by 1.3 m high.

The surveillance setup was housed in the Kim Engineering building to track a single target in the parking lot, which was divided up into four regions. A failure of surveillance occurs anytime the target moves out of the field of view of the slave camera [16]. Testing the surveillance setup in real-time (12 fps) at 405 × 305 resolution to track a target also showed excellent alignment capabilities as seen in **Figure 10**. A false positive in the experiment is defined as a feature that is detected which is not the target. They are a result of the master camera adjusting its setting to bring the target back within its region of interest. An average of two false positives were detected in 10 different adjustments of the master camera. These false positives can be minimized and/or eliminated by increasing the number of learning images needed to detect a background so that new objects within the scene are not considered as moving foreground objects. Increasing the number of images to find a background, however, does increase the latency in tracking the target with the slave camera.

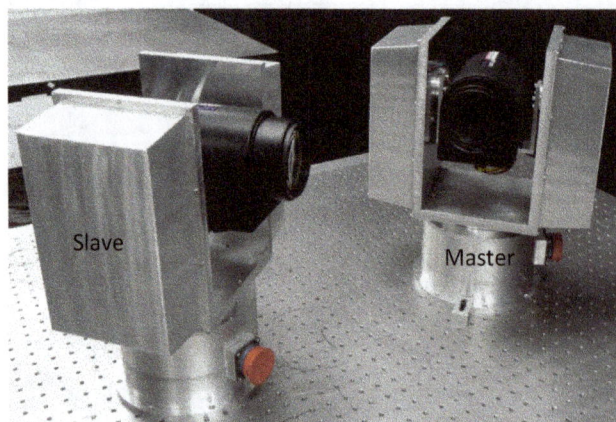

Figure 7. Master-slave camera surveillance setup (refer to gimbals graphically like master and slave).

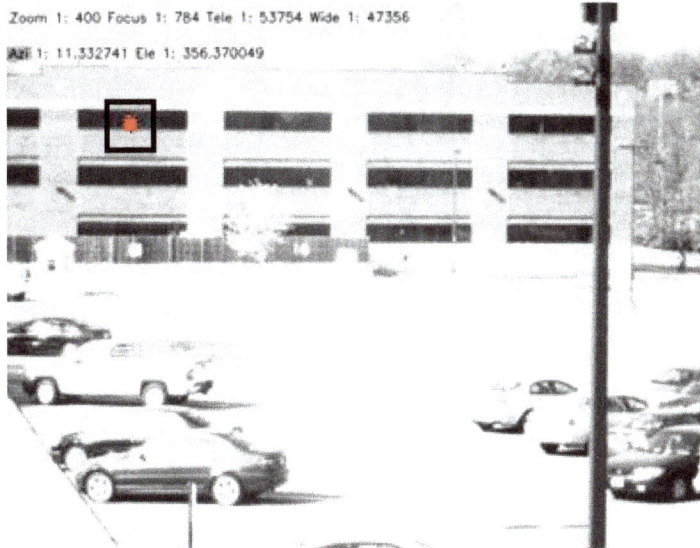

Zoom 1: 400 Focus 1: 784 Tele 1: 53754 Wide 1: 47356

Azi 1: 11.332741 Ele 1: 356.370049

(a)

Extracted Position Coordinate

X = 15.1078 m

Y = 3.8791

Z = 164.5592 m

Theoretical Position Coordinate

X = 14.2 m

Y = 4.2 m

Z = 170 m

Error In Coordinates

ΔX = 0.9078 m

ΔY = −0.3209 m

ΔZ = −5.4408 m

Zoom 2: 18288 Focus 2: 37936 Tele 2: 53754 Wide 2: 47356

Azi 2: 5.552559 Ele 2: 0.519104

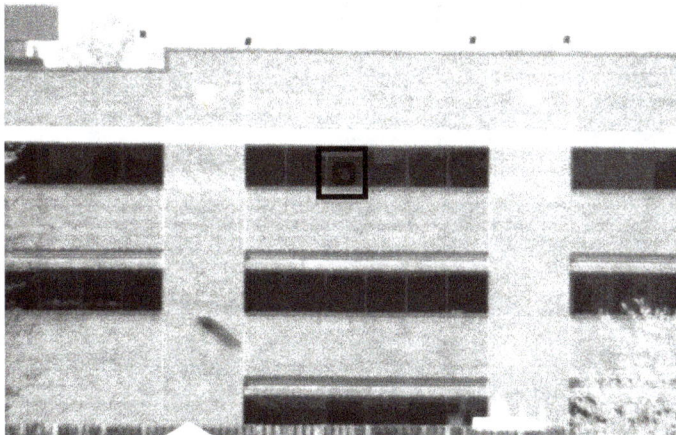

(b)

Figure 8. (a) Master Camera looking at a building with its point selected shown in red; (b) Slave camera centering that point within its image and computing the position relative to the master camera.

4. Conclusions/Future Work

We designed and developed a novel system with multiple dynamic cameras to track a target's 3D coordinate relative to the master camera in a master-slave relationship. As the target moved out of the region of interest in the master camera, the master camera moved to bring the target back into a certain predefined window. Calibrations between pan/tilt settings of the slave camera and pixel settings of the master camera are then updated based on the moves of the master camera to ensure the slave camera keeps the target within its center. The absolute encoders available on the optical system and the gimbals were then used in a stereo setup to find the 3D coordinate of the target relative to the master camera in real-time. To improve ranging accuracies, it was shown through simulation that the baseline of the system should be increased. This is relatively easy to incorporate within the master-slave system described.

To expand on the system currently running in real-time in our laboratory would require an implementation of

Figure 9. (a) Master Camera looking at a building with its point selected in a box; (b) Slave camera centering that point within its image and computing the position relative to the master camera.

image features to be used for correspondence. These vision algorithms are computationally expensive if they are to be run on the whole image, particularly when the video stream is in the form of uncompressed megapixel imagery data coming from machine vision cameras. Therefore, the master-slave relationship can act as an initialization of a region to correspond features. The larger baselines on the order of 7 m and above could then be tested to monitor the improvements of correspondence between the respective video streams. If the baseline is too large, the lighting coming into one camera could show a completely different image of the same scene between the two cameras and correspondence would fail. The initialization of two regions that should correspond to one another will help to alleviate this problem.

Our system was implemented with two cameras and a single target. A next step would be to use this setup as a single node within a much larger surveillance network. The network would communicate through a secondary control network to pass a target ID, namely the measured 3D coordinate and velocities from optic flow measurements, to other nodes for a longer track period. A lower data rate secondary channel would communicate small portions of data would allow the network to hand off the target from node to node in real-time. This would be a step towards the development of a fully cooperating, smart surveillance system.

(a)

(b)

(c)

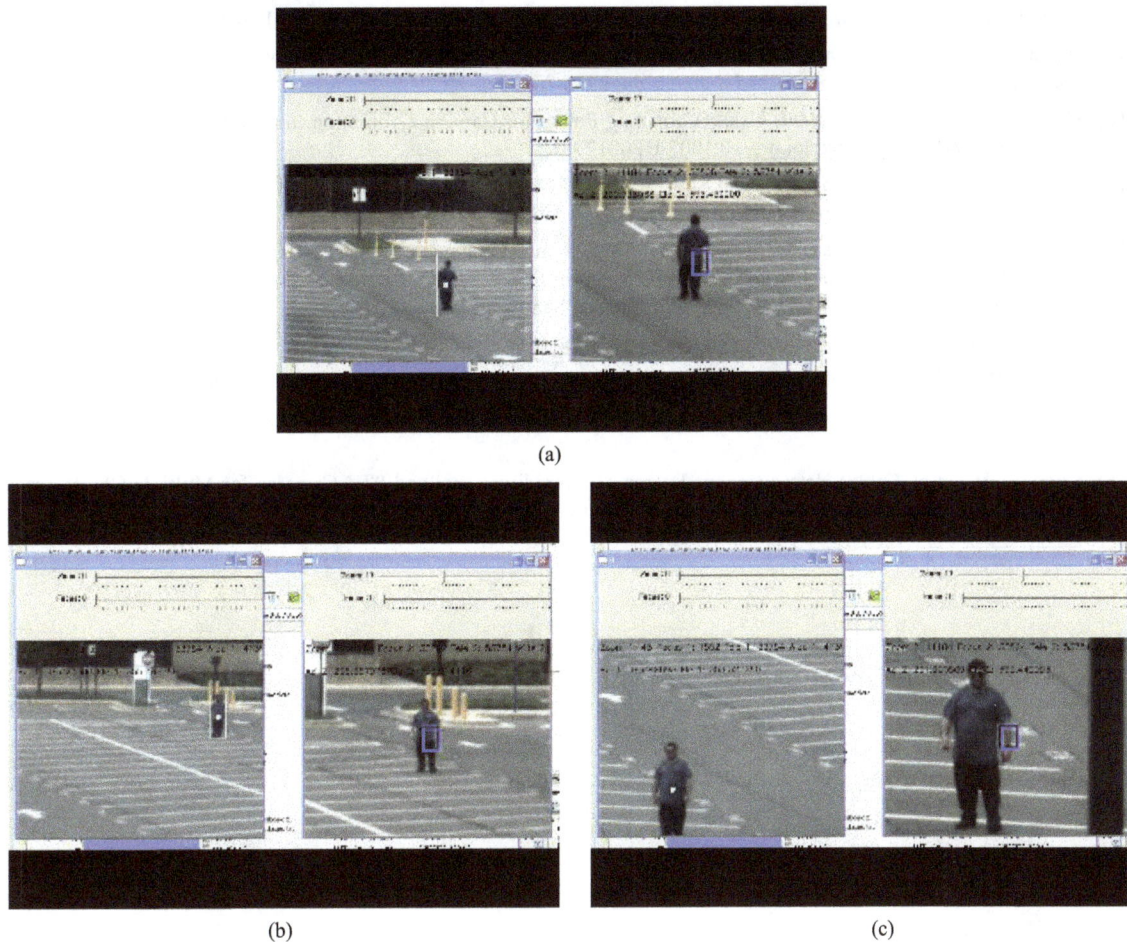

Figure 10. Surveillance system after calibration tracking a target moving from (a) Region 1; to (b) Region 2; (c) Region 3.

Support

This work was supported by the U.S. Department of Transportation and the Federal Highway Administration Exploratory Advanced Research Program contract. (DTFH6112C00015).

References

[1]   Zhou, J. and Wan, D. (2008) Stereo Vision Using Two PTZ Cameras. *Computer Vision and Image Understanding*, **112**, 184-194. http://dx.doi.org/10.1016/j.cviu.2008.02.005

[2]   Zhou, J., Wan, D. and Ying, W. (2010) The Chameleon Like Vision System. *IEEE Signal Processing Magazine*, **27**, 91-101. http://dx.doi.org/10.1109/MSP.2010.937310

[3]   Badri, J., Tilmant, C., Lavest, J.-M. and Pham, Q.-C. (2007) Camera-to-Camera Mapping for Hybrid Pan-Tilt-Zoom Sensors Calibration. *Lecture Notes in Computer Science*, **4522**, 132-141.

[4]   Horaud, R., Knossow, D. and Michaelis, M. (2006) Camera Cooperation for Achieving Visual Attention. *Machine Vision Applications*, **16**, 331-342. http://dx.doi.org/10.1007/s00138-005-0182-9

[5]   Khan, S., Javid, O. and Rasheed, Z. (2001) Human Tracking in Multiple Cameras. *Proceedings of IEEE International Conference of Computer Vision*, 331-336.

[6]   Senior, A., Hampapur, A. and Lu, M. (2005) Acquiring Multi-Scale Images by Pan-Tilt-Zoom Ontrola nd Automatic Multi-Camera Calibration. *IEEE Workshop on Applications on Computer Vision*, 433-438.

[7]   Sinha, S. and Pollefeys, M. (2004) Towards Calibrating a Pan-Tilt-Zoom Camera Network. *EECV Conference Workshop*.

[8]   Nelson, E.D. and Cockburn, J.C. (2007) Dual Camera Zoom Control: A Study of Zoom Tracking Stability. *Proceed-

ings of IEEE International Conference of Acoustics, Speech and Signal Processing, 941-944.

[9]   Bazin, J.-C. and Demonceaux, C. (2008) UAV Attitude Estimation by Vanishing Points in Catadioptric Images. *International Conference on Robotics and Automation*, 2743-2749.

[10] Caprile, B. and Torre, V. (1990) Using Vanishing Points for Camera Calibration. *International Journal of Computer Vision*, **4**, 127-140. http://dx.doi.org/10.1007/BF00127813

[11] Chen, Y.S., Hung, Y.P., Fuh, C.S. and Shih, S.W. (2000) Camera Calibration with a Motorized Zoom Lens. *International Conference on Pattern Recognition*, 495-498.

[12] Rzasa, J., Milner, S.D. and Davis, C.C. (2011) Design and Implementation of Pan-Tilt FSO Transceiver Gimbals for Real-Time Compensation of Platform Distrubances Using a Secondary Control Network. *SPIE Laser Communication and Propagation through the Atmosphere and Oceans*, San Diego.

[13] Bouget, J. http://www.vision.caltech.edu/bouguetj/calib_doc/

[14] Wilson, R.K. (1994) Modeling and Calibration of Automated Zoom Lenses. *Proceedings of SPIE*, 170-186.

[15] Hartley, R. and Zisserman, A. (2003) Multiple View Geometry in Computer Vision. 2nd Edition, Cambridge University Press, Cambridge.

[16] Chen, C.-H., *et al.* (2008) Heterogeneous Fusion of Omnidirectional and PTZ Cameras for Multiple Object Tracking. *IEEE Transactions on Circuits and Systems for Video Technology*, **18**, 1052-1063. http://dx.doi.org/10.1109/TCSVT.2008.928223

21

Temporal Prediction of Aircraft Loss-of-Control: A Dynamic Optimization Approach

Chaitanya Poolla[1], Abraham K. Ishihara[2]

[1]Electrical and Computer Engineering, Carnegie Mellon University (SV), Moffett Field, CA, USA
[2]Research Faculty, Electrical and Computer Engineering, Carnegie Mellon University (SV), Moffett Field, CA, USA
Email: chaitanya@cmu.edu, abe.ishihara@west.cmu.edu

Abstract

Loss of Control (LOC) is the primary factor responsible for the majority of fatal air accidents during past decade. LOC is characterized by the pilot's inability to control the aircraft and is typically associated with unpredictable behavior, potentially leading to loss of the aircraft and life. In this work, the minimum time dynamic optimization problem to LOC is treated using Pontryagin's Maximum Principle (PMP). The resulting two point boundary value problem is solved using stochastic shooting point methods via a differential evolution scheme (DE). The minimum time until LOC metric is computed for corresponding spatial control limits. Simulations are performed using a linearized longitudinal aircraft model to illustrate the concept.

Keywords

Pilot Assistance, Loss of Control, Aircrafts, Dynamic Optimization, Temporal Prediction, Pontryagin Maximum Principle, Differential Evolution, Stochastic Shooting Point Methods

1. Introduction

Air crash analyses during the past decade have concluded that about 40 percent of fatal air accidents in civil aviation occur due to Aircraft Loss of Control (LOC), the most contributing factor amongst others [1] [2]. During this time span, LOC related research has received increased attention in the aviation safety community [3]. Several investigations were carried out to understand the nature and characteristics of loss-of-control regimes [3]-[5]. In particular, a collaborative effort between Boeing and NASA Langley [6] provided a flight envelope based method to quantify LOC from air accident data. Similar envelopes have been used in this work to quantify LOC boundaries.

Quantifying LOC boundaries is a first step toward addressing the larger issue of LOC prevention. While there exist envelope protection features on an aircraft, they are of little use during LOC flight regimes due to degradation of normal control modes [7]. An alternative approach is to provide useful LOC information to pilots using flight states and pilot input data. Though LOC envelopes provide limits of operation of aircraft states and other auxiliary variables, there are not readily usable by the pilots. This is because, envelope data are provided in the flight state space whereas pilot decisions are executed in the control space. Furthermore, the mapping between the control-space inputs to state-space responses becomes unpredictable close to LOC regimes. This results in difficulty for human interpretation unlike flight regimes close to the trim conditions. However, it is possible to warn the pilot about potential LOC scenarios using intelligent algorithms by extracting accurate spatio-temporal information from the available LOC envelopes for direct pilot use. In this connection, recent experiments carried out at NASA Ames Research Center demonstrate favorable disposition of pilots to use pilot-friendly LOC tools [8]. A data based predictive control (DBPC) algorithm [9] was used to compute spatial control bounds for pilot use. However, in that work the time associated with the spatial bounds was considered fixed. This work complements the DBPC based spatial bounds by providing temporal bound information in framework of optimal control theory using Pontryagin's Maximum Principle.

This remainder of the paper is structured as follows. Section 2 provides an overview of the minimum time problem. Section 3 treats the optimal control problem using Pontryagin Maximum Principle and describes the resulting two point boundary value problem (TP-BVP). The solution to the TPBVP using differential evolution (DE) based methods is described in Section 4. Simulation results based on linear longitudinal model from [4] are provided in Section 5 followed by discussion and concluding remarks in Section 6.

2. Problem Formulation

We consider the problem of obtaining the minimum time to exit the flight envelope for a linearized longitudinal aircraft model. Let the operating envelope be defined in the (V, γ) state space for control limits in (T, δ_e) space as specified below [4].

$$\mathbb{E} := \left\{ (V, \gamma) \middle| 90 \le V \le 240, -22 \le \gamma \le 22 \right\}$$

$$\mathbb{U} := \left\{ (T, \delta_e) \middle| 0 \le T \le 40, -40 \le \delta_e \le 20 \right\}$$

$$\partial \mathbb{E} := \left\{ (V, \gamma) \middle| V = \{90, 240\}, \gamma = \{-22, 22\} \right\}$$

where, $\mathbb{E}$ denotes the flight envelope in (fps, deg) and $\mathbb{U}$ denotes the envelope for the control bound in (lbf, deg). The boundary of the envelope is denoted by $\partial \mathbb{E}$. The system is assumed to follow the dynamics [4] as shown below:

$$\begin{bmatrix} \Delta \dot{V} \\ \Delta \dot{\gamma} \end{bmatrix} = [A] \begin{bmatrix} \Delta V \\ \Delta \gamma \end{bmatrix} + [B] \begin{bmatrix} \Delta T \\ \Delta \delta_e \end{bmatrix} \tag{1}$$

where, $\Delta V = V - V_{trim}$, $\Delta \gamma = \gamma - \gamma_{trim}$, $\Delta T = T - T_{trim}$ and $\Delta \delta_e = \delta_e - \delta_{e_{trim}}$. For convenience of notation, let $x(t)$ denote the state of the system trajectory at time t starting at $\bar{x}(t_0)$ under the action of the control input $u(\cdot)$ on $[t_0, t)$.

The optimal control problem is posed as follows: *Given an initial point in state space* $x(t_0)$ *at* t_0, *find the control input* $u^*(t) \in \mathbb{U}$ *that minimizes the transfer time* $(t - t_0)$ *to reach any point on* $\partial \mathbb{E}$. The control u^* then becomes the optimal control input associated with minimum transfer time $(t^* - t)$. We shall compute the optimal trajectory using the Pontryagin Maximum Principle (PMP). The readers are referred to [10] for a proof of existence of optimal controls in case of linear systems. In this work, the boundary $\partial \mathbb{E}$ is approximated with a finite number of points. Let x_{b_i} denote the ith point on the boundary by $\partial \mathbb{E}$. Thus the optimal control law $u^*(t)$ needs to be computed such that the point $x(t_0)$ is transferred to $x_{b_i}(t_f)$ in minimum time $t^* = (t_f - t_0)$.

3. Minimum Time Problem

Consider the linear dynamical system given by:

$$\begin{bmatrix} \dot{\Delta V} \\ \dot{\Delta \gamma} \end{bmatrix} = [A] \begin{bmatrix} \Delta V \\ \Delta \gamma \end{bmatrix} + [B] \begin{bmatrix} \Delta T \\ \Delta \delta_e \end{bmatrix} \tag{2}$$

where, $A : \mathbb{R}^2 \to \mathbb{R}^2$ (state transition matrix) and $B : \mathbb{R}^2 \to \mathbb{R}^2$ (control matrix) are given by:

$$A = \begin{pmatrix} -0.1860 & -33.2125 \\ 0.0088 & 0.0150 \end{pmatrix}, \quad B = \begin{pmatrix} 0.4492 & 62.3033 \\ 0.0042 & -0.7512 \end{pmatrix}$$

The cost functional for the minimum time problem then becomes,

$$J(u) = \int_{t_0}^{t_f} 1 dt \tag{3}$$

where, t_f is the free terminal time to be optimized. The Hamiltonian for this dynamic optimization problem is given by:

$$H(t, x, u, p) = p^{\mathrm{T}} (Ax + Bu) - 1 \tag{4}$$

In the framework of PMP, the existence of an optimal control also mandates the existence of co-states (denoted by p), whose dynamics are governed by:

$$\dot{p} = -\frac{\partial H}{\partial x} = -A^{\mathrm{T}} x$$

The optimal control law is the one that maximizes the Hamiltonian at every time step from t_0 to t_f. It is evident that such a control law would have a bang-bang structure since the Hamiltonian is affine in u. Thus,

$$u^* = \arg \max_{u \in \mathbb{U}} H(t, x, u, p)$$

and so, the control law can be expressed as:

$$\Delta T = \begin{cases} T_{\max} - T_{trim}, & \text{if } \operatorname{sgn}(pB(1)) = 1 \\ T_{\min} - T_{trim}, & \text{otherwise} \end{cases} \tag{5}$$

$$\Delta \delta_e = \begin{cases} \delta_{e_{\max}} - \delta_{e_{trim}}, & \text{if } \operatorname{sgn}(pB(2)) = 1 \\ \delta_{e_{\min}} - \delta_{e_{trim}}, & \text{otherwise} \end{cases} \tag{6}$$

where, $p^{\mathrm{T}} B = \begin{bmatrix} pB(1) & pB(2) \end{bmatrix}$. This results in the TPBVP with the known initial and final states, $x(t_0)$ and $x(t_f)$ respectively as shown.

$$\begin{bmatrix} \dot{\Delta V} \\ \dot{\Delta \gamma} \\ \dot{p}_1 \\ \dot{p}_2 \end{bmatrix} = \begin{bmatrix} A_{11} & A_{12} & 0 & 0 \\ A_{21} & A_{22} & 0 & 0 \\ 0 & 0 & -A_{11} & -A_{21} \\ 0 & 0 & -A_{12} & -A_{22} \end{bmatrix} \begin{bmatrix} V \\ \gamma \\ p_1 \\ p_2 \end{bmatrix} + \begin{bmatrix} B_{11} & B_{12} \\ B_{21} & B_{22} \\ 0 & 0 \\ 0 & 0 \end{bmatrix} \begin{bmatrix} u_1 \\ u_2 \end{bmatrix} \tag{7}$$

subject to the boundary conditions:

$$x(t_0) = \begin{bmatrix} V_0 \\ \gamma_0 \end{bmatrix}$$

$$x(t_f) = \begin{bmatrix} V_f \\ \gamma_f \end{bmatrix}$$

$$H(t_f, x^*, u^*, p) = 0 \tag{8}$$

where, $H(t_f) = 0$ is necessary condition based on PMP to obtain the minimum final time t_f. The two point boundary value problem (TPBVP) is solved by matching the initial conditions of the unknown variables $p(t_0)$ and $\hat{t}_f$ so as to lead to the known final conditions described in Equation 8. This match between unknown guess and known final solutions provides the accurate map M shown below:

$$
M : \begin{bmatrix} p_1(t_0) \\ p_2(t_0) \\ \hat{t}_f \end{bmatrix} \rightarrow \begin{bmatrix} \dfrac{x_1(\hat{t}_f) - \hat{x}_1(\hat{t}_f)}{x_{1_{max}} - x_{1_{min}}} \\ \dfrac{x_2(\hat{t}_f) - \hat{x}_2(\hat{t}_f)}{x_{2_{max}} - x_{2_{min}}} \\ 0 - H(\hat{t}_f) \end{bmatrix} \tag{9}
$$

In other words, the solution to the TPBVP is obtained by solving for the zeros of M described in Equation 9. Due to the nonlinear nature of the composition maps to obtain M, analytical solutions are difficult to realize even for the case of linear systems (assuming they exist). Numerically, for the gradient based methods, step sizing for the update laws and good initial guesses are important factors for convergence. In such optimal control problems where the control law is bang-bang, small changes in the initial guess values could induce large changes in the gradients. For example, this scenario could arise when perturbations in $\hat{t}_f$ or $p(t_0)$ change the number of switches in the control law compared to the unperturbed case. In such cases, it could lead to jump discontinuities in the gradients and hence limit the gradient based solution techniques. In order to overcome this problem, we adopt a non-gradient marching technique based on stochastic swarm optimization using the differential evolution (DE) algorithm.

4. Differential Evolution for TPBVP

Differential Evolution (DE) is a metaheuristic iterative optimization strategy that tries to improve candidate solutions over generations. Since its inception in the mid 90s, there has been a growing interest in using DE due to its simple yet powerful approach to solve several engineering optimization problems [11]-[13]. In the framework of DE, each generation consists of population of candidate solutions, also known as agents. During every generation, each agent is moved a different position in the search space by a combination of update operations. If the new position of the agent is deemed better (based on a fitness measure), it is shortlisted to be a part of the next generation, else the original agent is retained. In this manner, the fitness of the solution candidates improves over generations. In this work, the DE algorithm was implemented as described below.

During the first generation, the population of guess vectors was initialized. Mutation was applied to each agent to generate the mutants. During the mutation step, a "local best" candidate was used along with the "global best" and "random" candidates to update the search direction, similar to that of Particle Swarm Optimization (PSO). These variants were crossed over with the existing agents based on a cross over probability (CR). In this implementation, a binomial crossover was performed to generate trial solutions which were compared to their counterparts from the original population to populate the next generation.

The objective function to be minimized is the final error, which depends on the initial conditions $\begin{bmatrix} p_1(t_0) & p_2(t_0) & \hat{t}_f \end{bmatrix}$, and hence approximating the roots of M. Let the kth agent $\vec{G} \in \vec{G}$ during generation g be denoted by $\vec{G}_k^g$. Also, let the kth agent's best (minimum error) historical position as of current generation be represented by $\overline{G\_PB}_k^g$ and let $\overline{G\_GB}$ denote the solution that results in least error $\|M\|$ (as defined in Equation 9) compared to all other candidates in population across all generations. If F, CR, N, g_{max} denote the parameters corresponding to mutation weight, cross over probability, number of agents in population, maximum allowable generations respectively, then the flow chart depicting differential evolution solution is shown in **Figure 1**. In this work, the population was set to $N = 100$ candidates along with a crossover probability of $CR = 0.4$ and mutation weight $F = 0.9$.

5. Results and Discussion

The solution to the TPBVP yields the initial conditions (IC) for forward simulation of the minimum time trajectories. The optimal control problems from an initial state $x_0(t_0)$ to various points along the boundary $\partial \mathbb{E}$ are solved and presented here. **Figure 2** depicts optimal trajectories from $x_0(t_0)$ to the corners of the rectangular envelope. The minimum time to envelope is approximated by computing minimum times to a collection of uniformly distributed points along the boundary as shown in **Figure 4**. The points along the boundary ("A" ... "L") are color coded based on magnitude of transfer time from "O" to boundary point (BP) along the optimal path. Further, it can be inferred that the optimal trajectory computations for various initial conditions (ICs) could be simplified progressively by applying the principle of optimality. Thus, the optimal trajectory computation from an IC to BP (IC$_1$-BP) overlapping with the optimal trajectory of another IC (IC$_2$-BP) leverages information from the optimal trajectory computation of the latter.

It can also be observed from **Figure 4** that some trajectories violate the bounds before the end point. In such cases, the minimum time to the boundary is estimated based on the first intersection with the boundary. Thus, the minimum time to envelope can be computed as the infimum of minimum times of all exit points under consideration. **Table 1** depicts the minimum exit times for each trajectory. The infimum obtained corresponds to 0.1 sec along path OI (**Figure 4**). Though the low minimum time information is not readily useful, it can be argued that by restricting control bounds $\mathbb{U}$ and applying the above approach, larger minimum times may be obtained.

The convergence of the optimal solution using the Differential Evolution algorithm is shown in **Figure 3**, wherein the errors converge to zero in a finite number of generations. The corresponding optimal controls—namely thrust and elevator deflection and their bang-bang structure (due to affine nature of Hamiltonian w.r.t control inputs) are shown in **Figure 5** and **Figure 6** respectively.

6. Conclusion

This work investigated the issue of LOC prediction using tools from optimal control theory to develop spatio-

Table 1. Minimum time to end points in **Figure 4**.

Point	A	B	C	D	E	F	G	H	I	J	K	L	C_1	C_2
Min Time	1.87	1.89	2.06	2.41	1.68	1.19	0.69	0.24	0.10	0.21	0.79	1.33	0.20	2.08

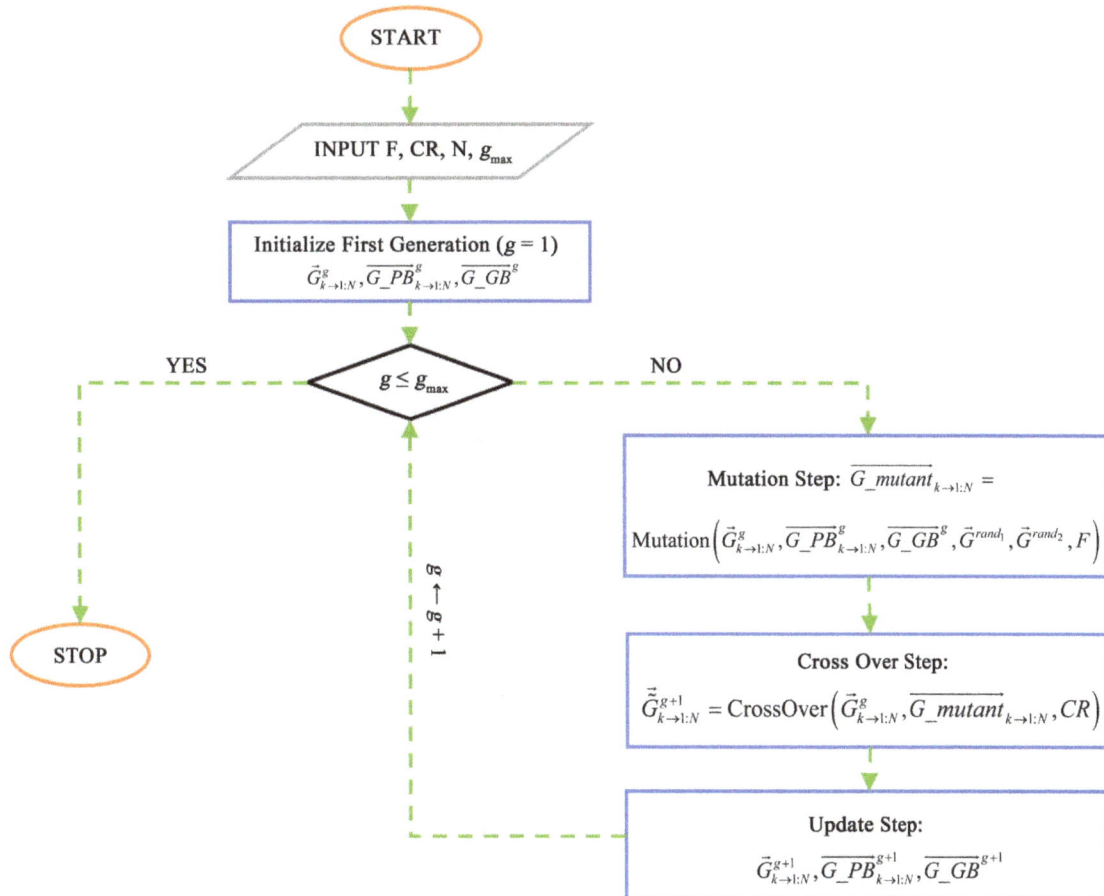

Figure 1. Differential evolution for solving TPBVP.

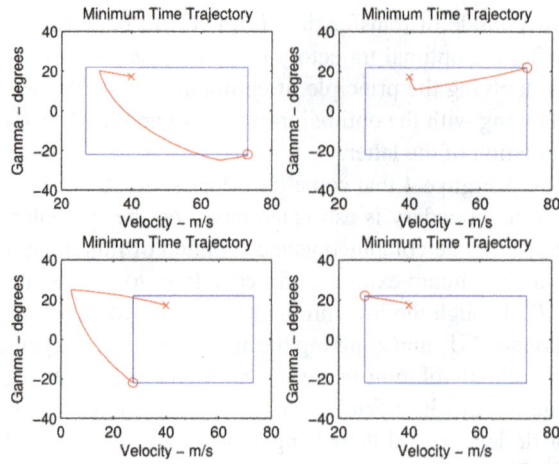

Figure 2. Minimum time paths to corners.

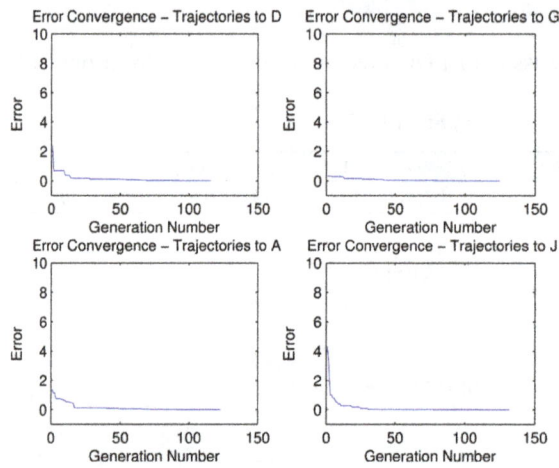

Figure 3. Error convergence profile.

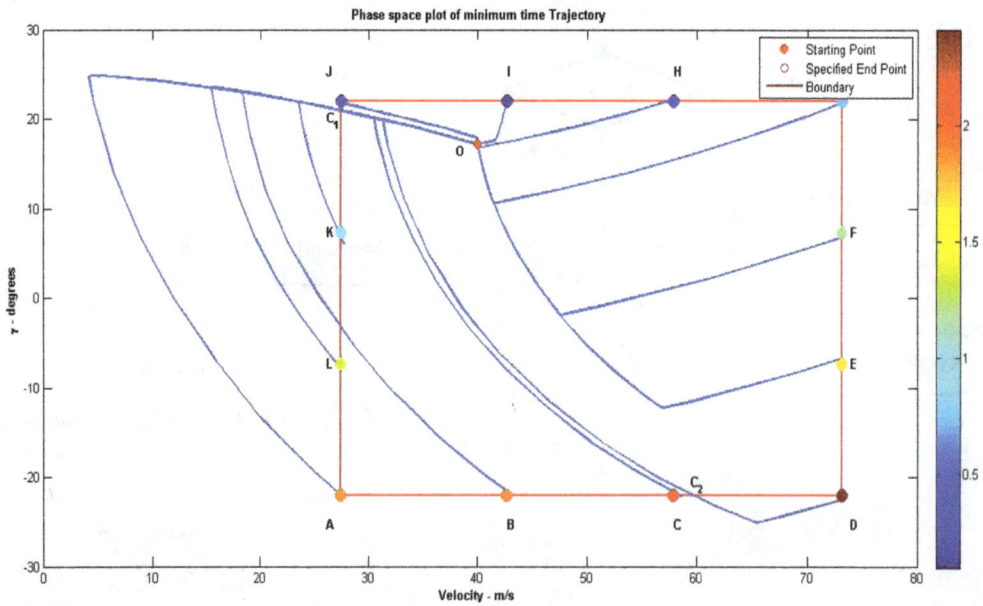

Figure 4. Minimum time trajectories to corners.

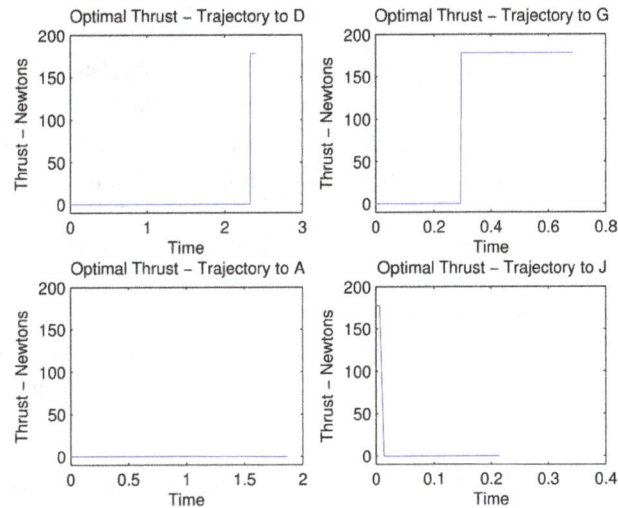

Figure 5. Optimal thrust.

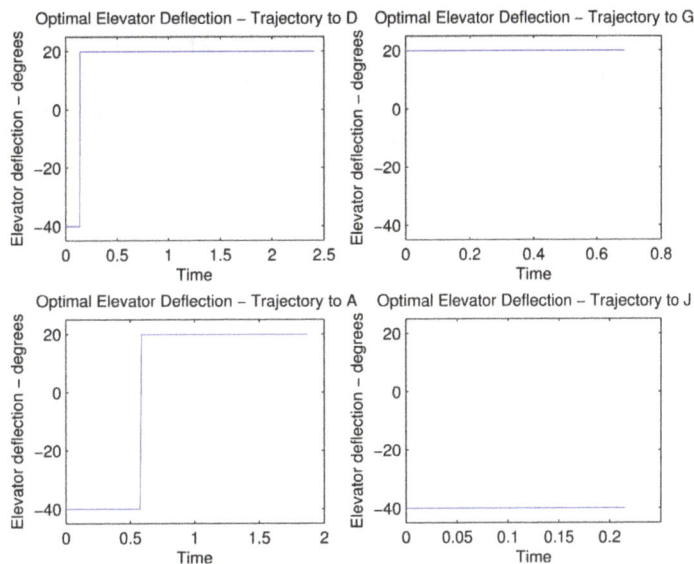

Figure 6. Optimal elevator deflection.

temporal pilot aids. The time optimal problem to violate the Loss-of-control boundary was considered. The minimum time to reach any point on the boundary was computed using PMP. The resulting TPBVP was solved using Differential Evolution. Simulation of the linear longitudinal model was carried out in MATLAB and the optimal trajectories were found to be not necessarily the minimum phase space distance paths. The minimum time to envelope was computed as the infimum of minimum times to various boundary points. Future work could investigate minimum time trajectory generation over a space of initial conditions (IC), which would expedite optimal trajectory generation by leveraging the principle of optimality. Alternative strategies for solving TPBVP like hybrid optimization schemes could be explored to reduce computational loads and restricted control bounds could be considered to obtain practically viable minimum times. Further, the solution methodology employed here (using PMP-DE) can be readily extended to nonlinear models, which better characterize dynamics near LOC boundaries away from local trim conditions. In conclusion, this work provides an initial step to augment spatial pilot aids with minimum time temporal information aimed at LOC prevention.

Acknowledgements

The authors thank the NASA Ames research center for their support.

References

[1]   Worldwide Operations (2012) Statistical Summary of Commercial Jet Airplane Accidents. Technical Report, Boeing.

[2]   Authority, Civil Aviation (2013) Global Fatal Accident Review 2002-2011. Tso.

[3]   Jacobson, S. and Edwards, C.A. (2010) Aircraft Loss of Control Study. NASA Internal Report.

[4]   Kwatny, H.G., *et al.* (2012) Nonlinear Analysis of Aircraft Loss of Control. *Journal of Guidance, Control, and Dynamics*, **36**, 149-162.

[5]   Michales, A.S. (2012) Contributing Factors among Fatal Loss of Control Accidents in Multiengine Turbine Aircraft.

[6]   Wilborn, J.E. and Foster, J.V. (2004) Defining Commercial Transport Loss-of-Control: A Quantitative Approach. *AIAA Atmospheric Flight Mechanics Conference and Exhibit.*

[7]   Randall, L. (2012) Brooks. LOC-I Training Foundations and Solutions. Technical Report, Boeing.

[8]   Krishnakumar, K., *et al.* (2014) Initial Evaluations of LoC Prediction Algorithms using the NASA Vertical Motion Simulator. SciTech 2014.

[9]   Barlow, V., Stepanyan, J. and Kalmanje, K. (2012) Estimating Loss-of-Control: A Data Based Predictive Approach.

[10] Pontryagin, L.S., *et al.* (1962) The Mathematical Theory of Optimal Processes (International Series of Monographs in Pure and Applied Mathematics. Interscience, New York.

[11] Qin, A.K., Huang, V.L. and Suganthan, P.N. (2009) Differential Evolution Algorithm with Strategy Adaptation for Global Numerical Optimization. *IEEE Transactions on Evolutionary Computation*, **13**, 398-417.

[12] Okdem, S. (2004) A Simple and Global Optimization Algorithm for Engineering Problems: Differential Evolution Algorithm. *Turkish Journal of Electrical Engineering and Computer Sciences*, **12**.

[13] Bersini, H., *et al.* (1996) Results of the First International Contest on Evolutionary Optimisation (1st ICEO). *Proceedings of IEEE International Conference on Evolutionary Computation*, 20-22 May 1996, 611-615.

22

Gap Navigation Trees for Discovering Unknown Environments

Reem Nasir, Ashraf Elnagar

Computational Intelligence Center, Department of Computer Science, University of Sharjah, Sharjah, UAE
Email: ashraf@sharjah.ac.ae

Abstract

We propose a motion planning gap-based algorithms for mobile robots in an unknown environment for exploration purposes. The results are locally optimal and sufficient to navigate and explore the environment. In contrast with the traditional roadmap-based algorithms, our proposed algorithm is designed to use minimal sensory data instead of costly ones. Therefore, we adopt a dynamic data structure called Gap Navigation Trees (GNT), which keeps track of the depth discontinuities (gaps) of the local environment. It is incrementally constructed as the robot which navigates the environment. Upon exploring the whole environment, the resulting final data structure exemplifies the roadmap required for further processing. To avoid infinite cycles, we propose to use landmarks. Similar to traditional roadmap techniques, the resulting algorithm can serve key applications such as exploration and target finding. The simulation results endorse this conclusion. However, our solution is cost effective, when compared to traditional roadmap systems, which makes it more attractive to use in some applications such as search and rescue in hazardous environments.

Keywords

Motion Planning, Gap-Navigation Trees, Roadmap, Robotics, Local Environments

1. Introduction

The problem of "exploring the environment surrounding a robot" can be divided into different sub-categories based on three parameters. The first parameter describes the environment itself and it is divided into simple and multiply connected. Simple environments are environments that have no holes (*i.e.*, obstacles) or discontinuity in its walls while multiply connected environments include holes. The second parameter is status of the obstacles whether it is static or dynamic. As more information is made available to the robot, the problem becomes more challenging [1]-[3]. The third parameter is knowledge about the robot's environment which is classified into known, partially known or unknown to a robot [1] [3].

When a point robot is placed in a given (known) polygonal region, the focus of path planning research would be finding the path a robot would follow. The tasks required of a robot would determine which path should be considered, the fastest, shortest, safest, or less mechanical movements from the robot. When the task involves search and rescue, then the shortest path is considered, and in known environments, computing shortest paths is a straightforward task. The most common approach is to compute a visibility graph which is accomplished in $O(n \log n)$ time by a radial sweeping algorithm [4] where n is the number of visible points for the robot.

When the environment is unknown (or partially unknown), sensors are used to explore and map the environment to develop navigation strategies. In this case, the robot is equipped with sensors that measure distances, able to identify walls from objects and track them. The information reported from these sensors is used by the robot to build a map for the environment that will be used later to navigate through. This approach needs the robot to build an exact model of the environment with all of its walls, edges and objects with their exact coordination and measurements. This leads to raise the question of whether this is practical or not and how accurate these maps are. That leads to many proposed solutions [1], which of course will keep increasing as new issues with the current algorithms arise [5]. On the other hand, some research has been conducted on how probabilistic techniques would be applied to the current algorithms [6] and how that would affect their performance [7].

Our work focused on a relatively new approach where researchers studied the sensory data of a robot [8]. We studied which of these data were essential to the robot's ability to explore and navigate an environment [9]. We further detailed the algorithms required to explore an unknown environment using minimal information stored in the Gap Navigation Trees (GNT) data structure.

2. Problem Statement

The problem tackled in this paper is of two fold. The first is to explore a local environment cluttered with static obstacles. The second is to navigate such environment for carrying out a specific task.

Given an unknown, multiply connected environment with static obstacles, and a robot; the goal is (for the robot) to explore such environment and build a data structure (roadmap) while using the least amount of sensors possible. This data structure shall to be sufficient to achieve the robot's goals (finding an object, tracking an object or even just learning the environment).

We assume that the robot is modeled as a point moving in static, unknown and multiply connected environment, to achieve its goal. The robot is equipped with depth discontinuities sensor. The robot is expected to use the sensor's feedback to build a data structure which is used along with building local grid map as virtual landmarks, to learn and navigate among obstacles in the environment.

3. Proposed Algorithm

3.1. Gap-Navigation Trees (GNT)

GNT is a dynamic data structure constructed over direction of depth discontinuities [8]. The robot is modeled as a point moving in an unknown planar environment. The robot is assumed to have an abstract sensor (in the sense of [10]) that reports the order of depth discontinuities of the boundary, from the current position of the robot. These discontinuities are called gaps, and the abstract sensor may be implemented in a number of ways, using a directional camera or a low-cost laser scanner. Once GNT is constructed, it encodes paths from the current position of the robot to any place in the environment. As the robot moves, the GNT is updated to maintain shortest-path information from the current position of the robot. These paths are globally optimal in Euclidean distance. GNT showed that the robot can perform optimal navigation without trying to resolve ambiguities derived from different environments which result in having same GNT [8].

More research have been done on GNT in simple, closed, piecewise-smooth curved environment [11].Then it has been extended to multiply-connected environments (where the environment has holes, obstacles, jagged holes and edges) and new modification been added to handle the new events that have occurred [12] [13].

We assume that the robot is equipped with a depth sensor. This sensor has the ability to report discontinuities which are referred to as gaps.

Figure 1 shows how a robot detects gaps, where the nodes A and B represent starting points of the visibility edges of the robot's view. Beyond these edges, the robot has no knowledge of the environment; hence these

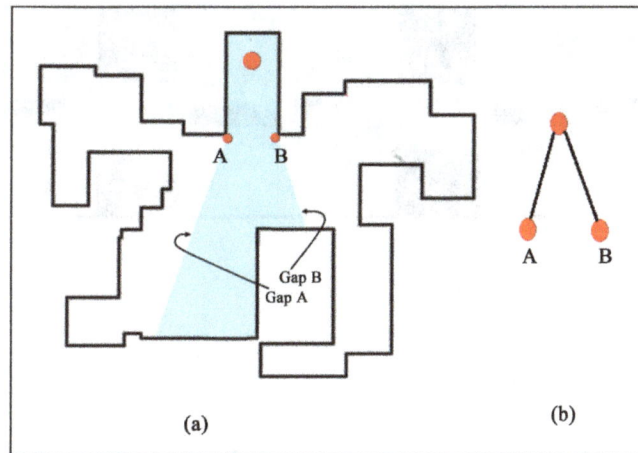

Figure 1. (a) Visibility region of a robot with depth sensor, (b) GNT.

nodes are added as children of the robot's current position. These nodes are called gaps. By adding children to each current position of the robot, and exploring them, the robot navigates and explores the whole environment until the whole environment is visible from these gaps.

There are three events that can occur while the robot investigates gaps:

- Addition of New Gaps: Each discontinuity in the robot's visibility field from its current position will be added as child gap node of the robot's node. This takes place while preserving the cyclic ordering from the gap sensor.
- Merger of Gaps: Two or more gaps could be merged into one gap, if they are the children of the same parent robot position and cover same area when investigated.
- Deletion of Gaps: If a gap becomes redundant, by being covered in the visibility range of the robot while examining another gap, and not a child for the same robot position parent node, then it will be deleted. When gaps belong to the same parent, but are in different directions (cannot be merged), the current gap will be kept, and the ones been seen (in the visibility range of the current gap) but not visited yet will be deleted.

Figure 2 is a detailed example of a robot that is building the GNT of a simple environment at different time steps. The highlighted area shows the visibility region from the current position of the robot. Edges instituted at the vertices are produced by the depth-discontinuity sensor (indicate gaps). Each gap shall be identified with a unique tag (a, b, a.1 ...) for the purposes of illustration. **Figure 2(a)**, shows three gaps, first one is (a), second one is (b), and the last one is (c). The robot will investigate the gaps in the order of detection (counterclockwise). Each black solid disc represents the position of the robot (Rc).

The GNT starts by constructing the Rc (the root is the initial robot's position) and 3 edges to the 3 newly identified gaps. **Figure 9(b)** demonstrates that while the robot is chasing the (a) gap, it encountered five new gaps, which are added as (a.1, a.2, a.3, a.4 and a.5). Notice that (a.1) and (a.2) share the same area but are represented as 2 gaps, same goes for (c), (a.3), and (a.4). There is also the merger between (a) and (b) in (a). That is because (b) is in the visibility range of (a) and they share the same parent. The robot has to explore these gaps in order to recognize that both cover the same area. That is because of its limited sensory data and its partial knowledge of the environment.

In (c) as the robot navigates toward (a.1), 2 events occur. First one is merging (a.1) and (a.2) as both children belong to the same parent. The second event is the deletion of (a.4) as it is completely in the robot's visibility range and therefore become redundant; (a.4) is deleted from GNT.

Applying the depth-first order, next gap to be investigated is (a.3). Once the robot reaches its new position, both of (c) and (a.5) are in the new Rc's visibility range. Therefore, both are deleted. As there are no further gaps to inspect in GNT, the robot stops exploration and declares that GNT is complete.

3.2. Landmarks Strategy

GNT works perfectly in simple planner environments and gives optimal results, but when it is used in more complex environments, it is not guaranteed to work. The depth discontinuity sensor is no longer sufficient; at

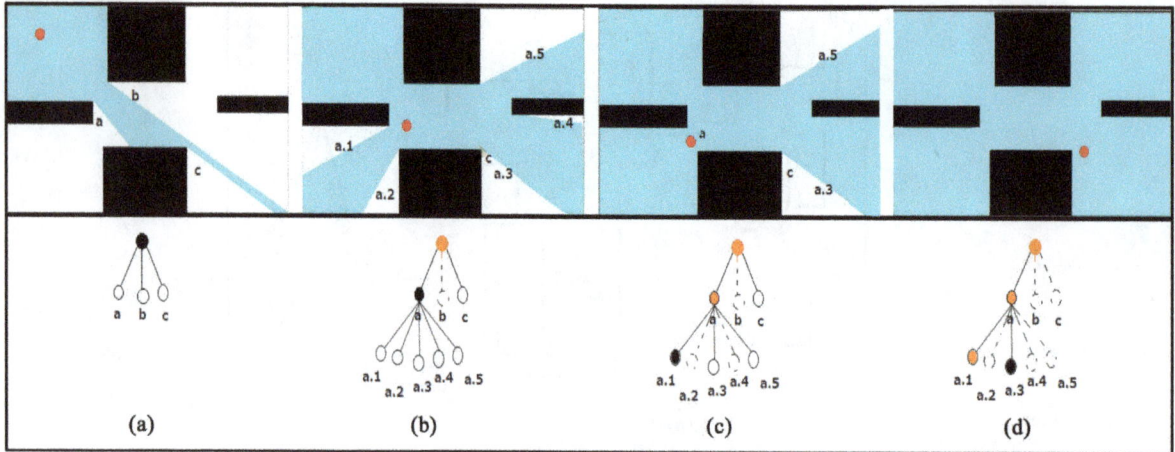

Figure 2. A robot navigating in simple environment and GNT building.

least another piece of data is required; because the robot needs to be able to recognize a gap that was previously visited. In simple environments each gap can be reached from one way only while in multiply connected environments, a gap could be reached from different sides. Therefore, the need to label such gaps becomes necessary. Previous works and suggestions had been made about combining GNT and landmarks, such as simple color landmark models [14], range measurements with respect to distinct landmarks, such as color transitions, corners, junctions, line intersections [15], scale-invariant image features as natural landmarks [16], and matching technology-RANSAC (Random Sample Consensus) [17].

3.3. Exploration Strategy

Before we present the proposed algorithm, we explain our landmark technique. The robot's initial configuration is assumed at location $(0, 0)$ and as a result the environment is divided into 4 quadrants. As the robot moves, we keep track of new positions (x, y) as well as the number of steps.

The robot will always keep track of its original position, and all of the movements it made, with respect to its origin and how it will know (x, y) from any other gap, at any position or time. Our proposed landmark technique, the robot will locally build grid map incrementally for (x, y) values, this is only used to assign different values for each gap. Which will be used to identify each of them, therefore, the robot will know if it's the same gap or not and it will not get stuck around an obstacle, since each gap will be identified by the localized grid. The robot will know which gaps to merge and which to delete. Each gap is identified by the virtual gird map configuration. Next we describe the proposed algorithm and the classical A* search algorithm.

Algorithm 1: Modified Gap Navigation Tree
Input: sensory data (discontinuity in the environment from the depth sensor)
Output: complete T

---

1. $T = \phi$
2. $T = \{G_0\}$
3. Let $G_i = G_0$
4. Update gridmap
5. **While** (G_i is not *cleared*)
6. $\forall_j$ where j<i
7.   **If** (any G_i is redundant)
8. **then** delete G_i
9.   **If** (G_i is not *visited*)
10. **then CHASE**(G_i)
11.   **else**
12.   **if** (C_{Gi} is not cleared)
13. **then** G_i = next in order of unvisited children nodes
14.   **else**
15.   mark G_i cleared
16.   make G_i a parent node.
17. **end**
18. report T

Algorithm 2: *CHASE(G_i)*
Input: a pointer to G_i in T
Output: perform clearing, deletion or addition

1. Mark G_i *visited*.
2. **If** (there are discontinuities)
3.   **For** (each G_i _ discontinuities)
4.   **If** (can't be seen by other previous gaps)
5.   **then** add it and update gridmap
6.   **else**
7.   **If** (current discontinuities and a previous gap have same parent)
8.   **then** merge it
9.   **else**
10.   delete it
11. G_i = first discontinuities added as a child
12. **else**
13.   gap cleared
14. G_i = parent node

A* search algorithm will be used for navigation applications, where the robot will demonstrate how it can move from one point to another using the constructed GNT which was sufficient to represent the environment. A* is a complete search algorithm and will always find a solution if one exists. If the heuristic function h is admissible, meaning that it never overestimates the actual minimal cost of reaching the goal, then A* is itself admissible (or optimal) if we do not use a closed set. If a closed set is used, then h must also be monotonic (or consistent) for A* to be optimal. This means that for any pair of adjacent nodes x and y, where $d(x, y)$ denotes the length of the edge between them, we must have: $h(x) \le d(x, y) + h(y)$.

Algorithm2: *A* search algorithm*
Input: *SN, GN, Q*.
Output: best path from start node to goal node, if any

1. initialize the OL
2. initialize the CL
3. add *SN* to *OL* (you can leave its f at zero)
4. **while** (*OL* not empty)
5.   find the node with the least f on the open list, call it "q"
6.   pop q off *OL*
7.   set q's *N* parents to q
8.   **for** (*N_i*)
9.   **if** (*N_i* is the *GN*)
10.   **then** stop the search
11.   *N_i.g* = q.g + distance between *N_i* and q
12.   *N_i.h* = distance from *GN* to *N_i*
13.   *N_i.f* = *N_i.g* + *N_i.*h
14.   **if** (*Q_i*'s position = *N_i* is in OL and   *Q_i.f*<*N_i.f*)
15.   **then** skip *N_i*
16.   **if** (*Q_i*'s position = *N_i* is in CL and   *Q_i.f*<*N_i.f*)
17.   **then** skip *N_i*
18.   **else** add *N_i* to *OL*
19.   **end**
20.   push q in the *CL*
30. **end**

3.4. Complexity Analysis

The following is an analysis for the time and space complexities of the proposed algorithm. The complexity stems from two main phases, which are the buildup phase (constructing GNT) and the query phase (applying A* search algorithm). The time complexity of the first phase is heavily affected by the chasing phase, were the robot needs to check each gap, which takes $O(n)$. Then for each of these gaps, the robot checks the GNT before deciding to add new gaps, which is also another $O(n)$ task. The result is:

$$\sum_{i=1}^{n} i = \frac{n(n+1)}{2} \approx O(n^2)$$

The storage space required to maintain the vertices of the resulted GNT is equal to the number of gaps n. Each gap will also maintain pointers to its parent and its children too. Therefore, there will be no need to store the information of the edges as well. This leads to $O(n)$ storage complexity.

On the other hand, the time complexity of A* depends on the heuristic used. In the worst case, the number of nodes expanded is exponential in the length of the solution (the shortest path), but it is polynomial when the search space is a tree, there is a single goal state, and the heuristic function h meets the following condition:

$$\left|h(x) - h^*(x)\right| = O\left(\log h^*(x)\right).$$

where h^* is the optimal heuristic, the exact cost to get from x to the goal. In other words, the error of h will not grow faster than the logarithm of the "perfect heuristic" h^* that returns the true distance from x to the goal.

4. Results and Analysis

In our simulations we classified the environments into six categories based on the size of obstacles and their distribution. The first four classifications are determined from the obstacles' sizes and the areas they occupy in the environment. The simulation started with sparse environments (*i.e.*, large free space) with uniform-size obstacles followed by sparse environments with variable-size obstacles. We repeated both simulations but in cluttered environments. At last, we simulated two popular problematic environments in the literature of the path planning field, namely, the Narrow-Passages problem and Indoor Environments. The following observations help appreciate the simulation results:

- Blue rectangles represent the obstacles in the environment.
- Red discs signify the initial robot position (Rs) or the current robot's position in the environment (Ri).
- Light orange discs indicate already visited gaps.
- Yellow discs represent the newly encountered gaps, which were added as children from the current robot position.
- Green discs indicate the next gap to be inspected.
- Green/black lines demonstrate the roadmap.
- **Maximum number of Gaps** (MNG): This refers to the theoretical maximum number of gaps, in the given multiply connected environment. The upper limit of the number of gaps is a function of the number of obstacles and their locations. For example, one rectangular obstacle would produce four gaps. If we consider having two obstacles in the environment, there will be a maximum number of eight gaps because the obstacles could be sharing a gap or more. Therefore we might end up with less than eight gaps, but it will not exceed eight. This info is reported by the simulator.
- **Unique Number of Gaps** (UNG): For validation purposes, this is obtained manually, by counting the number of gaps in a given environment. This number is less or equal to the maximum number of gaps. As the number of obstacles increase, in a given environment, this measure tends to be far less than the MNG.
- **Algorithm-computed Number of Gaps** (ANG): This refers to the total number of gaps encountered while constructing the GNT by the proposed algorithm. In an optimal scenario, ANG is the same as UNG. However, in a worse-case scenario it is close to MNG.
- **Gap Redundancy Rate Reduction** (GRR): This refers to the gap reduction in redundant gaps. The best-case scenario would yield testing UNG gaps only. Therefore, redundancy would be MNG-UNG. Therefore, we compute two reduction rates: optimal gap redundancy rate (OGRR) and algorithm-computed gap redundancy rate (AGRR), which are measured by:
  - o   OGRR= UNG/MNG.
  - o   AGRR= ANG/MNG.

Our objective is to minimize (AGRR-OGRR). That is the closer AGRR to OGRR is, the less is the redundancy in gaps.

We will start our simulation results with sparse environments which would have obstacles that occupy a small area of the environment. Here is an example of a uniform-size obstacles distribution. The environment has four small obstacles. Each 2 adjacent obstacles share a common gap.

As shown in **Figure 3**, the robot was able to identify each unique gap. Redundant gaps are well dealt with. The red node is the original robot's position and the root of the GNT (Rs). The light orange nodes represent the other gaps in the GNT. In this example, all of the gaps are the children of the root, which are sufficient to cover the whole environment. There were no further gaps to be added from any of its children.

Table 1 summarizes the observations on **Figure 3**. There were four obstacles in the environment. The final GNT consists of 6 nodes. In this example AGRR was 62.5%, which signifies a very good reduction of redundant gaps; it is close to the OGRR as well (68.7%).

Now we show how a robot explores a cluttered environment, in which the obstacles cover a large area of the environment. This example has 20 obstacles of uniform-size and uniform distribution. These obstacles cover a large area of the environment and are aligned as a grid. Rs is represented in red color.

Figure 4 depicts the resulting GNT. The light orange vertices correspond to the visited gaps. The four gaps in the first row of obstacles cannot cover the whole area and therefore another set of gaps has been added among the second row of obstacles. Although there are many obstacles, the final GNT was small because obstacles share significant part of the free-space, which is inversely proportional to the number of the required gaps in GNT.

Table 1. Observations on **Figure 2**.

Performance criterion	Measurement
Number of obstacles in the environment	4
MNG	16
UNG	5
ANG	6
OGRR	68.7%
AGRR	62.5%

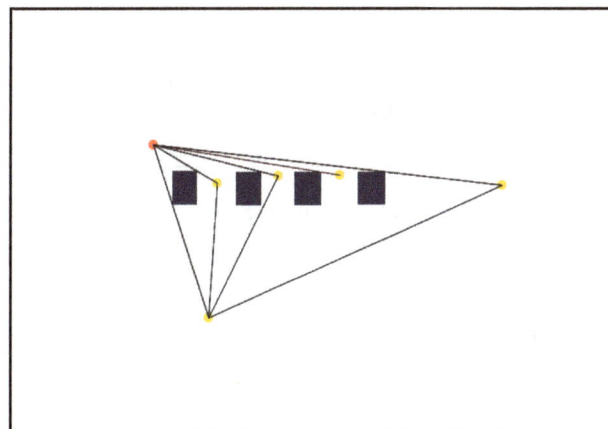

Figure 3. Sparse environment with four obstacles, complete GNT.

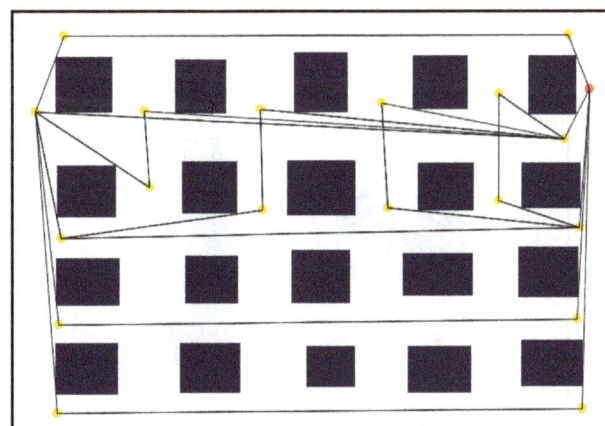

Figure 4. Cluttered environment with 20 obstacles, complete GNT.

Table 2 summarizes the observations on **Figure 4**. There are 20 obstacles in the environment with 80 unique vertices but only 19 gaps were added to the GNT. The final GNT was sufficient for this environment. Although the redundancy reduction rate (76.3%) is high, GNT provides a full coverage of the environment. The measure AGRR is not far from OGRR.

We also tested the performance in environments with variable-size obstacles. The obstacles are few; the free-space is sparse. The next example has only two obstacles.

The performance of the proposed algorithm is expected as when the number of obstacles decreases, the overlapping gaps are few and therefore GNT would yield a higher number of gaps. In practice, such environments are scarce.

Obstacle sizes and distribution in any environment are key parameters to construct the GNT. The algorithm computed a close number of gaps to the MNG. However, the actual number of gaps that are sufficient for this environment (3 gaps, see **Table 3**). This is the result of a global-based computation of the environment.

The following examples have environments with different obstacles' sizes that consume a large area of the environments. Some of these obstacles hide part of the environment (such part is not visible/shared with other obstacles) which increases ANG. Of course, it would decrease the AGRR as well.

Figure 6 represents an example of 5 obstacles that cover a large area of the environment and they differ in their sizes and distribution. One of these obstacles is a polygonal shape of 20 vertices.

Table 2. Observations on **Figure 4**.

Performance criterion	Measurement
Number of obstacles in the environment	20
MNG	80
UNG	11
ANG	19
OGRR	86.3%
AGRR	76.3%

Table 3. Observations on **Figure 5**.

Performance criterion	Measurement
Number of obstacles in the environment	2
MNG	8
UNG	3
ANG	7
OGRR	62.5%
AGRR	12.5%

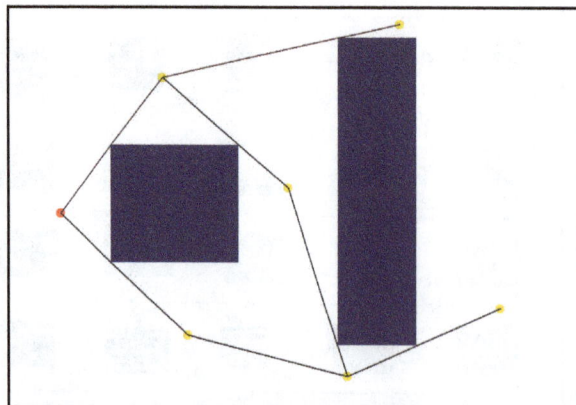

Figure 5. Sparse environment with four obstacles, complete GNT.

The polygonal-shape obstacle covers most of the middle area of the environment. The GNT structure grew surrounding this obstacle in order to cover the whole environment. The gap reduction rate of 55.5% is an encouraging result (**Table 4**).

The proposed algorithm can handle the popular narrow passages problem as well. It is a challenging environment that has been frequently reported in the literature. **Figure 7** shows an example of such environment with 5 obstacles that cover most of the workspace. There are two narrow passes in this environment.

The resulting GNT completely covers the free space. It is noticeable that the larger the obstacles get, the less gaps they share with other obstacles which leads to a less redundancy rate (**Table 5**).

Table 4. Observations on **Figure 6**.

Performance criterion	Measurement
Number of obstacles in the environment	5
MNG	36
UNG	11
ANG	16
OGRR	69.4%
AGRR	55.5%

Table 5. Observations on **Figure 7**.

Performance criterion	Measurement
Number of obstacles in the environment	5
MNG	20
UNG	13
ANG	14
OGRR	35%
AGRR	30%

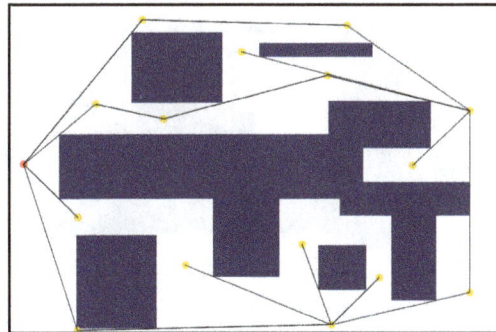

Figure 6. Polygonal-shape obstacle in a cluttered environment.

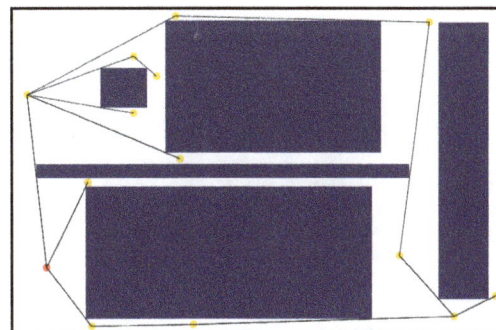

Figure 7. The narrow-passage problem.

Figure 8 combines a narrow-passage problem and a trap-type obstacle. The resulting GNT covers the whole environment's free-space eliminating the effect of such troublesome obstacles. There are eight obstacles. The robot successfully explored the environment and built the GNT. The performance of the algorithm is satisfactory as summarized in **Table 6**.

In summary, the vast set of simulations shed light on the performance of the proposed algorithm to construct a roadmap to cover the free-space. The algorithm is complete as it handles all types of environments including the challenging ones such as the narrow-passage or indoor environments. The resulting GNT is a function of Rs, obstacle size and placement. Although, the algorithm reduces redundant gaps, it does not eliminate all redundancy. This is very much attributed to the limited sensory data we use.

The following bar chart (**Figure 9**) describes the performance of the proposed algorithm with respect to gap redundancy reduction rate. The x-axis represents 13 simulations where **Figures 2-7** are represented as examples 2, 3, 6, 8, 9 and 11 respectively while the y-axis demonstrates the redundancy rate. This chart is a comparison between OGRR and AGRR.

Table 6. Observations on **Figure 8**.

Performance criterion	Measurement
Number of obstacles in the environment	8
MNG	40
UNG	16
ANG	26
OGRR	60%
AGRR	35%

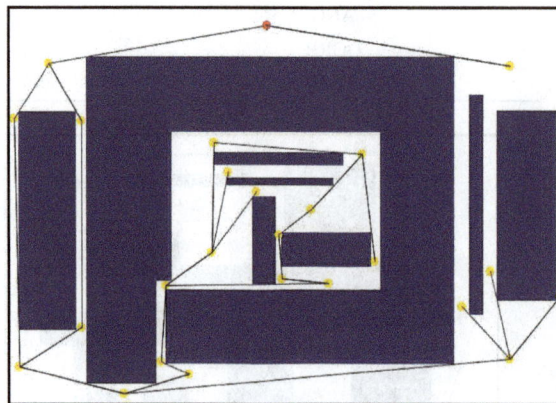

Figure 8. Trap-shape obstacles in an "indoor environment".

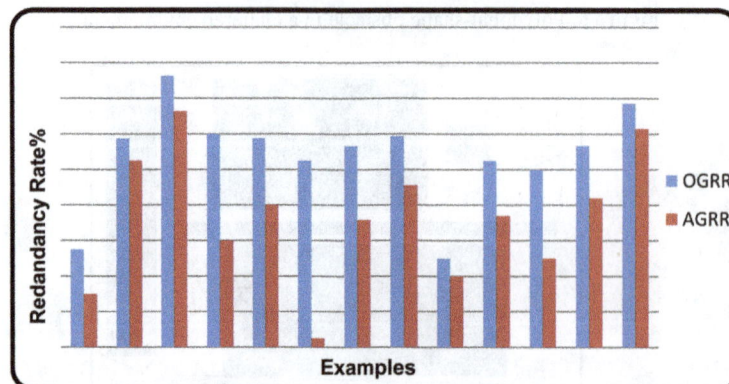

Figure 9. Bar-chart of the gap redundancy reduction rate.

Finding the minimum (unique) number of gaps is considered NP-hard problem. However, the performance of the algorithm (which uses local sensory data) is very much satisfactory when compared to the optimal one (NP-hard and based on global data). It should be noted that Example 6 (**Figure 5**) is a worst-case scenario where we had two obstacles; a scarce case.

5. Conclusions

In this paper, we have proposed using GNT while enabling the robot to count and record the steps it takes (as building its own grid) in order to enable GNT to work in multiply, connected environments. This solution would require using more RAM storage. However, the GNT is a versatile data structure that can serve a variety of applications.

The proposed solution is a cost-effective one which makes it practical to produce it in bigger quantities for specific application that require multiple robots such as search and rescue. A typical major assumption made to solve the robot motion planning using GNT is that obstacles are uniquely identified. Our solution does not require such assumption. Now, the robot is able to explore unknown multiply connected environments on its own with one very affordable sensor. Our solution eliminates the need for a landmark based sensor such as camera and it provides a cost effective prototype of a robot.

References

[1]   Canny, J. and Reif, J. (1987) New Lower Bound Techniques for Robot Motion Planning Problems. Proceedings *IEEE Symposium on Foundations of Computer Science*, Los Angeles, 12-14 October 1987, 49-60. http://dx.doi.org/10.1109/sfcs.1987.42

[2]   Murphy, L. and Newman, P (2008) Using Incomplete Online Metric Maps for Topological Exploration with the Gap Navigation Tree. *IEEE International Conference on Robotics and Automation*, Pasadena, 19-23 May 2008, 2792-2797. http://dx.doi.org/10.1109/robot.2008.4543633

[3]   Craig, J. (2004) Introduction to Robotics: Mechanics and Control. 3rd Edition, Prentice Hall, Upper Saddle River.

[4]   LaValle, S.M. and Hinrichsen, J. (2001) Visibility-Based Pursuit-Evasion: The Case of Curved Environments. *IEEE Transactions on Robotics and Automation*, **17**, 196-201. http://dx.doi.org/10.1109/70.928565

[5]   Thrun, S., Burgard, W. and Fox, D. (2005) Probabilistic Robotics. MIT Press, Cambridge.

[6]   Thrun, S., Burgard, W. and Fox, D. (1998) Probabilistic Mapping of an Environment by a Mobile Robot. *IEEE International Conference on Robotics and Automation*, Levene, 16-20 May 1998, 1546-1551.

[7]   Elnagar, A. and Lulu, L. (2005) An Art Gallery-Based Approach to Autonomous Robot Motion Planning in Global Environments. *IEEE/RSJ International Conference on Intelligent Robots and Systems*, 2-6 August 2005, 2079-2084. http://dx.doi.org/10.1109/iros.2005.1545170

[8]   Tovar, B., Murrieta-Cid, R. and LaValle, S.M. (2007) Distance-Optimal Navigation in an Unknown Environment without Sensing Distances. *IEEE Transactions on Robotics*, **23**, 506-518. http://dx.doi.org/10.1109/TRO.2007.898962

[9]   Nasir, R. and Elnagar, A. (2014) Exploration of Unknown Multiply Connected Environments Using Minimal Sensory Data. In: Zhang, X., *et al.*, Eds., *ICIRA* 2014, Part I, LNAI 8917, 479-490. http://dx.doi.org/10.1007/978-3-319-13966-1_47

[10]   Erdmann, M. (1995) Understanding Action and Sensing by Designing Action-Based Sensors. *The International Journal of Robotics Research*, **14**, 483-509. http://dx.doi.org/10.1177/027836499501400506

[11]   Tovar, B., LaValle, S.M. and Murrieta, R. (2003) Optimal Navigation and Object Finding without Geometric Maps or Localization. *IEEE International Conference on Robotics and Automation*, **1**, 464-470.

[12]   Tovar, B., LaValle, S.M. and Murrieta, R. (2003) Locally-Optimal Navigation in Multiply-Connected Environments without Geometric Maps. *IEEE/RSJ International Conference on Intelligent Robots and Systems*, **4**, 3491-3497. http://dx.doi.org/10.1109/iros.2003.1249696

[13]   Tovar, B., Guilamo, L. and LaValle, S.M. (2005) Gap Navigation Trees: Minimal Representation for Visibility-Based Tasks. *Algorithmic Foundations of Robotics VI*, Springer Berlin Heidelberg, 425-440.

[14]   Yoon, K-J. and Kweon, I. (2002) Landmark Design and Real-Time Landmark Tracking for Mobile Robot Localization. *Electrical Engineering*, **4573**, 219-226.

[15]   Bais, A. and Sablatnig, R. (2006) Landmark Based Global Self-Localization of Mobile Soccer Robots. *Computer Vision-ACCV*, **3852**, 842-851. http://dx.doi.org/10.1007/11612704_84

[16]   Se, S., Lowe, D. and Little, J. (2002) Mobile Robot Localization and Mapping with Uncertainty Using Scale-Invariant

Visual Landmarks. *The International Journal of Robotics Research*, **21**, 735-758.
http://dx.doi.org/10.1177/027836402761412467

[17] Zhao, L., Li, R., Zang, T., Sun, L. and Fan, X. (2008) A Method of Landmark Visual Tracking for Mobile Robot. *Lecture Notes in Computer Science*, **5314**, 901-910. http://dx.doi.org/10.1007/978-3-540-88513-9_97

Permissions

All chapters in this book were first published in ICA, by Scientific Research Publishing; hereby published with permission under the Creative Commons Attribution License or equivalent. Every chapter published in this book has been scrutinized by our experts. Their significance has been extensively debated. The topics covered herein carry significant findings which will fuel the growth of the discipline. They may even be implemented as practical applications or may be referred to as a beginning point for another development.

The contributors of this book come from diverse backgrounds, making this book a truly international effort. This book will bring forth new frontiers with its revolutionizing research information and detailed analysis of the nascent developments around the world.

We would like to thank all the contributing authors for lending their expertise to make the book truly unique. They have played a crucial role in the development of this book. Without their invaluable contributions this book wouldn't have been possible. They have made vital efforts to compile up to date information on the varied aspects of this subject to make this book a valuable addition to the collection of many professionals and students.

This book was conceptualized with the vision of imparting up-to-date information and advanced data in this field. To ensure the same, a matchless editorial board was set up. Every individual on the board went through rigorous rounds of assessment to prove their worth. After which they invested a large part of their time researching and compiling the most relevant data for our readers.

The editorial board has been involved in producing this book since its inception. They have spent rigorous hours researching and exploring the diverse topics which have resulted in the successful publishing of this book. They have passed on their knowledge of decades through this book. To expedite this challenging task, the publisher supported the team at every step. A small team of assistant editors was also appointed to further simplify the editing procedure and attain best results for the readers.

Apart from the editorial board, the designing team has also invested a significant amount of their time in understanding the subject and creating the most relevant covers. They scrutinized every image to scout for the most suitable representation of the subject and create an appropriate cover for the book.

The publishing team has been an ardent support to the editorial, designing and production team. Their endless efforts to recruit the best for this project, has resulted in the accomplishment of this book. They are a veteran in the field of academics and their pool of knowledge is as vast as their experience in printing. Their expertise and guidance has proved useful at every step. Their uncompromising quality standards have made this book an exceptional effort. Their encouragement from time to time has been an inspiration for everyone.

The publisher and the editorial board hope that this book will prove to be a valuable piece of knowledge for researchers, students, practitioners and scholars across the globe.

List of Contributors

Juan Yan and Huibin Yang
College of Mechanical Engineering, Shanghai University of Engineering Science, Shanghai, China

Mohammed Z. Al-Faiz
College of Information Engineering, Al-Nahrain University, Baghdad, Iraq

Ahmed F. Shanta
Department of Computer Engineering, Al-Nahrain University, Baghdad, Iraq

Vishal Goyal, Vinay Kumar Deolia and Tripti Nath Sharma
Department of Electronics and Communication Engineering, G. L. A. University, Mathura, India

Huibin Yang and Juan Yan
College of Mechanical Engineering, Shanghai University of Engineering Science, Shanghai, China

Arman Sargolzaei and Kang K. Yen
Department of Electrical and Computer Engineering, Florida International University, Miami, FL, USA

Mohamed N. Abdelghani
Department of Mathematics and Statistics, University of Alberta, Edmonton, Canada

Abolfazl Mehbodniya
Graduate School of Engineering-Sendai, Tohoku University, Sendai, Japan

Saman Sargolzaei
Department of Electrical and Computer Engineering, Wentworth Institute of Technology, Boston, MA, USA

Hosny Abbas and Mohammed Amin
Department of Electrical Engineering, Assiut University, Assiut, Egypt

Samir Shaheen
Department of Computer Engineering, Cairo University, Giza, Egypt

Basil Hamed
Electrical Engineering Department, Islamic University of Gaza, Gaza, Palestine

Abd Al Karim Abu Ras
Electrical Engineer, Biomedical Equipment Co., Gaza, Palestine

B. Aykent, D. Paillot, F. Merienne and C. Guillet
CNRS Le2i Arts et Metiers ParisTech, Chalon sur Saone, France

A. Kemeny
CNRS Le2i Arts et Metiers ParisTech, Chalon sur Saone, France
Technical Centre for Simulation, Renault, Guyancourt, France

Nizar J. Ahmad, Ebraheem K. Sultan, Mohammed Q. Qasem, Hameed K. Ebraheem
Faculty of Electronic Engineering Technology, College of Technological Studies, The Public Authority for Applied Education and Training (PAAET), Kuwait City, Kuwait

Jasem M. Alostad
Faculty of Computer Science, College of Basic Education, The Public Authority for Applied Education and Training (PAAET), Kuwait City, Kuwait

Sha Zhu, Yuexiang Li and Yu Wang, Yonghui Wang
School of Petroleum Engineering, Harbin Institute of Petroleum, Harbin, China

Aisha Jilani and Sadia Murawwat
Electrical Engineering Department, Lahore College for Women University, Lahore, Pakistan

Syed Omar Jilani
Electrical Engineering Department, University of Lahore, Lahore, Pakistan

Pierre Tety
Institut National Polytechnique Houphouët Boigny (INPHB), Yamoussoukro, Côte d'Ivoire

Adama Konaté, Olivier Asseu, Etienne Soro and Pamela Yoboué
Ecole Supérieure Africaine des Technologies de l'Information et de la Communication (ESATIC), Abidjan, Côte d'Ivoire

Sihem Saidani and Moez Ghariani
Laboratory of Electronics and Information Technology (LETI), Electric Vehicle and Power Electronics Group (VEEP), National School of Engineers of Sfax, University of Sfax, Sfax, Tunisia

Negar Honarmand
Faculty of Engineering and Applied Science, University of Ontario Institute of Technology, Oshawa, Canada

Ahmed M. Othman
Faculty of Engineering and Applied Science, University of Ontario Institute of Technology, Oshawa, Canada0
Faculty of Engineering, Electrical Power & Machine Department, Zagazig University, Zagazig, Egypt

Hossam A. Gabbar
Faculty of Engineering and Applied Science, University of Ontario Institute of Technology, Oshawa, Canada
Faculty of Energy Systems & Nuclear Science, University of Ontario Institute of Technology, Oshawa, Canada

Javanshir Mammadov, Tarana Tagiyeva, Akhmedova Sveta and Aliyeva Arzu
Department of "Information Technology and Programming" of Sumgait State University, Sumqayit, Azerbaijan

Nizar J. Ahmad, Mahmud J. Alnaser, Ebraheem Sultan and Khuloud A. Alhendi
Faculty of Electronic Engineering Technology, College of Technological Studies, The Public Authority for Applied Education and Training (PAAET), Kuwait City, Kuwait

Tariq Aldowaisan and Ali Allahverdi
Department of Industrial and Management Systems Engineering, Kuwait University, Kuwait City, Kuwait

Mohammed Bani Younis
Faculty of Engineering, Philadelphia University, Amman, Jordan

Mohammad Al-Shabi
Department of Mechatronics Engineering, Philadelphia University, Amman, Jordan

Mohammed A. Eslami, John R. Rzasa, Stuart D. Milner, Christopher C. Davis
Department of Electrical and Computer Engineering, University of Maryland, College Park, USA

Chaitanya Poolla
Electrical and Computer Engineering, Carnegie Mellon University (SV), Moffett Field, CA, USA

Abraham K. Ishihara
Research Faculty, Electrical and Computer Engineering, Carnegie Mellon University (SV), Moffett Field, CA, USA

Reem Nasir and Ashraf Elnagar
Computational Intelligence Center, Department of Computer Science, University of Sharjah, Sharjah, UAE

www.ingramcontent.com/pod-product-compliance
Lightning Source LLC
Chambersburg PA
CBHW080933240326

41458CB00143B/4108